Springer Theses

Recognizing Outstanding Ph.D. Research

Aims and Scope

The series "Springer Theses" brings together a selection of the very best Ph.D. theses from around the world and across the physical sciences. Nominated and endorsed by two recognized specialists, each published volume has been selected for its scientific excellence and the high impact of its contents for the pertinent field of research. For greater accessibility to non-specialists, the published versions include an extended introduction, as well as a foreword by the student's supervisor explaining the special relevance of the work for the field. As a whole, the series will provide a valuable resource both for newcomers to the research fields described, and for other scientists seeking detailed background information on special questions. Finally, it provides an accredited documentation of the valuable contributions made by today's younger generation of scientists.

Theses may be nominated for publication in this series by heads of department at internationally leading universities or institutes and should fulfill all of the following criteria

- They must be written in good English.
- The topic should fall within the confines of Chemistry, Physics, Earth Sciences, Engineering and related interdisciplinary fields such as Materials, Nanoscience, Chemical Engineering, Complex Systems and Biophysics.
- The work reported in the thesis must represent a significant scientific advance.
- If the thesis includes previously published material, permission to reproduce this must be gained from the respective copyright holder (a maximum 30% of the thesis should be a verbatim reproduction from the author's previous publications).
- They must have been examined and passed during the 12 months prior to nomination.
- Each thesis should include a foreword by the supervisor outlining the significance of its content.
- The theses should have a clearly defined structure including an introduction accessible to new PhD students and scientists not expert in the relevant field.

Indexed by zbMATH.

Masaki Donoshita

Design of Crystal Structures Using Hydrogen Bonds on Molecular-Layered Cocrystals and Proton–Electron Mixed Conductor

Doctoral Thesis accepted by
Kyoto University, Kyoto, Japan

Springer

Author
Dr. Masaki Donoshita
Institute for Materials Chemistry
and Engineering
Kyushu University
Fukuoka, Japan

Supervisor
Prof. Hiroshi Kitagawa
Graduate School of Science
Kyoto University
Kyoto, Japan

ISSN 2190-5053 ISSN 2190-5061 (electronic)
Springer Theses
ISBN 978-981-99-7061-2 ISBN 978-981-99-7062-9 (eBook)
https://doi.org/10.1007/978-981-99-7062-9

This Springer imprint is published by the registered company Springer Nature Singapore Pte Ltd.
The registered company address is: 152 Beach Road, #21-01/04 Gateway East, Singapore 189721, Singapore

Supervisor's Foreword

The Division of Chemistry, Graduate School of Science, Kyoto University, nominated this Ph.D. thesis for a Springer Thesis because the division regarded it as a remarkable work carried out with a quite high level of independence. Mr. Masaki Donoshita joined my research group in 2016 as an undergraduate student. My research group has been working on hydrogen-bonded molecular solids aiming at the realization of novel physical properties. He had worked on the research theme using hydrogen-bonding cocrystals and metal complexes. In the research process, he found several intriguing systems/phenomena and summarized them as a thesis, focusing on the designability of crystal structures based on hydrogen bonds.

This thesis consists of three central themes. First, a one-dimensional proton–electron mixed conductor is rationally synthesized, and its electronic state and physical properties are studied. Second, structural phase transitions accompanying the drastic rearrangement of hydrogen-bonded assemblies in a molecular cocrystal are investigated. Third, the stacking patterns of the hydrogen-bond-based molecular sheets in hydrogen-boned cocrystals are rationally controlled by a chemical modification. Some of these interesting results were published in leading chemistry journals including the *Journal of the American Chemical Society* and the *Angewandte Chemie International Edition*, and their novelty and significance have been highly evaluated by many peers. This proves that this thesis is worthy of a Springer Thesis.

Kyoto, Japan
April 2023

Prof. Hiroshi Kitagawa

Parts of this thesis have been published in the following journal articles:

1. **Drastic Rearrangement of Self-Assembled Hydrogen-Bonded Tapes in A Molecular Crystal**
   M. Donoshita, M. Hayashi, R. Ikeda, Y. Yoshida, S. Morikawa, K. Sugimoto, H. Kitagawa, *Chem. Commun.* **2018**, *54*, 8571–8574.
2. **Various Stacking Patterns of Two-Dimensional Molecular Assemblies in Hydrogen-Bonded Cocrystals: Insight on Competitive Intermolecular Interactions and Control of Stacking Patterns**
   M. Donoshita, Y. Yoshida, M. Hayashi, R. Ikeda, S. Tanaka, Y. Yamamura, K. Saito, S. Kawaguchi, K. Sugimoto, H. Kitagawa, *Angew. Chem. Int. Ed.* **2021**, *60*, 22839–22848.
3. **Rational Construction of Molecular Electron-Conducting Nanowires Encapsulated in Proton-Conducting Matrix in a Charge-Transfer Salt**
   M. Donoshita, Y. Yoshida, M. Maesato, H. Kitagawa, *J. Am. Chem. Soc.* **2022**, *144*, 17149–17155.

Acknowledgements

The presented thesis is the summary of the author's studies from April 2016 to March 2022 under the supervision of Prof. Hiroshi Kitagawa (Division of Chemistry, Graduate School of Science, Kyoto University).

The author would like to express his great acknowledgment to his supervisor, Prof. Hiroshi Kitagawa, for leading him to research. The author thanks his indispensable suggestions to the overall research and considerable support throughout the laboratory life.

The author sincerely acknowledges Dr. Yukihiro Yoshida (Kyoto University) for valuable discussion and continuous encouragement. This research was carried out based on a lot of discussion with him.

The author wishes to express his gratitude to Dr. Mikihiro Hayashi (Kyoto University, now at Nagasaki University) for his guidance. With his advice, the author could accomplish the earlier part of this research.

The author would like to express his indebtedness to Prof. Ryuichi Ikeda (Emeritus Professor of University of Tsukuba) for helpful discussion and continuous encouragement.

The author is deeply thankful to Dr. Mitsuhiko Maesato (Kyoto University) for his help in conductivity measurements and discussion on magnetism.

The author would like to thank Prof. Susumu Tanaka (National Institute of Technology, Yonago College) for his help in NMR experiments.

The author is deeply thankful to Dr. Sugimoto Kunihisa (JASRI) and Dr. Shogo Kawaguchi (JASRI) for their support for X-ray diffraction experiments at BL02B1 and BL02B2, respectively, in SPring-8.

The author wishes to express his gratitude to Prof. Kazuya Saito (University of Tsukuba) and Dr. Yasuhisa Yamamura (University of Tsukuba) for their discussion on thermodynamics.

The author expresses sincere appreciation to Prof. Hideki Yamochi (Kyoto University) and Dr. Akihiro Otsuka (Kyoto University) for their permission to use their ESR apparatus.

The author is thankful for the financial support of Research Fellowship for Young Scientists (No. 19J23309) from the Japan Society of Promotion of Science.

Great thanks are given to all members including staff, students, and secretaries, of the research group of Prof. Hiroshi Kitagawa for their support.

Finally, the author would like to thank every person who raised him including his family, sincerely.

Masaki Donoshita
March 2022

Contents

Chapter 1
General Introduction

Abstract Molecular packing in crystal structures can significantly influence the physical and chemical properties of molecular materials. Hydrogen bonds are promising intermolecular interactions that control molecular packings owing to their sufficient strength and directionality. This thesis addresses the rational design of functionalities and higher-order control of molecular packing, with a focus on intermolecular hydrogen bonds.

Keywords Hydrogen bond · Structure–property relationship · Intermolecular interactions · Crystal engineering

1.1 Structure–Property Relationship in Molecular Compounds

Various functional molecular materials are used in daily human life, such as those in medicines, plastics, and cosmetics. Their properties depend not only on the properties of the constituent molecules but also on their assembling manner. For example, the solubilities of drugs depend on their crystal structures [1]. Such differences in the properties are important for practical applications. Thus, fundamental science aimed at understanding structure–property relationships and controlling the assembling manner of molecules is of significant importance.

Crystalline materials offer an advantage in such studies because their structures can be investigated in detail. Extensive studies have been conducted on the relationships between the structures and properties, including conductivity, luminescence, magnetism, and ferroelectricity [2–6]. For example, an electron donor, bis(ethylenedithio)tetrathiafulvalene (BEDT-TTF), and an electron acceptor, tetracyanoquinodimethane (TCNQ), can form two types of 1:1 salts (Fig. 1.1): a segregated stack, where the donor and acceptor stack separately [2] and a mixed stack, where the donor and acceptor stack alternately [3]. The specific crystallization conditions determine the structure formed. The resistivity of the segregate stack ($\rho_{RT} \approx 10^{-1}$ Ω cm) is seven orders of magnitude smaller than that

M. Donoshita, *Design of Crystal Structures Using Hydrogen Bonds on Molecular-Layered Cocrystals and Proton–Electron Mixed Conductor*, Recognizing Outstanding Ph.D. Research, https://doi.org/10.1007/978-981-99-7062-9_1

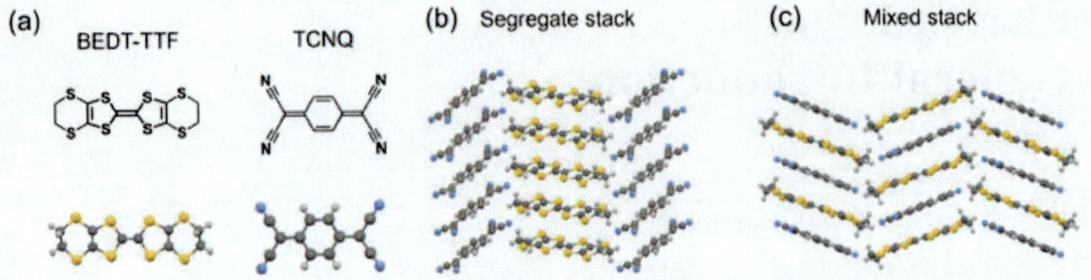

Fig. 1.1 **a** Molecular structures of bis(ethylenedithio)tetrathiafulvalene (BEDT-TTF; left) and tetracyanoquinodimethane (TCNQ; right). Side views of **b** segregate stack [2] and **c** mixed stack [3], respectively

of the mixed stack ($\rho_{RT} \approx 10^6\ \Omega$ cm), which can be explained by the difference in the homogeneity of the energy levels of molecular orbitals involved in the conduction pathway. Another example is the luminescence behavior of phenyl(3,5-dimethylphenyl isocyanide)gold(I) (Fig. 1.2) [4]. This compound undergoes a structural phase transition between the two crystal structures in response to mechanical stimuli and thermal treatment. The difference in the emission color was discussed considering the intermolecular aurophilic interactions.

In this thesis, the author focuses on the control of crystal structures of molecular solids to invoke the knowledge of structure–property relationships.

1.2 Intermolecular Interactions and Crystal Engineering

Design and control of the crystal structures are an effective approach for rationally controlling the physical and chemical properties of molecular solids. Therefore, the concept of "crystal engineering" is crucial. In the 1980s, researchers established the idea of recognizing crystal structures as analogous to molecular structures [7]. The molecular structure is composed of atoms assembled via covalent bonds, whereas the crystal structure is composed of molecules assembled via intermolecular interactions. Thus, G. R. Desiraju's renowned definition of "crystal engineering" is "*the understanding of intermolecular interactions in the context of crystal packing and the utilization of such understanding in the design of new solids with desired physical and chemical properties* [8]."

Among the various types of intermolecular interactions, those exhibiting well-defined directionality, sufficient strength, and a high frequency of occurrence are promising for crystal engineering [9]. Therefore, for example, the London dispersion force is not suitable because of its lack of directionality and weak interaction energy, despite its occurrence at high frequencies. Similarly, other interactions have drawbacks as well. For example, π–π interactions [10], halogen bonding [11, 12], and π-hole interactions [13, 14] are directional interactions with relatively strong interaction energies (< ~50 kJ mol^{-1} for π–π, [15] <~150 kJ mol^{-1} for halogen

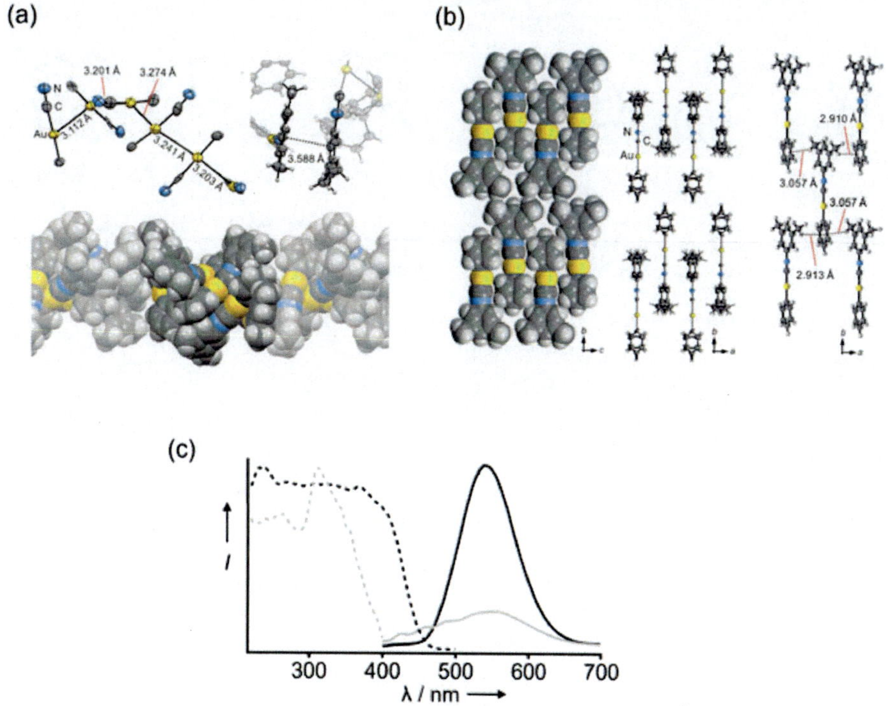

Fig. 1.2 Molecular packing of phenyl(3,5-dimethylphenyl isocyanide)gold(I) in **a** metastable and **b** thermodynamically stable polymorphs; **c** excitation and emission (dotted and solid lines, respectively) spectra for metastable and stable (gray and black, respectively) polymorphs. Reprinted with permission from Ref. [4]. Copyright 2013 John Wiley and Sons

bonding [11], and <~120 kJ mol^{-1} for π-hole [13]). However, their frequency of occurrence is not sufficiently high, making them less reliable.

Hydrogen bonds can be regarded as the most promising intermolecular interactions because they satisfy all the requirements, i.e., well-defined directionality, sufficient strength, and a high frequency of occurrence. Most traditional structural motifs used to construct the desired structures, called supramolecular synthons, involve hydrogen bonds [16] (Fig. 1.3).

1.3 Design of Crystal Structures (Crystal Engineering) Using Hydrogen Bonds

A hydrogen bond is an intra- or intermolecular interaction that forms between a hydrogen-donating chemical species composed of a hydrogen atom and an electronegative element (such as oxygen and nitrogen) and a hydrogen-accepting chemical species composed of an electronegative atom with a lone pair of electrons. The

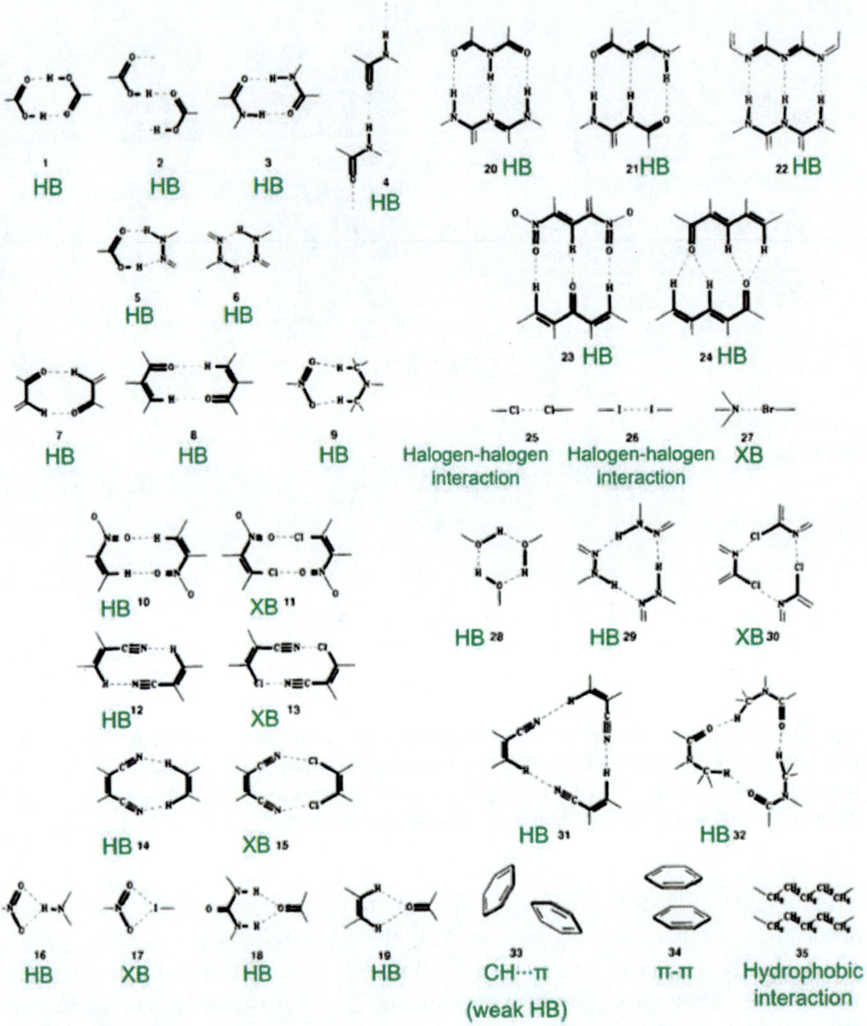

Fig. 1.3 Representative supramolecular synthons. Adapted with permission from Ref. [16]. Copyright 2003 John Wiley and Sons. The words in green were added by the author and represent the principal intermolecular interactions that maintain the structures. HB: hydrogen bond, XB: halogen bond

interaction energies span a wide range: ~160 kJ mol^{-1} ($[F–H–F]^-$) to ~1 kJ mol^{-1} (CH_4···F–CH_3) [17]. Notably, the strength of strong hydrogen bonds is relatively high compared to that of other interactions, such as π–π interaction (< ~50 kJ mol^{-1}) and van der Waals interactions (< ~5 kJ mol^{-1}) [15]. In addition, moderate or strong hydrogen bonds (according to G. A. Jeffrey's definition [18]) adopt well-defined geometries (Fig. 1.4). In X–H···Y type hydrogen bonds, the typical H···Y distances range from 1.2 to 2.2 Å and X···Y distances range from 2.2 to 3.2 Å. The typical

Fig. 1.4 Molecular geometry pertaining to moderate to strong hydrogen bonds

angles between the line connecting X and H and the line connecting Y and H are in the range 130–180° [17].

These favorable properties of hydrogen bonds allow designing functionalities through the design of crystal structures. For example, hydrogen-bonded organic frameworks (HOF) have been extensively studied using crystal engineering based on hydrogen bonds to construct porous structures and the resulting gas sorption behaviors [19–21]. For example, triptycenetrisbenzimidazolone (TTBI), shown in Fig. 1.5, exhibits a crystal structure with one-dimensional pores maintained by one-dimensional hydrogen-bonding networks, affording an exceptionally high surface area of 2796 m^2 g^{-1} (Brunauer–Emmett–Teller (BET) surface area) [21]. Magnetism can be another potential target; to achieve ferromagnetism using molecules, aligning the molecules while considering the spin polarization of each molecule is necessary. Matsushita et al. achieved ferromagnetism with $T_c = 0.5$ K using 2-(2',5'-dihydroxyphenyl)-4,4,5,5-tetramethyl-4,5-dihydro-1*H*-imidazolyl-1-oxyl-3-oxide (HQNN), a molecule comprising nitronyl nitroxide and hydroquinone moieties. In the crystal of an α-polymorph, spin polarization is transmitted via hydrogen bonds in the preferred manner for ferromagnetism (Fig. 1.6) [22, 23]. Furthermore, the ferroelectric hydrogen-bonded network systems are of particular interest. Several systems comprising proton donors and acceptors with similar pK_a and pK_b, respectively, have been studied [24, 25]. For example, the chloranilic acid/bromanilic acid–phenazine system exhibits a well-predicted one-dimensional tape structure, where the bistability of the proton position within a hydrogen bond and the resulting bistable electric dipole leads to ferroelectricity (Fig. 1.7) [25]. In this thesis, systems that could provide new insights into the design of functionalities based on hydrogen bonds were studied (Chaps. 2 and 3).

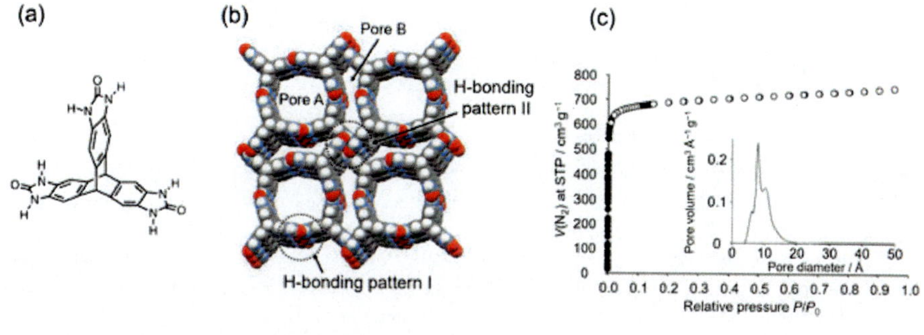

Fig. 1.5 **a** Molecular structure of TTBI. **b** Single-crystal X-ray structure of TTBI. **c** Nitrogen sorption isotherm of activated TTBI at 77 K. The inset shows the pore-size distribution. Adapted with permission from Ref. [21]. Copyright 2012 John Wiley and Sons

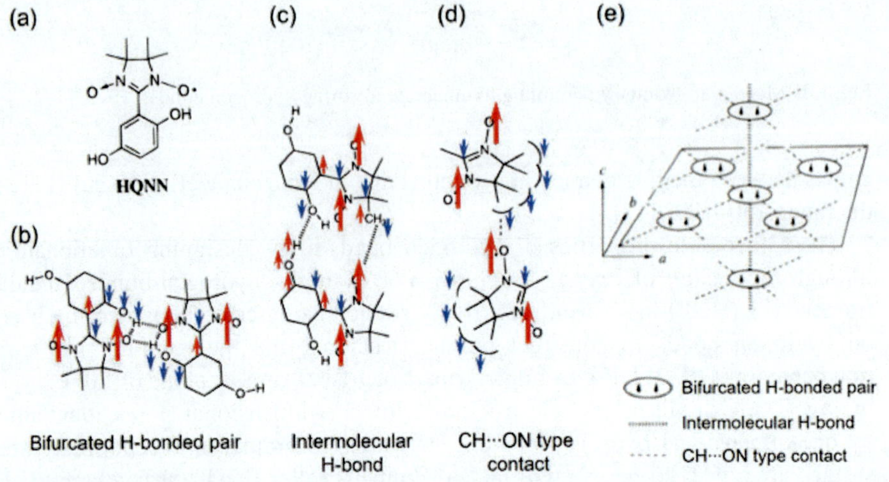

Fig. 1.6 Spin transmission along the hydrogen bond in α-HQNN. **a** Molecular structure of HQNN. **b** Ferromagnetic interaction along the hydrogen-bonded chain. The CH···ON type contact along this direction also contributes to the ferromagnetic interaction. **c** Ferromagnetic interaction between the doubly bifurcated hydrogen-bonded pair. **d** Ferromagnetic interaction through a CH···ON type contact within the ab-plane. **e** Schematic of the spin system in α-HQNN. Ferromagnetically coupled dimeric pair surrounded by six neighboring pairs. Adapted with permission from Ref. [23]. Copyright 1997 American Chemical Society

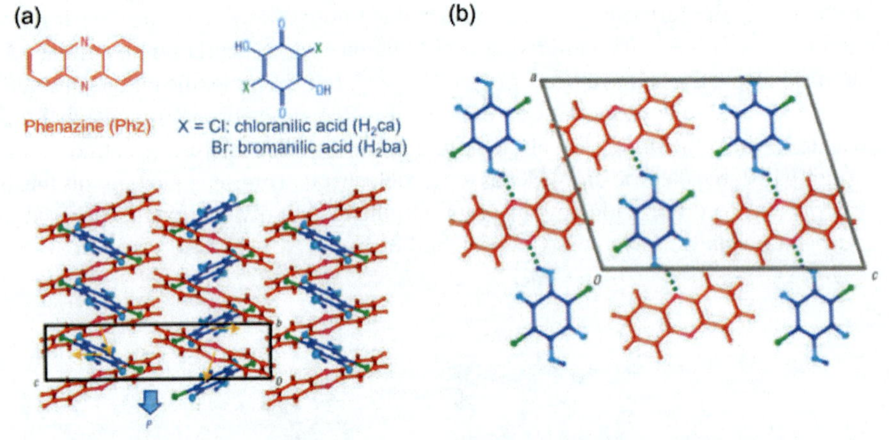

Fig. 1.7 Chemical formulae and crystal structures of cocrystals of phenazine and 2,5-dihalo-3,6-dihydroxy-p-benzoquinones (Phz-H_2xa). **a** The molecular stacking viewed along the crystallographic a direction. The orange arrows indicate the directions of molecular displacement at 160 K ($< T_c$) from the paraelectric symmetric structure for the Phz-H_2ca crystal. The resultant macroscopic polarization P is along the thick blue arrow. **b** The crystal structure viewed along the b (molecular stack) direction. The dotted lines indicate the intermolecular O–H···N hydrogen bonds. Reprinted with permission from Ref. [25]. Copyright 2005 Springer Nature

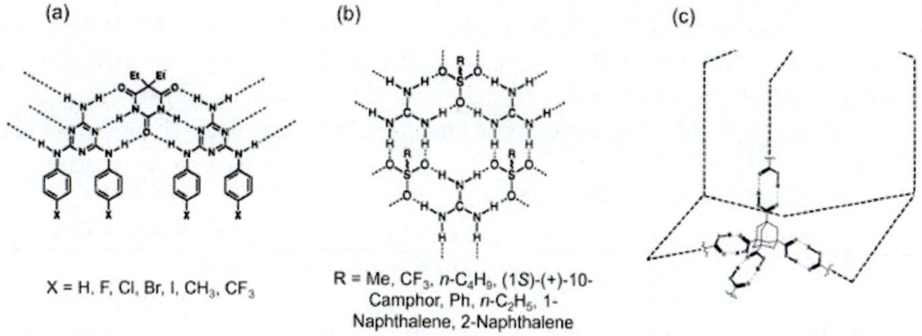

Fig. 1.8 **a** Hydrogen-bonded tapes in cyanuric acid-melamine complex. Adapted with permission from Ref. [37]. Copyright 1988 American Chemical Society. **b** Hydrogen-bonded sheets in guanidinium-sulfonates salts. Adapted with permission from Ref. [38]. Copyright 1988 American Chemical Society. **c** Hydrogen-bonded diamondoid frameworks in adamantane-1,3,5,7-tetracaboxylic acid. Adapted with permission from Ref. [40]. Copyright 1988 American Chemical Society

In addition, previous research in this field has yielded information regarding the relationship between the types and positions of the hydrogen-bonding groups of the component molecules and resulting local structure. Numerous hydrogen-bonded networks and frameworks have been explored to extend this knowledge for the extended structures [26–32]. Typical examples include tapes in cyanuric acid-melamine complexes [33–37], sheets in guanidinium sulfonate salts [38, 39], and diamondoid frameworks in adamantane-1,3,5,7-tetracaboxylic acid (Fig. 1.8) [40]. However, control over the assembling manner of such extended assemblies has not yet been achieved. To unambiguously control the whole crystal structures and thereby the properties of materials, it is necessary to control the assembling patterns of such assemblies, i.e., extended structure. Thus, in this study, control of the assembling pattern of hydrogen-bonded networks was focused upon (Chap. 4).

1.4 Structure of this Thesis

The purpose of this study is to advance research on the design of crystal structures (i.e., crystal engineering) using hydrogen bonds in two aspects: (1) utilization to provide functionalities (Chaps. 2 and 3) and (2) development of the concept by controlling the assembling manner of hydrogen-bonded assemblies, that is, higher-order structural control (Chap. 4). This section provides a brief overview of the chapters.

Chapter 2, titled "Rational Construction of Molecular Electron-conducting Nanowires Encapsulated in a Proton-conducting Matrix in a Charge-Transfer Salt," illustrates the utilization of crystal-structure design using hydrogen bonds in the rational construction of molecular electron-conducting wires encapsulated in a proton-conducting matrix in a charge-transfer salt. Based on the design of the molecular structure and, hence, the crystal structure, the target molecular assembly was synthesized, and its electronic state and physical properties were investigated.

In Chap. 3, titled "Drastic Rearrangement of Self-assembled Hydrogen-bonded Tapes in a Molecular Crystal," the hydrogen bond in a cocrystal was shown to provide a primary building block whose rearrangement, affording the structural phase transitions, leads to significant anisotropic changes in the crystal shape, which could potentially be advantageous for applications in mechanical devices.

Chapter 4, "Various Stacking Patterns of Two-Dimensional Molecular Assemblies in Hydrogen-Bonded Cocrystals: Insight into Competitive Intermolecular Interactions and Control of Stacking Patterns," contains description of the control over the higher-order structural control of hydrogen-bonded assembly. As described in Chap. 3, hydrogen bonds afford a robust two-dimensional sheet structure in a cocrystal system. Rational control of the stacking patterns of the two-dimensional sheets was demonstrated via the substitution of halogens in a component molecule of the cocrystal.

Finally, a general conclusion in Chap. 5 concludes the Ph.D. thesis.

References

1. Bernstein J (1997) Polymorphism in molecular crystals, 2nd edn. Oxford University Press, Oxford
2. Yamamoto HM, Hagiwara M, Kato R (2003) Synth Met 133–134:449–451
3. Mori T, Inokuchi H (1987) Bull Chem Soc Jpn 60:402–404
4. Seki T, Sakurada K, Ito H (2013) Angew Chem Int Ed 52:12828–12832
5. Nishihara S, Akutagawa T, Hasegawa T, Fujiyama S, Nakamura T, Nakamura T (2002) J Solid State Chem 168:661–667
6. Kobayashi K, Horiuchi S, Ishibashi S, Kagawa F, Murakami Y, Kumai R (2014) Chem Eur J 20:17515–17522
7. Desiraju GR (2007) Angew Chem Int Ed 46:8342–8356
8. Desiraju GR (1989) Crystal engineering. In: The design of organic solids, Elsevier, Amsterdam
9. Bishop R (2015) CrystEngComm 17:7448–7460
10. Hunter CA, Sanders JKM (1990) J Am Chem Soc 112:5525–5534
11. Politzer P, Murray JS, Clark T (2013) Phys Chem Chem Phys 15:11178–11189
12. Cavallo G, Metrangolo P, Milani R, Pilati T, Priimagi A, Resnati G, Terraneo G (2016) Chem Rev 116:2478–2601
13. Wang H, Wang W, Jin WJ (2016) Chem Rev 116:5072–5104
14. Politzer P, Murray JS (2018) J Comput Chem 39:464–471
15. Steed JW, Turner DR, Wallace K (eds) (2007) Core concepts in supramolecular chemistry and nanochemistry, Wiley, West Sussex
16. Desiraju GR (1995) Angew Chem Int Ed 34:2311–2327
17. Steiner T (2002) Angew Chem Int Ed 41:48–76
18. Jeffrey GA (1997) An introduction to hydrogen bonding. Oxford University Press, Oxford
19. Hisaki I, Xin C, Takahashi K, Nakamura T (2019) Angew Chem Int Ed 58:11160–11170
20. Lin RB, He Y, Li P, Wang H, Zhou W, Chen B (2019) Chem Soc Rev 48:1362–1389
21. Mastalerz M, Oppel IM (2012) Angew Chem Int Ed 51:5252–5255
22. Sugawara T, Matsushita MM, Izuoka A, Wada N, Takeda N, Ishikawa M (1994) J Chem Soc Chem Commun 1723–1724

23. Matsushita MM, Izuoka A, Sugawara T, Kobayashi T, Wada N, Takeda N, Ishikawa M (1997) J Am Chem Soc 119:4369–4379
24. Horiuchi S, Ishibashi S (2020) J Phys Soc Jpn 89:051009
25. Horiuchi S, Ishii F, Kumai R, Okimoto Y, Tachibana H, Nagaosa N, Tokura Y (2005) Nat Mater 4:163–166
26. MacDonald JC, Whitesides GM (1994) Chem Rev 94:2383–2420
27. Williams LJ, Jagadish B, Lyon SR, Kloster RA, Carducci MD, Mash EA (1999) Tetrahedron 55:14281–14300
28. Williams LJ, Jagadish B, Lansdown MG, Carducci MD, Mash EA (1999) Tetrahedron 55:14301–14322
29. Takasawa R, Yoshikawa I, Araki K (2004) Org Biomol Chem 2:1125–1132
30. Sada K, Inoue K, Tanaka T, Tanaka A, Epergyes A, Nagahama S, Matsumoto A, Miyata M (2004) J Am Chem Soc 126:1764–1771
31. Gong B, Zheng C, Skrzypczak-Jankun E, Yan Y, Zhang J (1998) J Am Chem Soc 120:11194–11195
32. Gong B, Zheng C, Zeng H, Zhu J (1999) J Am Chem Soc 121:9766–9767
33. Zerkowski JA, Seto CT, Wierda DA, Whitesides GM (1990) J Am Chem Soc 112:9025–9026
34. Zerkowski JA, Seto CT, Whitesides GM (1992) J Am Chem Soc 114:5473–5475
35. Zerkowski JA, Whitesides GM (1994) J Am Chem Soc 116:4298–4304
36. Zerkowski JA, Mathias JP, Whitesides GM (1994) J Am Chem Soc 116:4305–4315
37. Zerkowski JA, MacDonald JC, Seto CT, Wierda DA, Whitesides GM (1994) J Am Chem Soc 116:2382–2391
38. Russell VA, Etter MC, Ward MD (1994) J Am Chem Soc 116:1941–1952
39. Russell VA, Evans CC, Li W, Ward MD (1997) Science 276:575–579
40. Ermer O (1988) J Am Chem Soc 110:3747–3754

Chapter 2
Rational Construction of Molecular Electron-Conducting Nanowires Encapsulated in Proton-Conducting Matrix in a Charge-Transfer Salt

Abstract Insulated molecular wires have gained significant attention owing to their potential contribution to the fields of nanoelectronics and low-dimensional chemistry/physics. This study demonstrates, for the first time, the rational construction of molecular electron-conducting wires encapsulated in a proton-conducting matrix via the use of a molecular charge-transfer salt, which may pave the way for iono-electronics. As expected from the molecular structure of the newly designed complex anion (a propeller-shaped structure with hydrogen-bonding sites at four edges), a three-dimensional hydrogen-bonded framework was constructed within the crystal, which contained one-dimensional arrays of the electron donor, tetrathiafulvalene (TTF). Single-crystal crystallographic and spectroscopic studies clarified that non-stoichiometric deprotonation of anions and partial oxidation of TTF molecules occurred, whereas the anion was electronically inert. The moderate conductivities of electrons and protons were confirmed by DC- and AC-conductivity measurements. In addition, electronic isolation of the TTF wires was confirmed using magnetic susceptibility data.

Keywords Hydrogen bond · Mixed conductor · Charge-transfer salt

2.1 Introduction

Insulated molecular wires, in which an electron-conducting molecular wire is surrounded by insulating sheaths or embedded in insulating matrices, have gained significant attention owing to their potential in low-dimensional physics and chemistry and the miniaturization of electronic circuits by providing nanoscale wiring materials [1–3]. To date, electronic behavior has been investigated in various systems, including conjugated polymers sheathed with other compounds such as polymers [4] and macrocyclic molecules (e.g., cyclodextrin) [5, 6] or encapsulated in frameworks (e.g., zeolites [7] and metal–organic frameworks [8]), as well as charge-transfer salts comprising one-dimensionally aligned cation radicals surrounded by insulating matrices of counter anions [9].

M. Donoshita, *Design of Crystal Structures Using Hydrogen Bonds on Molecular-Layered Cocrystals and Proton–Electron Mixed Conductor*, Recognizing Outstanding Ph.D. Research, https://doi.org/10.1007/978-981-99-7062-9_2

Providing the insulating part of insulated molecular wires with ionic conductivity can result in the manifestation of novel properties derived from the cooperation between electrons and ions (a type of iono-electronics) [10]. Such materials can, for example, interface biological and electronic systems, in which ions and electrons are the dominant charge carriers, respectively. Thus far, several insulated ion–electron dual-channel systems based on columnar liquid crystals have been reported [11, 12], such as a phthalocyanine molecule bearing crown ethers with long alkyl chain moieties [11]. However, there is no report on a single-crystalline compound that allows the investigation of structure–property relationships to guide the rational design of such molecular mixed conductors. Although there are several reports on single-crystalline proton–electron-mixed-conducting ion radical salts [13–16], the radical cations in such systems exhibit quasi-one- or two-dimensional arrangements, unlike insulated molecular wire systems.

In general, it is challenging to predict the arrangement of electron-conducting π-conjugated molecules in a crystal prior to crystallization [17]. This is mainly because the molecular packing in molecular crystals is determined by the subtle balance of various weak intermolecular interactions; π–π interactions ($< \sim 50$ kJ mol^{-1}) [18], which dominate the arrangement of π-conjugated molecules, are not significantly stronger than other interactions such as dipole–dipole interactions ($< \sim 50$ kJ mol^{-1}) [18], halogen interactions ($< \sim 150$ kJ mol^{-1}) [19], and hydrogen-bonding interactions ($< \sim 160$ kJ mol^{-1}) [20].

In this study, we designed a molecular assembly by focusing on the different intermolecular interaction strengths; specifically, we formed a rigid framework with one-dimensional channels using strong and directional hydrogen bonds and subsequently arranged electronically active π-conjugated molecules in these channels using weak interactions (Fig. 2.1). We focused on the use of a propeller-shaped molecule with hydrogen-bonding sites at the edge of each blade (Fig. 2.1a, top). As a manifestation of the orthogonal hydrogen-bonding sites, the molecule is expected to form a square grid layer (Fig. 2.1b) that readily undergoes stacking via hydrogen bonds to form a one-dimensional hollow matrix (Fig. 2.1c, top). In addition, the connected hydrogen-bonding sites can serve as proton-conducting pathways. The introduction of electrically active π-conjugated molecules causes one-dimensional alignment in the channel to form electron-conductive wires (Fig. 2.1d). Here, we designed a new paddle-wheel-type dinuclear platinum complex, $Pt_2(pmdt)_4$ (pmdt: phosphonomethanedithioate), with four phosphonic acid moieties (Fig. 2.1a, bottom) as the required hydrogen-bonding compound. Hereinafter, this complex is referred to as **1** (also represented as **$1\text{-}H_x$** depending on the number of attached protons, x). Using its deprotonated anionic form as the counter anion, a charge-transfer salt was prepared using a typical electron donor molecule, tetrathiafulvalene (TTF, Fig. 2.1c, bottom).

Based on the aforementioned molecular design, we report the rational construction of a molecular assembly of electron-conducting nanowires encapsulated in a proton-conducting matrix.

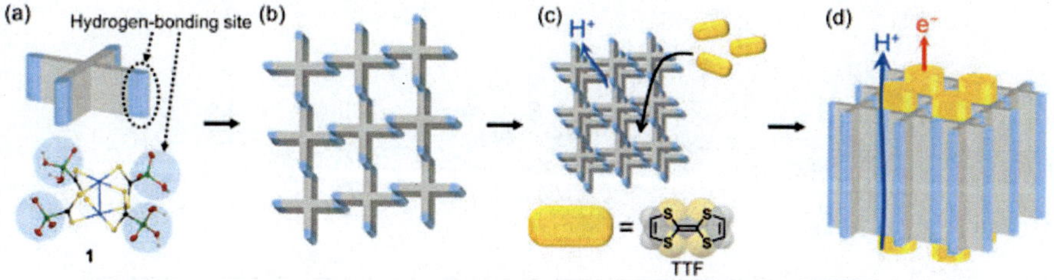

Fig. 2.1 Design of the electron-conducting molecular wires encapsulated in proton-conducting matrix. **a** (top) Schematic of a propeller-shaped molecule, and (bottom) molecular structure of complex 1. Molecular structure is adapted from the crystal structure of $(TTF)_2$(**1-H**$_{6+\delta}$). **b** Two-dimensional grid structure formed by the propeller-shaped molecules. **c** (top) Hollow matrix comprising the propeller-shaped molecules with yellow units representing π–conjugated molecules and (bottom) molecular structure of TTF. **d** Schematic of the complete structure of the molecular assembly, wherein the π–conjugated molecules form electron-conducting molecular wires

2.2 Experimental Section

Materials

Anhydrous methanol (Wako Chemical), methanol (Wako Chemical), anhydrous acetonitrile (Wako Chemical), acetonitrile (Wako Chemical), dichloromethane (Wako Chemical), chloroform (Wako Chemical), and trimethylsilyl trifluoromethanesulfonate (TMSOTf, Tokyo Kasei) were used as received. Sodium hydrosulfide (Wako Chemical) was recrystallized from an ethanol solution by adding diethyl ether. *trans*-$PtCl_2(PhCN)_2$, [21] $(TTF)_3(BF_4)_2$, [22] $TTFBr_{0.76}$, [23] and $TTFClO_4$ [24] were synthesized according to the literature methods.

Elemental analysis

Elemental analysis was performed for C, H, and N by the Elemental Analysis Center of the Faculty of Pharmaceutical Science at Kyoto University.

Synthesis of complex 2

The synthesis scheme of 2 is shown in Scheme 2.1. First, *S*-methyl diethoxyphosphinyldithioformate (**3**) was synthesized from sodium diethoxyphosphonate using a method similar to that for *S*-methyl diethoxyphosphinylthioformate described in the literature [25]. The purity of the product was confirmed using ^{1}H NMR spectroscopy. To an anhydrous methanolic solution (54 mL) of **3** (4.1069 g, 18.0 mmol), an anhydrous methanolic solution (54 mL) of sodium hydrosulfide (1.0050 g, 17.9 mmol) was added at room temperature. The color of the solution gradually changed from purple to dark orange. After stirring for 8 h, the reaction mixture was condensed in vacuo. Water (50 mL) was added, and the mixture was washed with CH_2Cl_2 (4 × 50 mL). The aqueous layer was evaporated to yield a dark orange oil (2.8763 g). The oil was added to anhydrous acetonitrile (200 mL) and filtered to obtain a red solution. This solution was then added dropwise (over approximately 1 h) to a solution of *trans*-$PtCl_2(PhCN)_2$ (2.8628 g, 6.09 mmol) in anhydrous acetonitrile (750 mL) at room temperature. After stirring for 30 min, the resulting green precipitate was

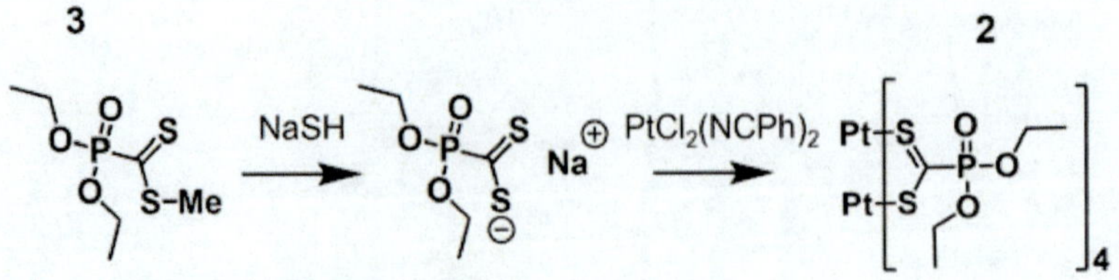

Scheme 2.1 Synthesis scheme of **2**

filtered and washed with methanol (50 mL). After drying in vacuo, a dark-green powder (949.7 mg) was obtained. Recrystallization from a chloroform solution by the slow addition of acetonitrile afforded yellow needles of product **2** (458.9 mg, 36.9 mmol) in a yield of 20% against **3**. The crystal structure was investigated using single-crystal X-ray diffraction.
Elemental analysis:
Calcd (%) for $Pt_2S_8C_{20}P_4O_{12}H_{40}$: C, 19.32; H, 3.24; N, 0.
Found (%): C, 19.08; H, 3.22; N, 0.00.

Synthesis of $(TTF)_2(\mathbf{1\text{-}H_{6+\delta}})$

The synthesis scheme of $(TTF)_2(\mathbf{1\text{-}H_{6+\delta}})$ is shown in Scheme 2.2. To the suspension of **2** (124.4 mg, 100 μmol) in anhydrous acetonitrile (200 mL) at 70 °C, TMSOTf (722.5 μL, 4.00 mmol) was added under an argon atmosphere, leading to the dissolution of the solids. After stirring overnight under the argon atmosphere, the reaction solution was cooled to room temperature, and water (0.66 mL) was added to the solution to afford a dark-green precipitate. The reaction mixture was then transferred to a pressure-resistant vial, and the supernatant was removed to leave the suspension (30 mL). A methanolic suspension (166 mL) of $(TTF)_3(BF_4)_2$ (90.0 mg, 114 μmol) was then added. The mixture was heated to 80 °C for 30 min, maintained at that temperature for 6 h, and cooled to room temperature for 6 h using a solvothermal method. The resulting product contained green crystals, cloudy brown precipitates, and black rod-like crystals. The black rod-like crystals were exclusively collected using different fall speeds for the three products in methanol. The supernatant was removed to leave 10 mL of the suspension. Methanol (140 mL) was then added to the suspension, which was shaken in a conventional chromatography tube ($\phi = 2.2$ cm). Shortly thereafter, the supernatant, containing mainly green crystals and brown precipitates, was removed. The process was repeated several times until the product, pure black rod-like crystals of $(TTF)_2(\mathbf{1\text{-}H_{6+\delta}})$ (56.6 mg, 39.8 μmol calculated as $(TTF)_2(\mathbf{1\text{-}H_6})$), were obtained in a yield of 40% against **2**.

Elemental analysis:
Calcd(%) for $Pt_2S_{16}C_{16}P_4O_{12}H_{12}$: C, 13.48; H, 0.99; N, 0.
Found(%): C, 13.30; H, 1.05; N, 0.00.

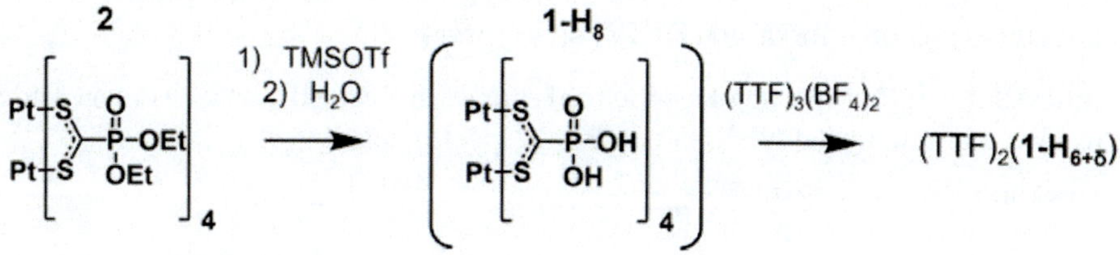

Scheme 2.2 Synthesis scheme of $(TTF)_2$(**1-H**$_{6+\delta}$)

Single-crystal X-ray diffraction

For the single-crystal X-ray diffraction (SCXRD) analysis of **2** and $(TTF)_2$(**1-H**$_{6+\delta}$), diffraction data were collected using a Rigaku XtaLAB PRO diffractometer with graphite-monochromatized Mo-Kα radiation and a Dectris PILATUS detector. The single crystal was mounted on a plastic loop (MicroLoops, MiTeGen) with a minimal amount of oil (Parabar 10,312, Hampton Research) and cooled under a stream of nitrogen gas. The structures were solved using direct methods (SHELXT) [26] and refined by full-matrix least-squares refinement of F^2 (SHELXL) [27] using the CrystalStructure software package [28]. All non-hydrogen atoms were refined anisotropically. For the hydrogen atoms belonging to **2** or the TTF molecules in $(TTF)_2$(**1-H**$_{6+\delta}$), the positional parameters were calculated using the sp^2 or sp^3 configuration of the bonding atoms. For the hydrogen atoms involved in the hydrogen bonds originating from the phosphonic acid groups of **1-H**$_{6+\delta}$, the atomic positions were refined based on the P–O distance. Namely, when the distances of two P–O bonds adjacent to an O···H···O hydrogen bond were similar, the position of the hydrogen atom was refined to be located in the middle of two oxygen atoms. On the other hand, when the distances between the two P–O bonds were significantly different, the hydrogen atom was placed on the oxygen atom with a longer P–O bond in the sp^2 configuration. In the refinement procedures, isotropic atomic displacement parameters with magnitudes that were 1.2-fold those of the equivalent isotropic atomic displacement parameters of the bonding atoms were applied to the hydrogen atoms.

Powder X-ray diffraction

Powder X-ray diffraction (PXRD) measurements were performed on a Dectris MYTHEN detector using synchrotron X-rays with a wavelength of 0.699628(1) Å at the BL02B2 beamline at SPring-8. The ground samples were sealed in a glass capillary and the temperature was changed using a stream of nitrogen gas.

Calculation of transfer integrals

The overlap integrals (s) between the HOMOs of the TTF in $(TTF)_2$(**1-H**$_{6+\delta}$) were calculated using the extended Hückel method. The parameters for the Slater-type atomic orbitals were obtained from the literature [29]. Transfer integrals (t) were estimated based on the approximation $t = Es$ ($E = -10$ eV). All calculations were performed using a program developed by Mori et al. [29]

Fourier-transform infrared (FT-IR) spectroscopy

Infrared absorption spectra of a pelletized sample in KBr (~0.5 wt%) were measured using a ThermoNicolet NEXUS 670 FT-IR spectrometer.

Ultraviolet–visible (UV–vis) spectroscopy

UV–vis spectra were obtained using a JASCO V-770 UV–Visible/NIR spectrometer in the range 190–2700 nm at room temperature. Pelletized samples diluted in KBr (~0.5 wt%) were used.

Electron spin resonance (ESR)

The ESR spectra of the polycrystalline $(TTF)_2(\mathbf{1\text{-}H_{6+\delta}})$ were recorded at room temperature and 4 K using a JEOL JES-TE200 X-band ESR spectrometer equipped with a TE_{011} cavity. The microwave frequency and modulation width were 9.071 GHz and 0.1 G, respectively. The obtained g-factors were corrected using a standard sample (Bruker strong pitch: $g = 2.0028$).

X-ray photoelectron spectroscopy

XPS spectra were collected on a Shimadzu ESCA-3400 X-ray photoelectron spectrometer (Shimadzu, Japan) using Mg $K\alpha$ anode. The binding energies were corrected using the Ag $3d_{5/2}$ peak (368.27 eV) of Ag particles (Wako Chemical) that were mixed with the samples as the inner reference. Fittings of the line shapes were performed with the iterative least-squares method using the XPSPEAK41 software program, wherein asymmetric Gaussian–Lorentzian mixed functions were used.

Raman spectroscopy

Raman spectra were recorded using a JASCO NRS-5100 with a microscope attachment at room temperature. A Nd-doped YAG laser (532 nm) was used as the excitation source, and the Raman shift was calibrated using a Ne-lamp signal. Pelletized samples diluted in KBr (~0.5 wt%) were used. For each sample, we integrated a spectrum obtained at ~100–200 different sample points separated by 5 μm. In addition to $(TTF)_2(\mathbf{1\text{-}H_{6+\delta}})$, the spectra of neutral TTF, $(TTF)_3(BF_4)_2$, $TTFBr_{0.76}$, and $TTFClO_4$ were recorded to obtain the frequency(ν_3)–valence relationship of TTF.

DC conductivity

DC resistivity of the crystalline rod-like sample was measured using a standard four-probe technique. Four gold wires ($\phi = 10$ μm) were attached to the sample with carbon paste. A constant electric current was then applied along the long axis of the sample corresponding to the stacking direction of the TTF molecules.

AC impedance

The proton conductivities of the compressed pellet (2.5 mm in diameter) sandwiched by Nafion membranes (Nafion® 117, Aldrich) were measured by the ac impedance technique with an applied voltage of 100 mV using the Solartron SI 1260 Impedance/

Gain-Phase Analyzer and the 1296 Dielectric Interface in the frequency range of 0.1–1 MHz. Measurements were performed in a humidity- and temperature-controlled chamber.

Magnetic susceptibility

A Quantum Design MPMS-XL superconducting quantum interference device (SQUID) magnetometer was used to collect the magnetic susceptibility data for polycrystalline $(TTF)_2$(**1-H$_{6+\delta}$**). A polywrap sheet was used as the sample holder. Variable-temperature measurements were performed between 2 and 298 K. Magnetic susceptibility was calculated from the magnetic moments at 20 and 30 kOe. The core diamagnetism value was estimated on the basis of the sum of Pascal's constants (-5.47×10^{-4} emu mol^{-1}).

In the Results and Discussion section (Sect. 2.3), we show the fitting results of the magnetic susceptibility data of $(TTF)_2$(**1-H$_{6+\delta}$**) using the alternating chain model with ignored t_3 and t_4 (please refer to Figs. 2.2a and 2.8 (Sect. 2.3)). In addition, we performed fitting using the two-leg ladder model with ignored t_2 and t_4 (refer to Figs. 2.2b and 2.8). The fitting function of the $S = 1/2$ two-leg ladder model is shown below [30]:

$$\chi_{spin} = \frac{N_{spins} g^2 \beta^2}{-2J_{rung}} \frac{e^{-\frac{\Delta^*_{fit}}{t}}}{4t} P^{(p)}_{(q)}(t),$$

where N_{spins} is the number of spins per 1 mol of formula $(TTF)_2$(**1-H$_6$**), g is the g-factor, β is the Bohr magneton, J_{rung} is the primary magnetic interaction proportional to the square of t_1, Δ^*_{fit} is the parameter related to spin gap, and t is the parameter related to temperature. $P^{(p)}_{(q)}(t)$ is the Padé approximant containing J_{leg}, a secondary magnetic interaction proportional to the square of t_3. The details of this equation are provided in Appendix 5 of Ref.[30]. Note that we defined J_{rung} and J_{leg} corresponding to $-J'/2$ and $-J/2$ in Ref. [30], respectively, to unify the definitions of the magnetic interactions in the two-leg ladder model and the alternating chain model.

Fitting using the two-leg ladder model assuming $g = 2.0079$ also reproduced the experimental data well, as depicted in Fig. 2.2b, with the obtained parameters of $N_{\text{spins}} = 1.97(2)$ mol, $J_{\text{rung}}/k = -56.7(3)$ K, and $J_{\text{leg}}/k = -8.1(8)$ K, where k is Boltzmann's constant. However, in the ground state, the two-leg ladder model affords the ferromagnetic spin configuration between the electrons connected by t_2, whereas the alternating chain model affords the ferromagnetic spin configuration between the electrons connected by t_3 (Fig. 2.3). Given that $|t_2| > |t_3|$ (see Sect. 2.3), the latter is more acceptable. Therefore, we used an alternating chain model to explain the magnetic behavior of $(TTF)_2$(**1-H$_{6+\delta}$**).

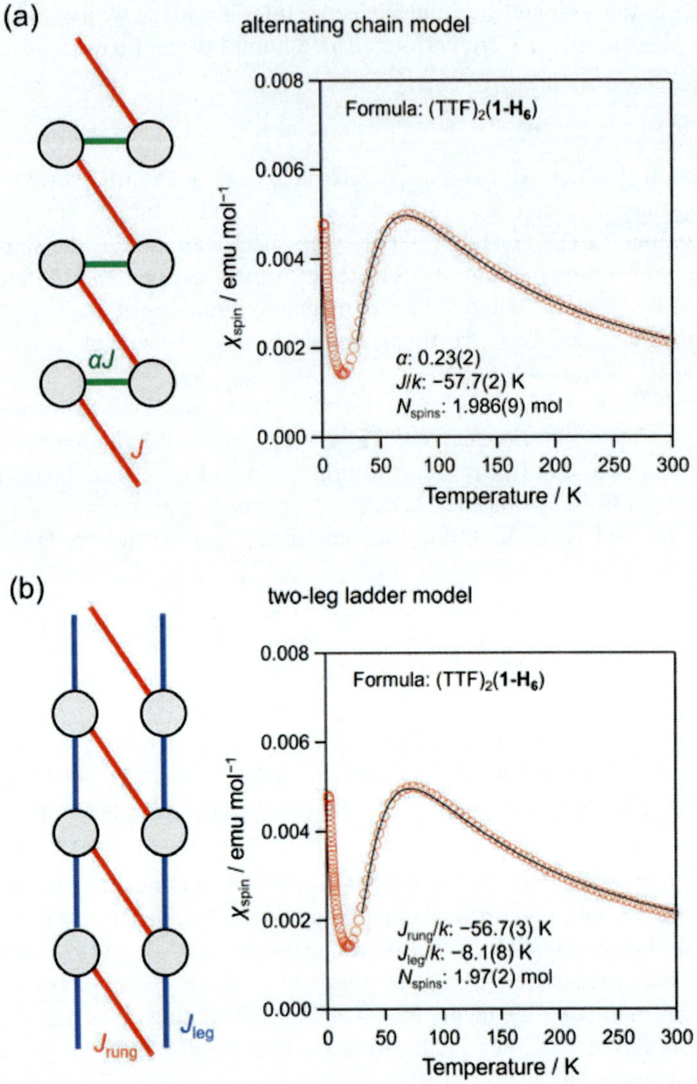

Fig. 2.2 Fitting results for the temperature dependence of the magnetic susceptibility (χ_{spin}) of $(\mathrm{TTF})_2(\mathbf{1\text{-}H_{6+\delta}})$. **a** Schematic of the alternating chain model (left) and the fitting result obtained by the model (right). **b** Schematic of the two-leg ladder model (left) and the fitting result obtained by the model (right)

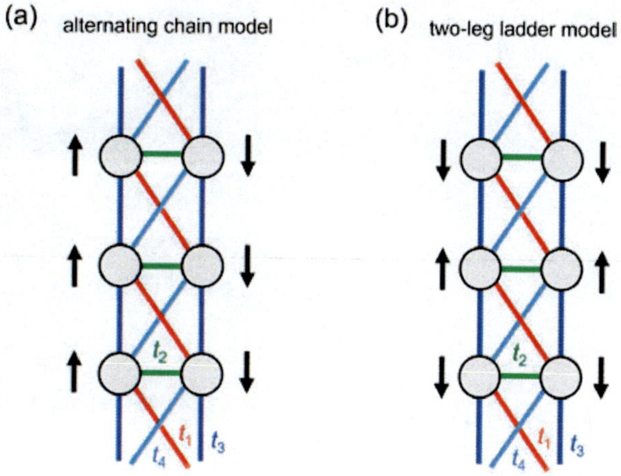

Fig. 2.3 Schematic of spin configurations in the ground states in **a** the alternating chain model and **b** the two-leg ladder model. Black arrows indicate the direction of electron spin

2.3 Results and Discussion

2.3.1 Synthesis and Crystal Structure

The dinuclear platinum complex **1** was synthesized from the ethyl ester derivative **2** (Scheme 2.3) using a deprotection procedure, which comprised silylation and subsequent hydrolysis. First, to a hot suspension of **2** in acetonitrile, trimethylsilyl trifluoromethanesulfonate (TMSOTf), a silylating agent, was added to dissolve **2**. Subsequent addition of water afforded a green precipitate of **1**. The supernatant was removed, and a methanolic suspension of $(TTF)_3(BF_4)_2$ was added. Solvothermal treatment of this mixture afforded rod-like black crystals in 40% yield (Fig. 2.4; refer to Sect. 2.2 for detailed synthetic procedure). A large percentage (>90%) of the obtained crystals were twinned [31], although some selected crystals afforded good diffraction spots in single-crystal X-ray diffraction (SCXRD), which could be properly solved. The powder X-ray diffraction (PXRD) pattern of the ground sample (Fig. 2.5), which was recorded to confirm the phase purity, matched well with the simulated pattern of the structure obtained via SCXRD. The reproducibility of the synthesis method was confirmed by fabricating different sample batches. The obtained samples showed no degradation during storage in air for several months.

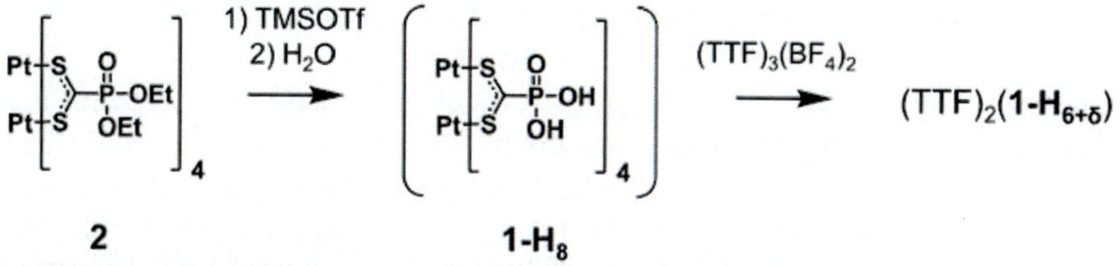

Scheme 2.3 Synthesis scheme of $(TTF)_2$(**1-H**$_{6+\delta}$)

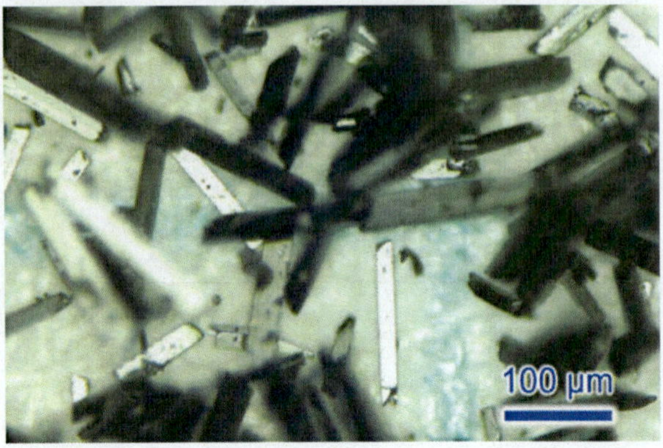

Fig. 2.4 Photograph of crystals of $(TTF)_2$(**2-H$_{6+\delta}$**)

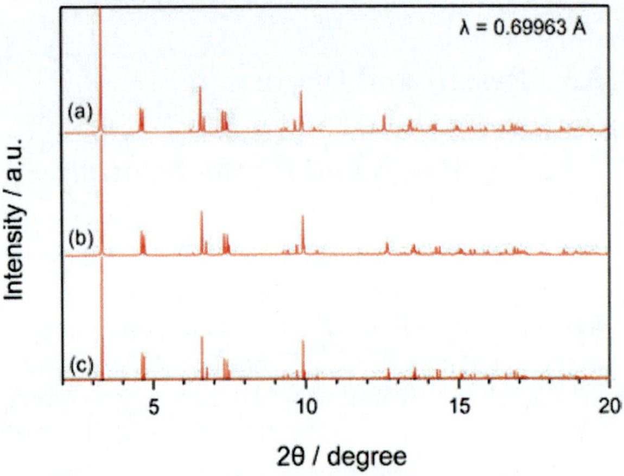

Fig. 2.5 Powder X-ray diffraction pattern of $(TTF)_2$(**2-H$_{6+\delta}$**) at **a** 298 K and **b** 100 K. **c** shows the pattern simulated from single-crystal X-ray diffraction result (100 K)

SCXRD measurements at 100 K revealed that the molecular ratio of TTF to complex **1** is 2:1. The crystal structure adopts a monoclinic crystal system (space group: *I*2/*a*), with $Z = 4$ (Table 2.1). The asymmetric unit contains one molecule of TTF and half a molecule of **1**. In the crystal structure, each phosphonic acid moiety forms hydrogen bonds with two other phosphonic acid moieties in the adjacent complex anions (Fig. 2.6a). There are two types of hydrogen bonds: complementary hydrogen bonds comprising O–H···O-type hydrogen bonds, and single O···H···O hydrogen bonds. The alternating alignment of these two hydrogen bonds results in a one-dimensional hydrogen-bonded network along the *a*-axis. Their bond lengths, which are similar to each other (O···O distance: 2.439(12) and 2.435(12) Å for the former and 2.401(6) Å for the latter), are relatively short and located near the boundary that determines the shape of the proton potential in the hydrogen bond, namely double well or single well [32]. The hydrogen in the former bond is located near one of the oxygen atoms, whereas the hydrogen in the latter bond is located at the center of two oxygen atoms. Accordingly, each phosphonic acid moiety bears one

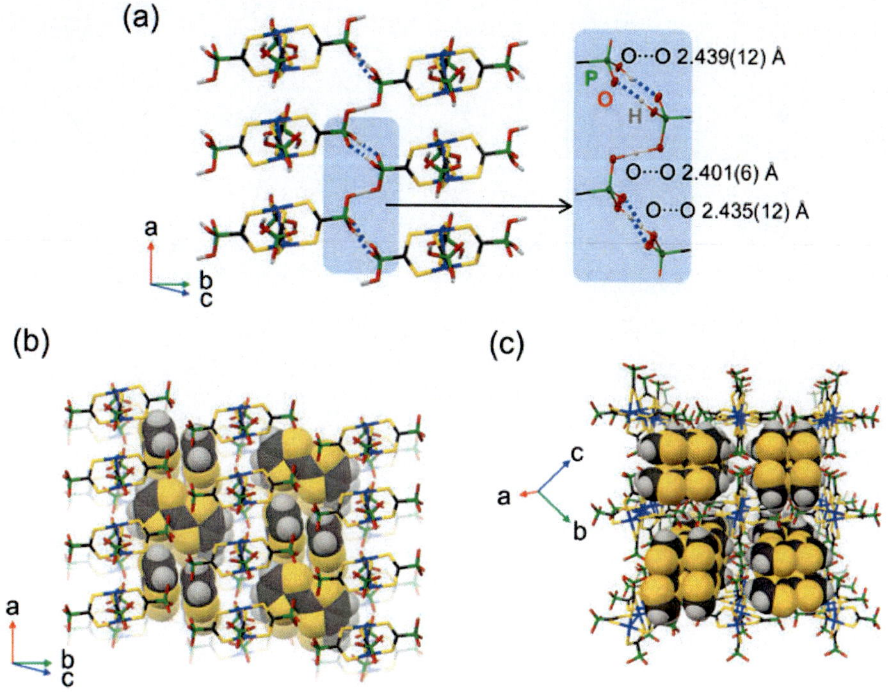

Fig. 2.6 Crystal structure of $(TTF)_2(\mathbf{1\text{-}H_{6+\delta}})$. **a** Side view of one-dimensional hydrogen-bonded network (left) and the detailed geometry of hydrogen bonds (right). For clarity, only $\mathbf{1\text{-}H_{6+\delta}}$ is shown. **b** Side view and **c** top view of the one-dimensional hydrogen-bonded networks and TTF arrays. Complex anion $\mathbf{1\text{-}H_{6+\delta}}$ and TTF are drawn as stick models and van der Waals radii models. Hydrogen atoms are omitted for clarity. Color code in **a**–**c**: C, black; H, white; O, red; P, green; Pt, blue; S, yellow

and a half hydrogen atoms (one attached and one shared); each complex anion therefore has six hydrogen atoms. However, the protonated state of **1** in salt is apparently non-stoichiometric, as discussed in the following section, and therefore, hereinafter, we represent the complex anion as $\mathbf{1\text{-}H_{6+\delta}}$ with the number of the attached hydrogen, $6 + \delta$. Thus, the formula for the salt is $(TTF)_2(\mathbf{1\text{-}H_{6+\delta}})$.

Because each complex anion $\mathbf{1\text{-}H_{6+\delta}}$ has four phosphonic acid moieties, it connects with eight neighboring $\mathbf{1\text{-}H_{6+\delta}}$ anions to form four hydrogen-bonded networks. In other words, the complex anion behaves as a node that connects parallelly aligned one-dimensional hydrogen-bonded networks to form a three-dimensional framework (Fig. 2.6b, c). This hydrogen-bonded three-dimensional framework bears one-dimensional channels (Fig. 2.6c) parallel to the one-dimensional hydrogen-bonded networks; the TTF molecules are confined in the channels (Fig. 2.6b and c). TTF molecules form face-to-face pairs with an interplanar distance of 3.77 Å (for top view of stacked two molecules; see Fig. 2.7), which is orthogonally arranged to construct a one-dimensional array. In the array, the intermolecular S···S distance is 3.70 Å at

Table 2.1 Cell parameters and crystallographic information for $(TTF)_2(\mathbf{1\text{-}H_{6+\delta}})$ and **2**

	$(TTF)_2(\mathbf{1\text{-}H_{6+\delta}})$	$(TTF)_2(\mathbf{1\text{-}H_{6+\delta}})$	$(TTF)_2(\mathbf{1\text{-}H_{6+\delta}})$	**2**
Formula	$Pt_2S_{16}C_{16}P_4O_{12}H_{14}$	$Pt_2S_{16}C_{16}P_4O_{12}H_{14}$	$Pt_2S_{16}C_{16}P_4O_{12}H_{14}$	$Pt_2S_8C_{20}P_4O_{12}H_{40}$
Formula weight	1425.31	1425.31	1425.31	1243.08
Temperature (K)	100	200	298	100
Wavelength (Å)	0.71075	0.71075	0.71075	0.71075
Crystal system	Monoclinic	Monoclinic	Monoclinic	Monoclinic
Space group	*I*2/*a*	*I*2/*a*	*I*2/*a*	*P*2/*n*
Color	Black	Black	Black	Yellow
a (Å)	12.8986(2)	12.9535(2)	12.9908(2)	27.5324(7)
b (Å)	17.3088(2)	17.3411(3)	17.3771(3)	11.7540(2)
c (Å)	17.3119(3)	17.3501(3)	17.3819(3)	27.6487(7)
α (°)	90	90	90	90
β (°)	100.3283(15)	100.4385(16)	100.5223(17)	118.122(3)
γ (°)	90	90	90	90
V (Å^3)	3802.43(10)	3832.82(11)	3857.86(12)	7891.3(4)
Z	4	4	4	8
Calcd density (g cm^{-3})	2.490	2.470	2.454	2.092
μ (mm^{-1})	8.451	8.384	8.329	7.719
F(000)	2712.00	2712.00	2712.00	4800.00
Crystal size (mm^3)	0.07 × 0.02 × 0.01	0.07 × 0.02 × 0.01	0.07 × 0.02 × 0.01	0.48 × 0.04 × 0.02
Total reflections	82,838	83,620	84,134	376,153

(continued)

Table 2.1 (continued)

	$(TTF)_2(\mathbf{1\text{-}H_{6+\delta}})$	$(TTF)_2(\mathbf{1\text{-}H_{6+\delta}})$	$(TTF)_2(\mathbf{1\text{-}H_{6+\delta}})$	**2**
Unique reflections	5357	5409	5468	22,660
Parameters used	231	231	231	852
R_{int}	0.0498	0.0553	0.0488	0.0728
Goodness-of-fit	1.256	1.249	1.212	1.048
R_1 $[I > 2\sigma(I)]$	0.0530	0.0591	0.0595	0.0504
wR_2 (all reflections)	0.1467	0.1651	0.1761	0.1107
max, min $\Delta\rho$ (e Å^{-3})	7.37, −6.52	7.92, −6.05	7.85, −6.47	5.90, −7.25
CCDC deposition nos	2,127,382	2,189,181	2,189,182	2,127,381

the shortest, and hence, there is no short intermolecular van der Waals contact (sum of van der Waals radii: 3.60 Å [33]).

In the TTF array, four types of molecular interactions (t_1, t_2, t_3, and t_4) were observed (Fig. 2.8). Calculations based on the extended Hückel method [29] gave moderate transfer integrals ($t_1 = -60.35$ meV, $t_2 = 35.17$ meV, $t_3 = -22.73$ meV, $t_4 = -2.284$ meV). Notably, the largest value was found not within the facing pairs (connected by the green line, t_2), but between the pairs (connected by the red line, t_1). Accordingly, the molecular assembly was composed of a one-dimensional array of π-conjugated molecules and an insulating matrix with a proton conduction pathway as designed (Fig. 2.1d).

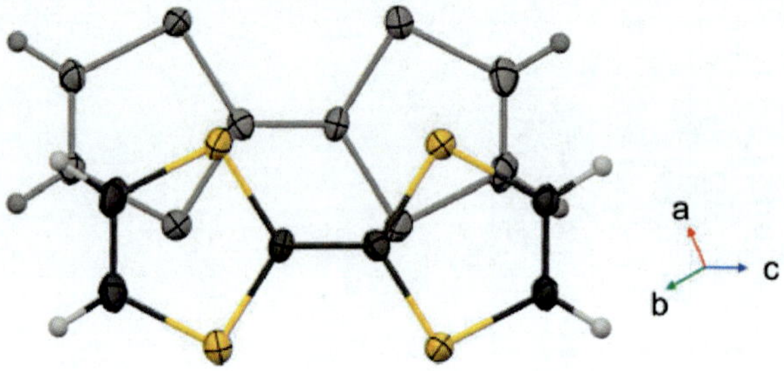

Fig. 2.7 Overlapping pattern of face-to-face TTF dimers in $(TTF)_2$(**1-H$_{6+\delta}$**). Gray and colored molecules are located at back and front, respectively

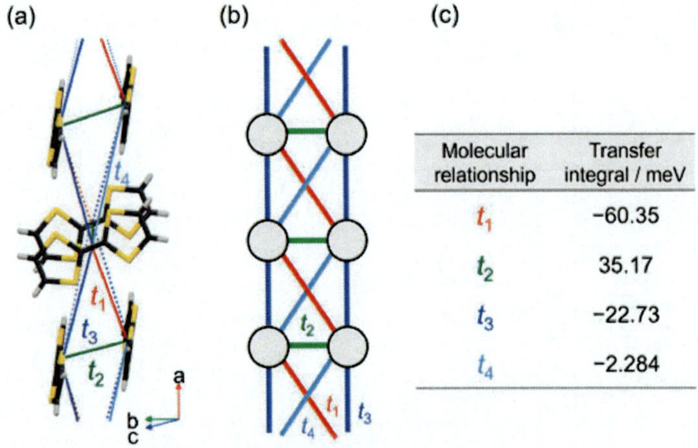

Fig. 2.8 Molecular arrangement and interactions in one-dimensional TTF array in $(TTF)_2$(**2-H$_{6+\delta}$**). Molecular interactions at the front and back are shown by solid and dotted lines, respectively

2.3.2 Valence States

To clarify the electronic state of the salt, ultraviolet–visible-near infrared (UV–vis-NIR) absorption spectroscopy was applied. The high-wavenumber region (> ~15,000 cm^{-1}) of the spectrum obtained for $(TTF)_2$(**1-H$_{6+\delta}$**) (Fig. 2.9a) was explained by combining the absorption bands attributed to intramolecular transitions of the TTF cation as seen in $(TTF)_3(BF_4)_2$ (Fig. 2.9c) and those attributed to intramolecular transitions derived from the central part of the Pt-dimer complex as seen in **2** (Fig. 2.9b). It should be noted that the absorption band at approximately 4000 cm^{-1} was derived from the electron transfer between components with different redox states [34, 35]. Here, two possibilities should be considered: electron transfer from divalent Pt to trivalent Pt, and from monovalent TTF to neutral TTF. Thus, it is apparent that either Pt or TTF is partially oxidized as a carrier source.

To determine the location of the carrier, we performed electron spin resonance (ESR) measurements. The spectra obtained at 4 and 298 K (Fig. 2.10) could be fitted using a single Lorentzian with *g*-factors of 2.0078 and 2.0080, respectively. Given that the isotropic *g*-factors of the oxidized TTF cations and Pt-dithiocarboxylate paddle-wheel complexes have been reported to be 2.00838 (for TTF·Cl [36]) and 2.129 (for $Pt_2(n\text{-}penCS_2)_4$ [37]), respectively; the obtained value was found to be relatively similar to that of oxidized TTF. Thus, it is expected that the carrier is generated in one-dimensional TTF arrays and that the Pt in the complex anions is divalent, that is, a closed shell. Notably, the bond length of intradimer Pt–Pt in **1-H$_{6+\delta}$** is 2.7567(7) Å, which is more similar to that of a similar divalent paddle-wheel complex (2.77 Å for $Pt^{II}_2(PhCH_2CS_2)_4$ [38, 39]) than that of a trivalent paddle-wheel complex (2.60 Å for

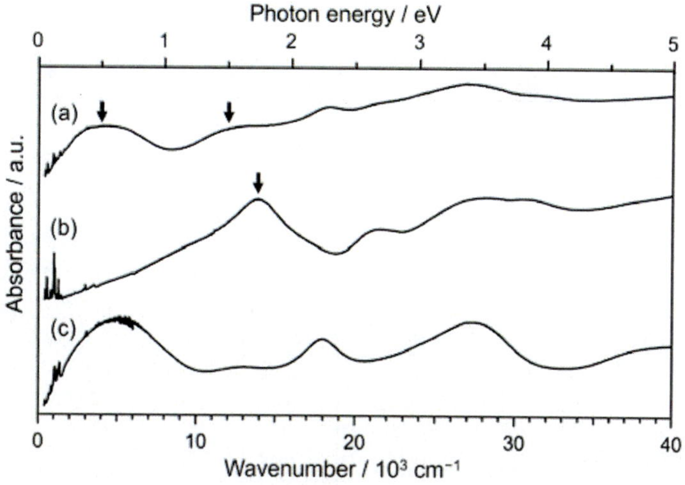

Fig. 2.9 UV–vis-NIR spectra of **a** $(TTF)_2$(**1-H$_{6+\delta}$**), **b** **2**, and **c** $(TTF)_3(BF_4)_2$. Black arrows indicate absorptions at approximately 4000, 12,000, and 14,000 cm^{-1} (refer to text)

$Pt^{III}{}_2(PhCH_2CS_2)_4$ [40]). In addition, the X-ray photoelectron spectrum (Pt 4*f*) could be explained by peaks originating from a single divalent component (Fig. 2.11).

To confirm the partially oxidized state of TTF, Raman spectrum was recorded (Fig. 2.12). It is known that the ν_3 mode is sensitive to the valence (γ) of the TTF (1516 cm^{-1} for neutral and 1420 cm^{-1} for monovalent [23]). We observed a distinct peak at 1441 cm^{-1} for $(TTF)_2(\mathbf{1\text{-}H_{6+\delta}})$, indicating that TTF is in a partially oxidized state. From the frequency–valence relationship derived by measuring several TTF salts with different valence states (Fig. 2.12b), we obtained the valence of TTF in $(TTF)_2(\mathbf{1\text{-}H_{6+\delta}})$ as $+\ 0.73 \pm 0.03$. Moreover, the valence estimated based on the intramolecular bond lengths (C–S bonds near the central C= C bond; Fig. 2.13) is $+\ 0.88 \pm 0.19$, which covers the value estimated by Raman spectroscopy [41]. Thus, it is obvious that the absorption band at approximately 4000 cm^{-1} in the UV–vis-NIR spectrum can be assigned to a charge-transfer band between TTF° and TTF^{+} molecules (so-called A-band) [34]. In addition, the band at approximately

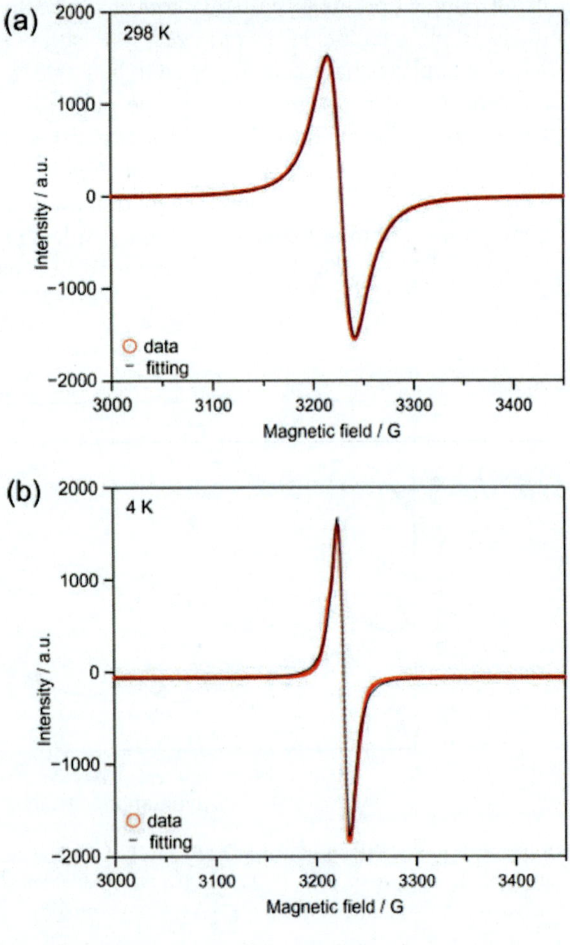

Fig. 2.10 Electron spin resonance spectrum of $(TTF)_2(\mathbf{1\text{-}H_{6+\delta}})$ at **a** 298 K and **b** 4 K. Red circles and black line are the experimental data and fitting curve, respectively. The *g*-factors at 298 K and 4 K are 2.0078 and 2.0080, respectively. Peak-to-peak line widths (ΔH_{pp}) at 298 K and 4 K are 26.37 and 10.9 G, respectively

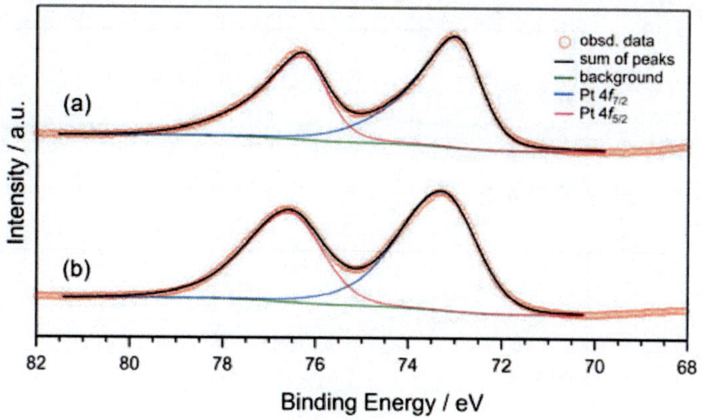

	Pt $4f_{5/2}$	Pt $4f_{7/2}$
$(TTF)_2$(**1-H$_{6+\delta}$**)	76.299 (1.244)	72.999 (1.244)
2	76.578 (1.696)	73.278 (1.696)

Fig. 2.11 (top) Pt 4*f* XPS spectra of **a** **2** and **b** $(TTF)_2$(**1-H$_{6+\delta}$**). (bottom) Binding energies (eV) for Pt 4*f*. Values in parentheses correspond to the full width at half-maximum (FWHM) of each peak

12,000 cm^{-1} can be attributed to a charge-transfer band between TTF^+ molecules (so-called B-band), which is observable in the range of $+0.5 < \gamma < +1$ [42], although the Frenkel exciton originating from one-dimensionally stacked paddle-wheel type Pt complexes [38, 39], which was clearly observed in **2** at ~14,000 cm^{-1} (refer to Fig. 2.14 for the molecular packing of **2**), could overlap with the absorption band.

Given that the Pt in the complex anion is divalent and the ratio of TTF to the complex anion is 2:1, there should be additional protons (δ) outside the hydrogen bonds to balance the charge neutrality; hence, the formula should be $[TTF^{(1-\delta/2)+}]_2$[**1-H$_{6+\delta}$**$^{(2-\delta)-}$] instead of $[TTF^+]_2$[**1-H$_6$**$^{2-}$]. Considering that γ lies in the range of $+0.5 < \gamma < +1$, δ must satisfy $0 < \delta < 1$. Notably, this assumption does not conflict with the elemental analysis results. The expected weight percentages of (C, H, N) for $(TTF)_2$(**1-H$_6$**) and $(TTF)_2$(**1-H$_7$**) are (13.48, 0.99, 0) and (13.45, 1.20, 0), respectively. Both values are consistent with the observed values (13.30, 1.05, 0).

The positions of the additional protons could not be determined via SCXRD, but it is possible that they are attached to the lone pairs of electrons on the oxygen atoms of the phosphonic acids, which are not used for hydrogen-bonding networks.

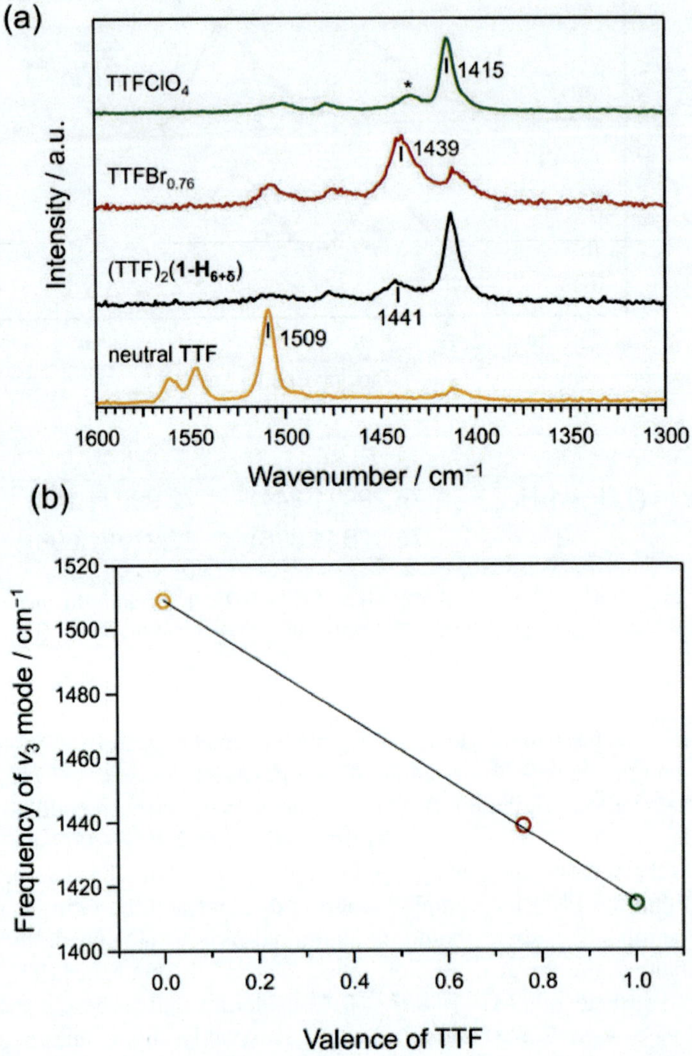

Fig. 2.12 **a** Raman spectra of $TTFClO_4$, $TTFBr_{0.76}$, $(TTF)_2(\mathbf{1\text{-}H_{6+\delta}})$, $(TTF)_3(BF_4)_2$, and neutral TTF (from top to bottom). Black short lines show the peak tops of ν_3 modes obtained by fitting. The numbers near the lines show the frequencies of the peak tops. Asterisk at the spectrum of $TTFClO_4$ is probably ascribable to the ν_3 mode of contaminated partially reduced TTF^+. **b** Relationship between the frequency of the ν_3 mode of TTF and the valence of TTF. Circles are experimental values obtained from **a**. Black line is the fitting line of $y = a + bx$, where $a = 1509.2 \pm 1.3\ cm^{-1}$ and $b = -93.8 \pm 1.9\ cm^{-1}$. The results are consistent with the literature [23]

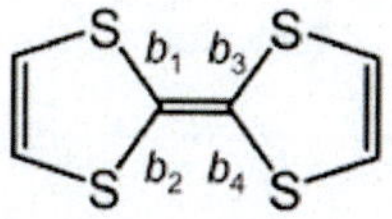

$$\text{Valence} = \frac{|1.757 - (b_1+b_2+b_3+b_4)/4|}{0.0385}$$

b_1 / Å	b_2 / Å	b_3 / Å	b_4 / Å	Valence
1.708(12)	1.729(13)	1.718(12)	1.735(12)	0.88(19)

Fig. 2.13 Estimation of valence of TTF in $(TTF)_2$(**1-H$_{6+\delta}$**). Valence was estimated from the b_1, b_2, b_3, and b_4 lengths (top left) using an equation from the literature [41] (top right). The values are summarized in the bottom table

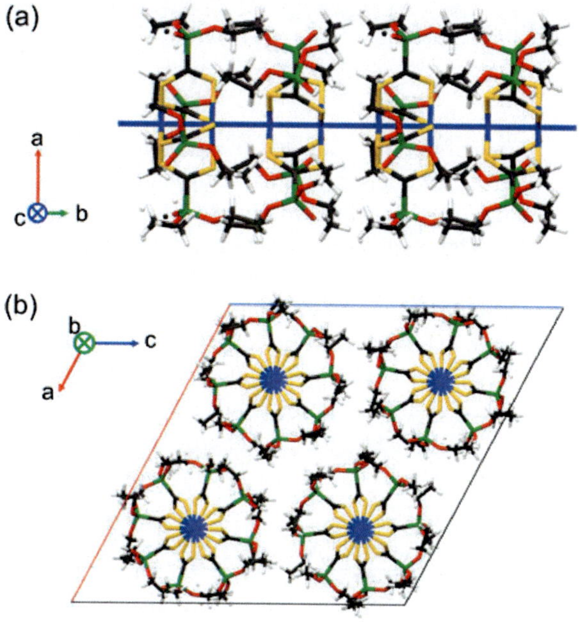

Fig. 2.14 Crystal structure of **2**, which is constructed from parallelly aligned infinite chain structures consisting of stacked molecules. **a** Side view of four molecules extracted from one chain structure. **b** Top view of packed infinite chain structures. Color code: C, black; H, white; O, red; P, green; Pt, blue; S, yellow

2.3.3 Physical Properties

To evaluate the itinerancy of electrons on the TTF molecules, we investigated the electronic conductivity. DC conductivity measurements were performed on the rod-like crystal using the standard four-probe method with the current flowing along the TTF column direction. The sample exhibited moderate room-temperature conductivity ($\sigma_e = 1.3 \times 10^{-4}$ S cm^{-1}) and Arrhenius-type activation behavior upon cooling (activation energy $E_a = 0.13$ eV, Fig. 2.15). This indicates that the electrons on the TTF are localized because of strong Coulomb interactions [43]. Indeed, we observed large paramagnetic susceptibility, as demonstrated below.

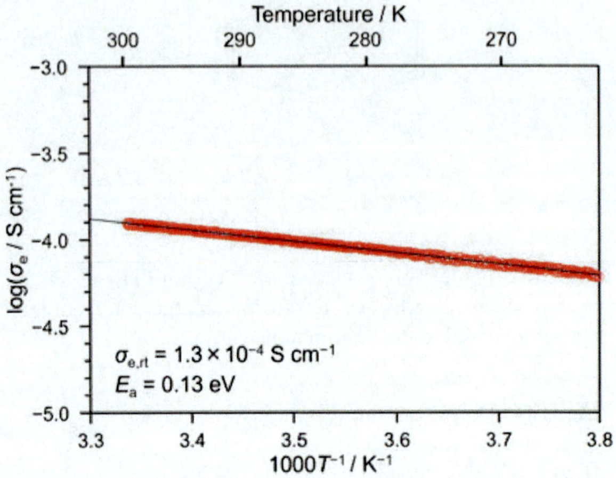

Fig. 2.15 Arrhenius plot of the temperature dependence of the electronic conductivity (σ_e) of $(TTF)_2(\mathbf{1\text{-}H_{6+\delta}})$

Thereafter, we evaluated the proton conductivity of $(TTF)_2(\mathbf{1\text{-}H_{6+\delta}})$ using the pelletized sample sandwiched with Nafion membranes as the electron-blocking electrode [13, 14, 44] (refer to Fig. 2.16 for the result of the experiment on only Nafion films). The Nyquist plots obtained from AC impedance measurements at different temperatures (25–45 °C; Fig. 2.17a) under 40%RH were well fitted by assuming a combination of the sample intrinsic (proton) conductivity and the sample/electrode interface. The obtained proton conductivity at 25 °C was moderate ($\sigma_H = 2.8 \times 10^{-6}$ S cm^{-1}) and the conductivity exhibited Arrhenius-type activation (Fig. 2.17b, $E_a = 0.63$ eV). This proton conductivity is readily attributed to the hydrogen-bonded networks formed by the phosphonic acid moieties.

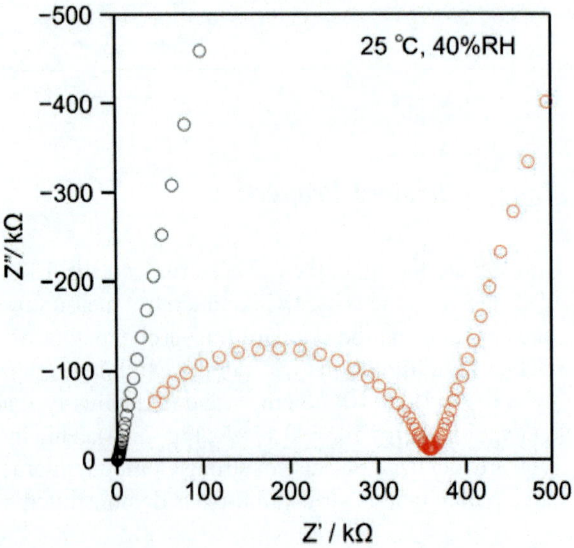

Fig. 2.16 Impedance spectra of $(TTF)_2(\mathbf{1\text{-}H_{6+\delta}})$ sandwiched by Nafion films (red circles) and Nafion films alone (black circles) at 25 °C under 40%RH

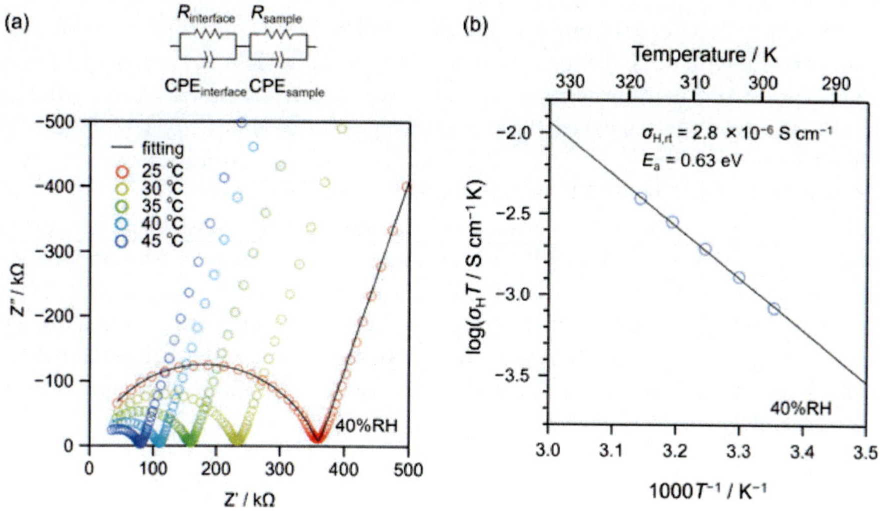

Fig. 2.17 **a** Nyquist plot of the impedance spectra of $(TTF)_2$(**1-H$_{6+\delta}$**) under 40%RH. The equivalent circuit used to model the plot is illustrated on top, where R_{sample} ($R_{interface}$) and CPE_{sample} ($CPE_{interface}$) are the intrinsic (sample/electrode interface) resistance and constant-phase element, respectively. **b** Arrhenius plot of the temperature dependence of the product of proton conductivity (σ_H) and temperature (T) under 40%RH

Notably, ceramics and organic polymers have been studied as typical mixed ionic–electronic conductors [45]. Among them, ceramics have a great advantage because their structure–property relationships can be investigated owing to their crystalline nature, which is also the case for $(TTF)_2$(**1-H$_{6+\delta}$**). At approximately room temperature, typical ceramic mixed conductors exhibit an ionic conductivity of approximately 10^{-5}–10^{-2} S cm^{-1} and an electronic conductivity of approximately 10^{-8}–10^{-5} S cm^{-1} [46–49]. The former is higher than the room-temperature proton conductivity of $(TTF)_2$(**1-H$_{6+\delta}$**) ($\sigma_H = 2.8 \times 10^{-6}$ S cm^{-1}), whereas the latter is lower than the room-temperature electronic conductivity of $(TTF)_2$(**1-H$_{6+\delta}$**) ($\sigma_e = 1.3 \times 10^{-4}$ S cm^{-1}). In addition, the design of the unique structure and conduction pathway addressed herein is based on molecular designability, which is not available for ceramics. It should be noted that the proton conductivity of $(TTF)_2$(**1-H$_{6+\delta}$**) ($\sigma_H = 2.8 \times 10^{-6}$ S cm^{-1} at room temperature) is relatively low when compared to the values for typical ceramic phosphate-based proton conductors such as $Zr(HPO_4)\cdot H_2O$ ($\sigma_H = 4 \times 10^{-5}$ S cm^{-1} at 20 °C) [50, 51]. One possible reason for the lower proton conductivity is the lower concentration of Brønsted acidic/basic moieties as proton hopping sites (i.e., phosphonic acid/phosphate moieties and water molecules); the concentrations of –OH moieties were calculated as 0.0062 and 0.022 Å^{-3} for $(TTF)_2$(**1-H$_{6+\delta}$**) and $Zr(HPO_4)\cdot H_2O$, respectively, based on their crystal structures [52].

Finally, the magnetic susceptibility was investigated using a superconducting quantum interference device (SQUID). The sample contained a very small amount of ferromagnetic impurity, the saturation magnetic moment of which was almost temperature-independent in the range of 2–298 K [53]. Thus, after confirming the linear relation between the magnetization and magnetic field in the range of 2 T and 3 T (Fig. 2.18), the intrinsic susceptibility was estimated based on the difference between magnetization data at 2 and 3 T, i.e., $\chi = [M(3\ \text{T}) - M(2\ \text{T})/1\ \text{T}]$. Figure 2.19 depicts the temperature dependence of molar spin susceptibility (χ_{spin}), which is defined for a formula of $(\text{TTF})_2(\mathbf{1\text{-}H_6})$, after subtracting the contribution of the core diamagnetism (-5.47×10^{-4} emu mol^{-1}) using Pascal's constants. The susceptibility was 2.14×10^{-3} emu mol^{-1} at 298 K and showed a broad maximum at approximately 70 K. Below approximately 20 K, a steep increase in χ_{spin}; that is, a Curie tail was observed. The temperature dependence of χ_{spin} originating from the sample was well fitted by the $S = 1/2$ antiferromagnetically coupled alternating chain model, which assumes alignment of the spins in one dimension while alternately coupling with coupling constants J and αJ ($\alpha < 1$) [54, 55];

$$\chi_{spin} = \frac{N_{spins} g^2 \beta^2}{3kT} \frac{A + B\left(\frac{|J|}{kT}\right) + C\left(\frac{|J|}{kT}\right)^2}{1 + D\left(\frac{|J|}{kT}\right) + E\left(\frac{|J|}{kT}\right)^2 + F\left(\frac{|J|}{kT}\right)^3},$$

where N_{spins} is the number of spins per 1 mol of formula $(\text{TTF})_2(\mathbf{1\text{-}H_6})$, g is the g-factor, β is the Bohr magneton, and k is Boltzmann's constant. A–F are constants containing the power of α. In the fitting, we fixed the g-factor as 2.0079 according to the ESR results. The obtained parameters are $N_{\text{spins}} = 1.986(9)$ mol, $J/k = -57.7(2)$ K, and $\alpha = 0.23(2)$.

The fact that N is nearly equal to 2 indicates that approximately one spin is localized on each TTF molecule and itinerant electrons are likely to be absent. In addition, the obtained α values could be explained by the crystal structure. The

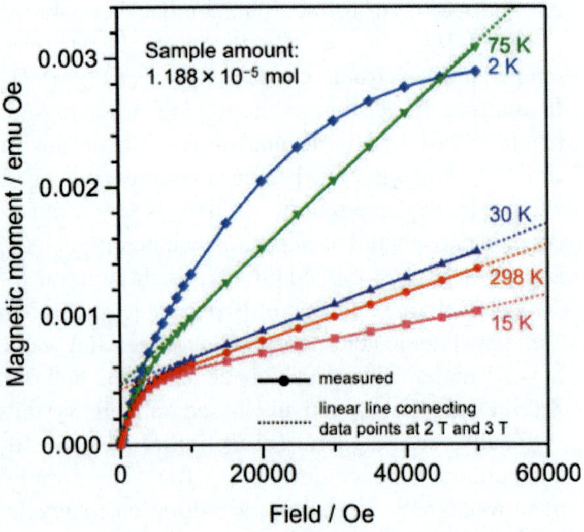

Fig. 2.18 *M–H* curves for $(\text{TTF})_2(\mathbf{1\text{-}H_{6+\delta}})$ at 2 K (blue), 15 K (pink), 30 K (purple), 75 K (green), and 298 K (red)

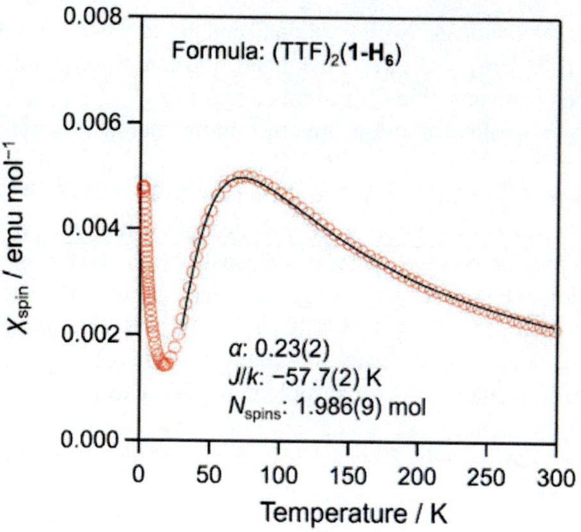

Fig. 2.19 Temperature dependence of the spin susceptibility of $(TTF)_2$(**1-H**$_{6+\delta}$). Open red circles: measured data. Solid black line: fitting line of $S = 1/2$ antiferromagnetically coupled alternating chain model (see text)

magnetic interaction of spins on molecules is proportional to t^2/U, where t is the transfer integral and U is the on-site Coulomb repulsion [56]. The ratio of the squares of the largest and the second largest t values obtained by our calculation, $t_2{}^2/t_1{}^2 = 0.34$, was comparable to $\alpha = 0.23$ obtained by the susceptibility experiment. This magnetic behavior demonstrates the one-dimensional nature of the TTF array.

2.4 Conclusion

In this study, we reported the rational construction of molecular electron-conducting nanowires encapsulated in a proton-conducting matrix in a charge-transfer salt. We described the synthesis and investigation of valence states and physical properties of a TTF cation radical salt with a propeller-shaped hydrogen-bonding complex anion.

SCXRD revealed that the molecular assembly comprises a three-dimensional hydrogen-bonded framework of the complex anion and one-dimensional TTF arrays (i.e., molecular wires) confined within it. From UV–vis-NIR absorption spectroscopy, ESR spectrometry, Raman spectroscopy, and intramolecular bond length analysis, it was clarified that the valence of Pt in the complex anions is divalent and that TTF is partially oxidized. Notably, the non-stoichiometric deprotonation compensates for charge neutrality; hence, the formula can be represented as $(TTF)_2$(**1-H**$_{6+\delta}$). The semiconducting transport behavior and singlet–triplet type magnetic behavior are well explained by the localized electrons in the one-dimensional TTF arrays. On the other hand, the hydrogen-bonding ability of the complex anions is responsible for proton conduction.

The single-crystalline mixed-conducting molecular wire system constructed in this study may pave the way for the development of iono-electronics. It should be noted that the expected electronic filling control via control of the amount of non-stoichiometric deprotection is advantageous in this context. In addition, the design principle adopted here, namely, the construction of a hollow framework with strong intermolecular interactions (e.g., hydrogen bonds) and the arrangement of π-conjugated molecules with weak intermolecular interactions (e.g., π–π interaction) within it, is an effective way to design the arrangement of a wide range of π-conjugated molecules, which can potentially lead to the realization of various desired electronic states.

This chapter was reproduced from Ref. [57] with permission from the American Chemical Society.

References

1. Frampton MJ, Anderson HL (2007) Angew Chem Int Ed 46:1028–1064
2. Cardin DJ (2002) Adv Mater 14:553–563
3. Yamamoto HM (2014) CrystEngComm 16
4. Li C, Numata M, Bae AH, Sakurai K, Shinkai S (2005) J Am Chem Soc 127:4548–4549
5. Taylor PN, O'Connell MJ, McNeill LA, Hall MJ, Aplin RT, Anderson HL (2000) Angew Chem Int Ed 39:3456–3460
6. Cacialli F, Wilson JS, Michels JJ, Daniel C, Silva C, Friend RH, Severin N, Samori P, Rabe JP, O'Connell MJ, Taylor PN, Anderson HL (2002) Nat Mater 1:160–164
7. Nguyen TQ, Wu J, Doan VV, Schwartz BJ, Tolbert SH (2000) Science 288:652–656
8. MacLean MW, Kitao T, Suga T, Mizuno M, Seki S, Uemura T, Kitagawa S (2016) Angew Chem Int Ed 55:708–713
9. Yamamoto HM, Kosaka Y, Maeda R, Yamaura J, Nakao A, Nakamura T, Kato R (2008) ACS Nano 2:143–155
10. Lehn J-M (1990) Angew Chem Int Ed 29:1304–1319
11. van Nostrum CF, Picken SJ, Schouten A-J, Nolte RJM (2002) J Am Chem Soc 117:9957–9965
12. Chen J, Sun Y, Zhao W, Liu J, Fang J, Xu T, Chen D (2021) J Mater Chem C 9:3871–3881
13. Akutagawa T, Hasegawa T, Nakamura T, Inabe T, Saito G (2002) Chem Eur J 8:4402–4411
14. Akutsu-Sato A, Akutsu H, Turner SS, Day P, Probert MR, Howard JAK, Akutagawa T, Takeda S, Nakamura T, Mori T (2005) Angew Chem Int Ed 117:296–299
15. Kobayashi Y, Yoshioka M, Saigo K, Hashizume D, Ogura T (2009) J Am Chem Soc 131:9995–10002
16. Kobayashi Y, Fujii T, Terasaki I, Kino H, Jin Y, Hibino T, Kobayashi T, Nishibori E, Sawa H, Yoshikawa H, Terauchi T, Sumi S (2013) J Mater Chem A 1:5089–5096
17. Ishiguro T, Yamaji K, Saito G (1998) In: Organic superconductors, 2nd ed. Springer, Heidelberg
18. Steed JW, Turner DR, Wallace K (eds) (2007) In: Core concepts in supramolecular chemistry and nanochemistry, Wiley, West Sussex
19. Cavallo G, Metrangolo P, Milani R, Pilati T, Priimagi A, Resnati G, Terraneo G (2016) Chem Rev 116:2478–2601
20. Steiner T (2002) Angew Chem Int Ed 41:48–76
21. Anderson GK, Lin M (1990) Inorg Synth 28:60–63

22. Wudl F (2002) J Am Chem Soc 97:1962–1963
23. Matsuzaki S, Moriyama T, Toyoda K (1980) Solid State Commun 34:857–859
24. Miles MG, Wilson JD (2002) Inorg Chem 14:2357–2360
25. Grisley JD (2003) J Org Chem 26:2544–2546
26. Sheldrick GM (2014) Acta Cryst A70:C1437. SHELXT Version 2014/5
27. Sheldrick GM (2008) Acta Cryst A64:112–122. SHELXL Version 2014/3
28. Rigaku (2018) Crystal structure. Version 4.3. Rigaku Corporation, Tokyo, Japan
29. Mori T, Kobayashi A, Sasaki Y, Kobayashi H, Saito G, Inokuchi H (1984) Bull Chem Soc Jpn 57:627–633
30. Johnston DC, Troyer M, Miyahara S, Lidsky D, Ueda K, Azuma M, Hiroi Z, Takano M, Isobe M, Ueda Y, Korotin MA, Anisimov VI, Mahajan AV, Miller LL (2000). arXiv:con-mat/0001147
31. X-ray structural analysis of a twinned crystal demonstrated that although each crystallite is properly aligned in the *a*-axis direction (direction of the TTF column), it is misaligned in the *b*- and *c*-axis directions (refer to the main text for the crystal structure and crystallographic axes)
32. Grabowski SJ (ed) (2006) In: Hydrogen bonding—new insights. Springer, Dordrecht
33. Bondi A (1964) J Phys Chem 68:441–451
34. Torrance JB, Scott BA, Welber B, Kaufman FB, Seiden PE (1979) Phys Rev B 19:730–741
35. Mitsumi M, Ueda H, Furukawa K, Ozawa Y, Toriumi K, Kurmoo M (2008) J Am Chem Soc 130:14102–14104
36. Wudl F, Smith GM, Hufnagel EJ (1970) J Chem Soc Chem Commun 1453–1454
37. Tanaka H, Kuroda S-I, Yamashita T, Mitsumi M, Toriumi K (2003) J Phys Soc Jpn 72:2169–2172
38. Bellitto C, Flamini A, Piovesana O, Zanazzi PF (1980) Inorg Chem 19:3632–3636
39. Bellitto C, Dessy G, Fares V, Flamini A (1981) J Chem Soc Chem Commun 409–411
40. Bellitto C, Bonamico M, Dessy G, Fares V, Flamini A (1986) J Chem Soc Dalton Trans 595–601
41. Clemente DA, Marzotto A (1996) J Mater Chem 6:941–946
42. Sugano T, Yakushi K, Kuroda H (1978) Bull Chem Soc Jpn 51:1041–1046
43. Saito G, Yoshida Y (2007) Bull Chem Soc Jpn 80:1–137
44. Kimura Y, Yoshida Y, Tanaka Y, Maesato M, Komatsu T, Kitagawa H (2022) Inorg Chem 61:4453–4458
45. Malti A, Edberg J, Granberg H, Khan ZU, Andreasen JW, Liu X, Zhao D, Zhang H, Yao Y, Brill JW, Engquist I, Fahlman M, Wagberg L, Crispin X, Berggren M (2016) Adv Sci 3:1500305
46. Kamaya N, Homma K, Yamakawa Y, Hirayama M, Kanno R, Yonemura M, Kamiyama T, Kato Y, Hama S, Kawamoto K, Mitsui A (2011) Nat Mater 10:682–686
47. Wada H, Amiel O, Sato A (1995) J Alloys Compd 219:55–58
48. Koebel M, Ibl N, Freit AM (1974) Electrochim Acta 19:287–295
49. Inaguma Y, Liquan C, Itoh M, Nakamura T, Uchida T, Ikuta H, Wakihara M (1993) Solid State Commun 86:689–693
50. Clearfield A, Smith GD (1969) Inorg Chem 8:431–436
51. Krogh Andersen E, Krogh Andersen IG, Knkkergardård Møller C, Simonsen KE, Skou E (1982) Solid State Ionics 7:301–306
52. The –OH concentration for $(TTF)_2$(**1-H**$_{6+\delta}$) was calculated assuming the formula $(TTF)_2$(**1-H**$_6$) and the cell volume at 298 K (Table 2.1)
53. The saturation magnetic moment of ferromagnetic component was ~40 emu Oe mol^{-1}, where one mole is defined as a gram-formula weight of $(TTF)_2$(**1-H**$_6$)
54. Hall JW, Marsh WE, Weller RR, Hatfield WE (1981) Inorg Chem 20:1033–1037
55. A comparison between the fitting results using the alternating chain model and the two-leg ladder model is presented in Section 2.2 (Figs. 2.2 and 2.3)
56. Mori T (2016) In: Electronic properties of organic conductors, Springer Japan, Tokyo, pp 198–204
57. Donoshita M, Yoshida Y, Maesato M, Kitagawa H (2022) J Am Chem Soc 144:17149–17155

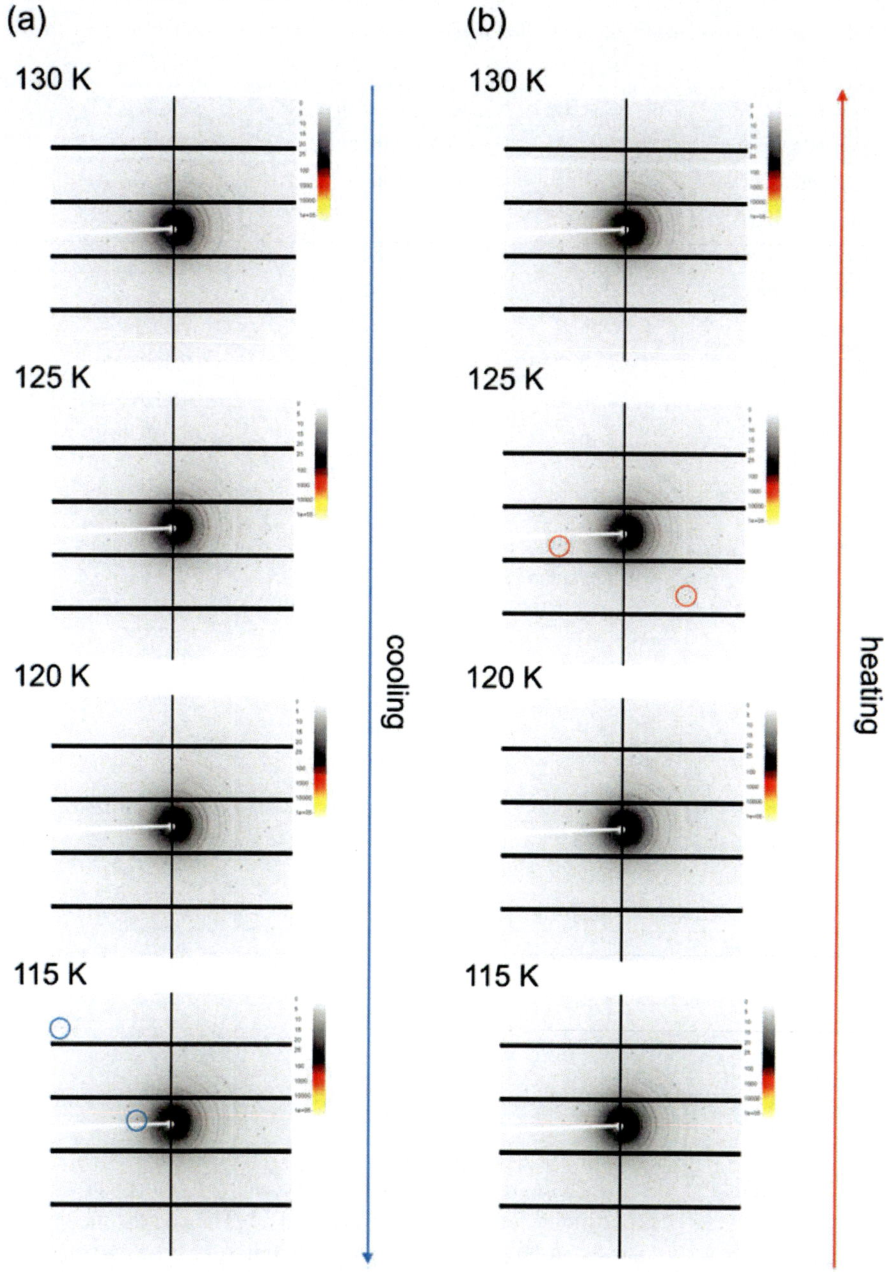

Fig. 4.6 Change in diffraction spots owing to the High-A ↔ Low-A transition of **PyCA** upon **a** cooling from 298 to 100 K and **b** subsequent heating from 100 to 298 K, showing thermal hysteresis. Blue and red circles indicate emergent spots of Low-A and High-A, respectively

Table 4.1 Cell parameters and crystallographic information for PyFA and Low-A of PyCA

	PyFA	PyFA	PyCA (Low-A)
Formula	C_{14} H_{16} F_2 N_2 O_6	C_{14} H_{16} F_2 N_2 O_6	C_{14} H_{16} Cl_2 N_2 O_6
Formula weight	346.29	346.29	379.20
Temperature (K)	298	100	298
Wavelength (Å)	0.71075	0.71075	0.7030
Crystal system	Triclinic	Triclinic	Triclinic
Space group	$P-1$	$P-1$	$P-1$
Color	Orange	Orange	Orange
a (Å)	5.07820(10)	5.0659(2)	7.3693(8)
b (Å)	8.5928(2)	8.4193(4)	10.3919(11)
c (Å)	9.4682(2)	9.1720(2)	11.7970(12)
α (°)	70.342(2)	72.275(3)	98.405(7)
β (°)	79.433(2)	81.444(3)	106.585(8)
γ (°)	81.633(2)	82.604(3)	108.916(8)
V (Å^3)	380.905(15)	367.03(2)	790.19(16)
Z	1	1	2
Calcd density (g cm^{-3})	1.510	1.567	1.594
μ (mm^{-1})	0.133	0.138	0.427
$F(000)$	180	180	392
Crystal size (mm^3)	0.26 × 0.24 × 0.10	0.26 × 0.24 × 0.10	0.09 × 0.04 × 0.01
Total reflections	16,804	16,225	10,442
Unique reflections	1953	1894	3594
Parameters used	110	120	219
R_{int}	0.0508	0.0677	0.0622
Goodness-of-fit	1.111	1.117	1.021
R_1 [$I > 2\sigma(I)$]	0.0598	0.0422	0.0475
wR_2 (all reflections)	0.1921	0.1333	0.1321
max, min $\Delta\rho$ (e Å^{-3})	0.42, −0.43	0.57, −0.60	0.58, −0.44
CCDC deposition nos	2,076,400	2,076,401	2,076,399

To understand the energetic relationships between the polymorphs, variable-temperature PXRD experiments were performed (Fig. 4.9). The coexistence of irreversible High-A → Low-B → High-B and reversible High-A ↔ Low-A transitions were observed upon cooling and subsequent heating (298 K → 100 K → 298 K). Notably, the transition temperature of the Low-B → High-B transition upon heating (190 K) was higher than that of the High-A → Low-B transition upon cooling (170 K), implying that the Gibbs energy–temperature curves for High-A and Low-B do not intersect. Given that High-B does not exhibit a phase transition upon cooling [48], an energy diagram of the four polymorphs can be drawn, as shown in Fig. 4.10.

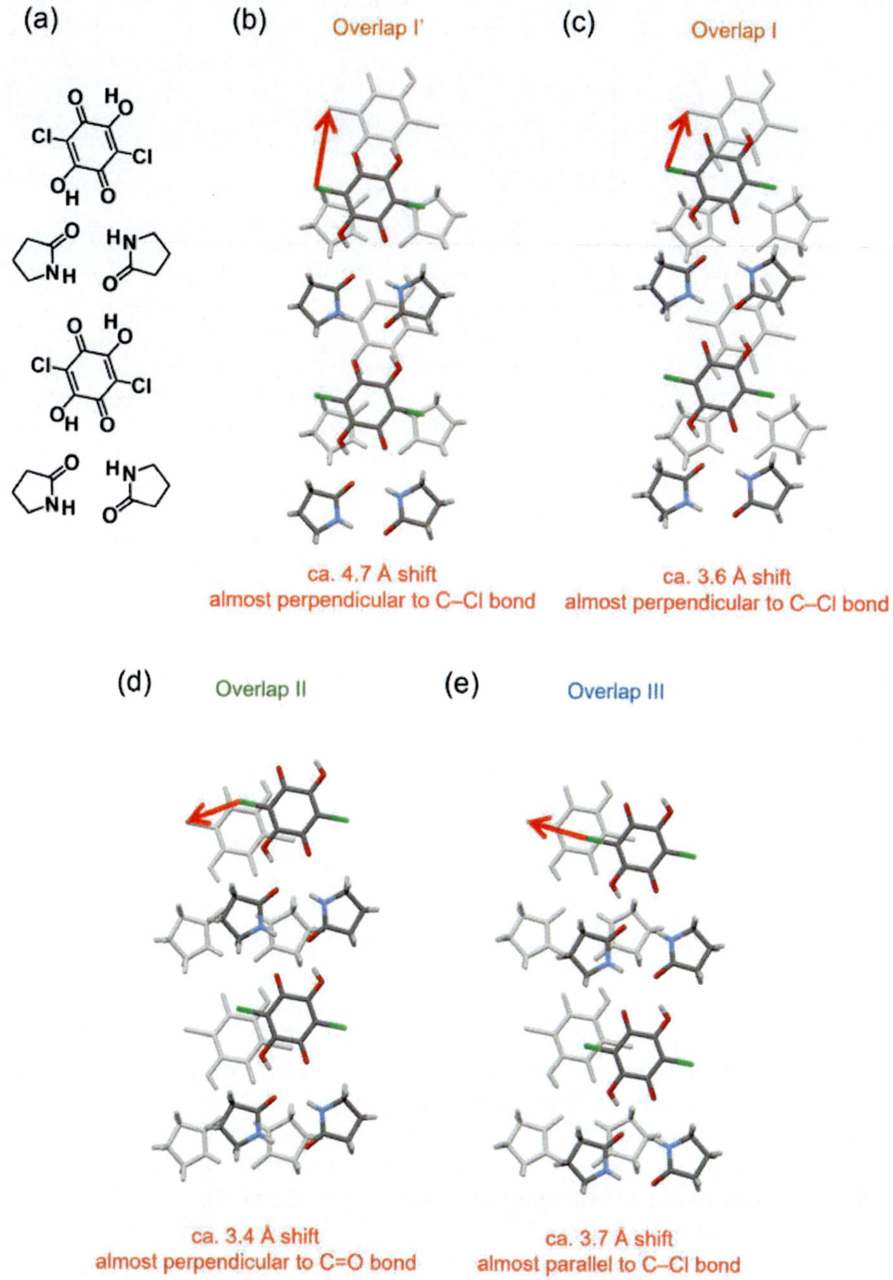

Fig. 4.7 **a** Schematic molecular arrangement in a tape and face views of two stacked tapes of **PyCA** in **b** Overlap I′ in High-A at 298 K, **c** Overlap I in Low-A at 100 K, **d** Overlap II in Low-B at 100 K, and **e** Overlap III in High-B at 298 K. Colored molecules (gray, C; white, H; purple, N; red, O; green, Cl) are placed in front of gray molecules. Red arrows in **b–e** show the translational shift between adjacent tapes

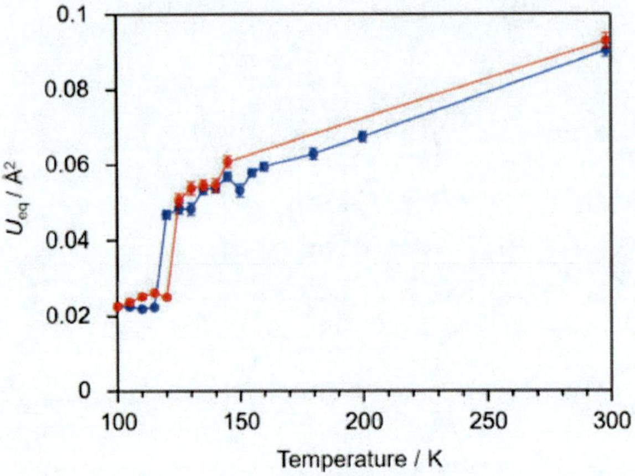

Fig. 4.8 Temperature dependence of equivalent isotropic atomic displacement parameter (U_{eq}) of 4-positioned carbon of Py molecules with conformational disorder upon High-A ↔ Low-A transition. Red and blue circles represent data obtained during cooling and heating processes, respectively. High-A → Low-A and Low-A → High-A transitions occur at approximately 120 K on cooling and heating, respectively

The disorder of Py can be regarded as one of the possible origins of the trapping of metastable High-A, which affords two series of transitions. As the disorder of Py disappears upon cooling, High-A can transform into the more thermodynamically stable Low-B, with a significant change in the stacking patterns from Overlaps I′ to II ((i) in Figs. 4.5 and 4.10). However, this change needs to be accompanied by a large translation of the sheets (~7 Å); therefore, some of the High-A crystals may fail to transform into Low-B, and instead transform into Low-A at the intersection of the energy curves of High-A and Low-A ((ii) in Figs. 4.5 and 4.10). The factor determining which crystals have a Low-A or Low-B structure at low temperatures is unclear at this stage, although the crystal quality may be an important factor because lattice imperfections, such as defects, might preferentially disturb the irreversible transition accompanied by the significant translation of 2D sheets.

Accordingly, we can conclude that High-A, Low-B, and Low-A are all metastable states and that the only stable phase is High-B.

4.3.1.2 Nature of the Intersheet Intermolecular Interactions

To investigate the nature of the intersheet interactions, we calculated the surface electrostatic potential of each component molecule at the MP2/aug-cc-pVDZ level of theory (Fig. 4.11, top). The central benzene ring in CA is positively charged (Fig. 4.11b). In contrast, the O and Cl atoms at the periphery were negatively charged, whereas the H atoms are positively charged, as expected from their atomic electronegativities. For Py (Fig. 4.11e), the O and N atoms are negatively charged, whereas the C and H atoms are positively charged. As mentioned above, SCXRD measurements

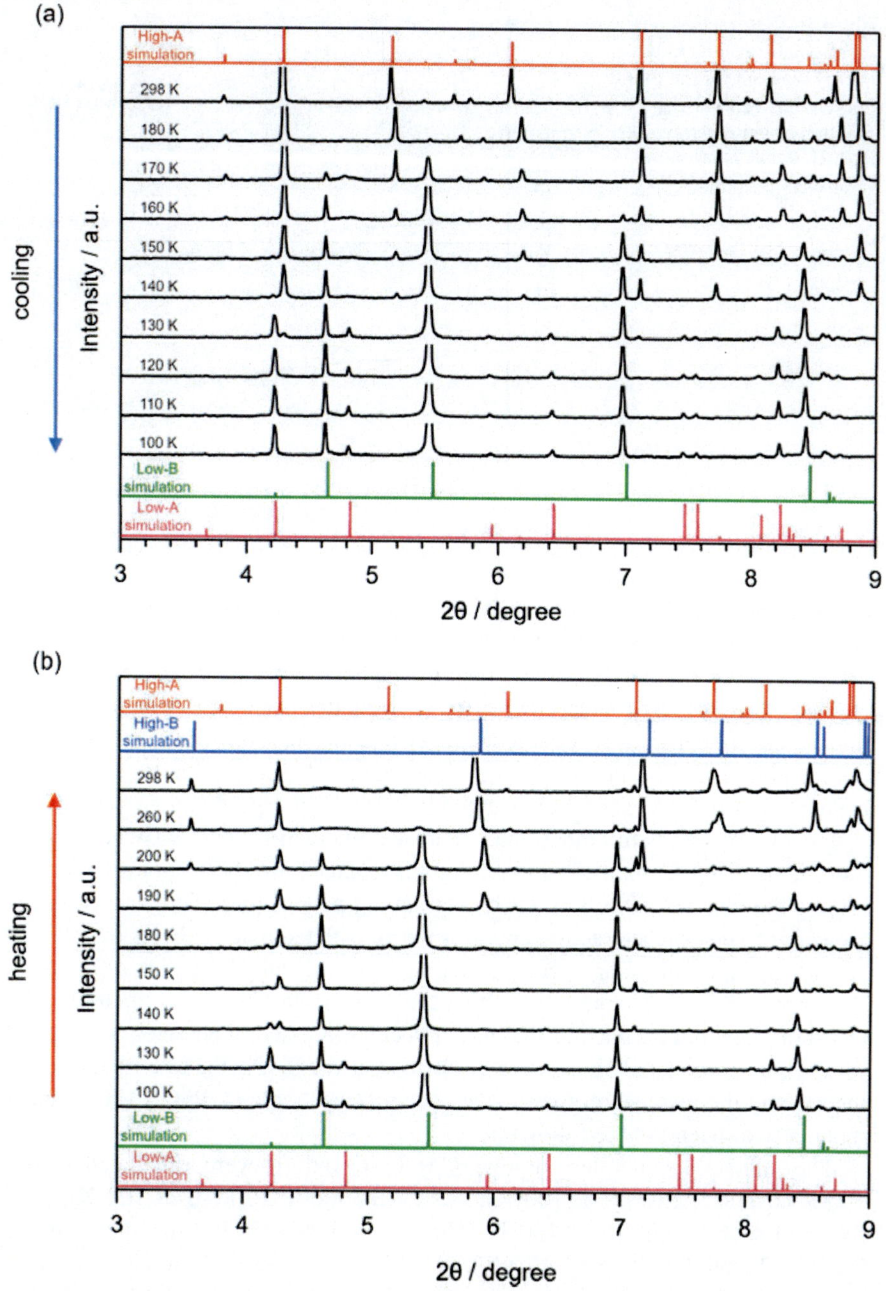

Fig. 4.9 Variable-temperature powder X-ray diffraction patterns of High-A sample upon **a** first cooling from 298 to 100 K and **b** subsequent heating from 100 to 298 K. Simulation patterns of High-A (298 K), Low-A (100 K), Low-B (100 K), and High-B (298 K) are calculated based on the crystal structures determined by single-crystal XRD

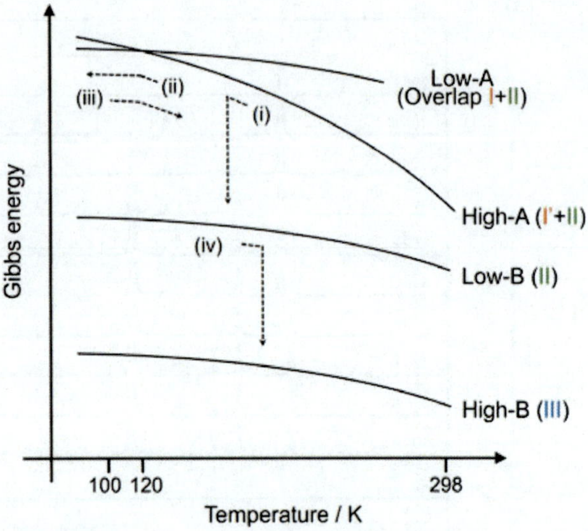

Fig. 4.10 Schematic representation of the Gibbs energy diagram for the four polymorphs of **PyCA** as a function of temperature. The dashed arrows represent the observed transitions

revealed that intersheet vdW contacts are present between Py and CA in Overlap I, CA and CA in Overlap II, and Py and Py in Overlap III, in which oppositely charged molecular sites approach each other owing to electrostatic interactions (Fig. 4.11f–h). Notably, the CA···CA contact in Overlap II originates from the lone pair···π interaction, wherein the negatively charged halogen and the positively charged benzene ring (i.e., π-hole) are attracted to each other. Interactions based on π-holes have recently been recognized and extensively studied in the field of crystal engineering [62, 63].

In addition to the aforementioned intersheet interactions with vdW contacts, it is most likely that pairs with no short interatomic contacts are also responsible for determining the relative stability of each overlapping pattern [64]. Thus, we considered the entire intersheet interaction energies for an isolated pair of sheets by taking the sum of the intermolecular interaction energies for the intersheet pairs of molecules. We employed all molecular pairs with a center-to-center distance of up to $d = 16$ Å based on the d dependence of the intermolecular interaction energies (see Sect. 4.2 for details). The calculations were performed at the SAPT0/jun-cc-pVDZ level of theory, and the results are given as the quantity per mole of **PyCA** (i.e., total one mole in two sheets) in the following.

The interaction energies for Overlaps I, II, and III were calculated to be −55.83 kJ mol^{-1} (Low-A at 100 K), −56.54 kJ mol^{-1} (Low-B at 100 K), and −59.34 kJ mol^{-1} (High-B at 100 K) (Fig. 4.12), respectively (intersheet distance: 3.27, 3.42, and 3.34 Å, respectively). Notably, the division of the entire energy in each pair, that is, Py–Py, Py–CA, and CA–CA, showed a marked difference in the nature of the stacking forces (Fig. 4.12). In Overlap I, the Py–CA term is the major component (81.0%) of the intersheet interactions, whereas in Overlap III, the Py–Py term is the major component (54.2%). In Overlap II, the contributions from Py–Py and CA–CA are comparable (45.4 and 34.4%, respectively) and constitute the majority of the intersheet interactions (79.9%).

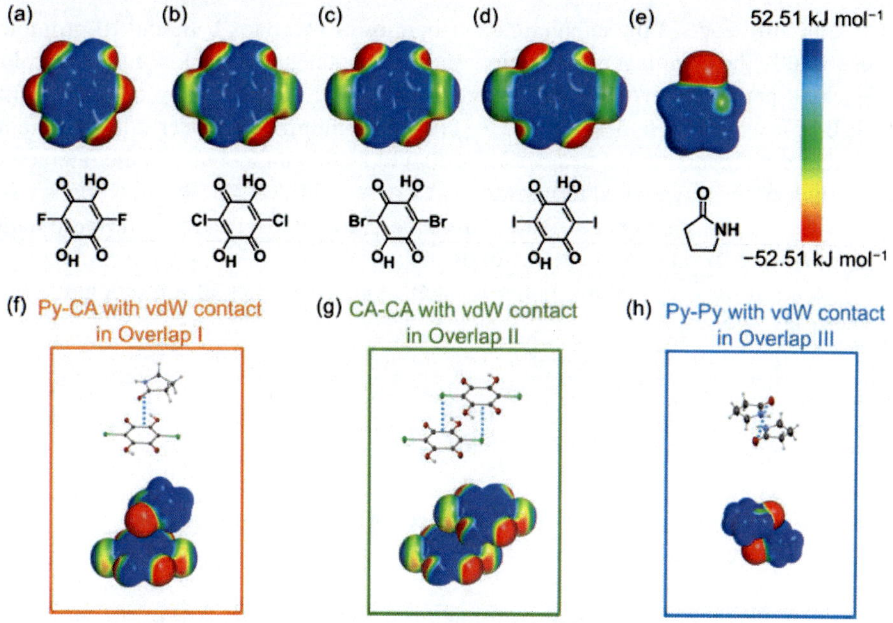

Fig. 4.11 Surface electrostatic potentials calculated at the MP2/aug-cc-pVDZ(-PP) level for **a** FA at 100 K, **b** CA in High-B at 100 K, **c** BA at 100 K, **d** IA at 100 K, and **e** Py in High-B at 100 K computed on the 0.005 electron bohr $^{-3}$ contour of the electronic density (blue, positive region; green, transition region; red, negative region). Color range (kJ mol $^{-1}$): blue, >> 52.51; red, << – 52.51. The molecular structures (top) and surface electrostatic potentials (bottom) of the molecular pair with intersheet vdW contacts consisting of **f** Py and CA in Low-A at 100 K, **g** CA and CA in Low-B at 100 K, and **h** Py and Py in High-B at 100 K. The bottom figures are drawn by combining surface electrostatic potentials calculated for the individual monomers. Computational methods are the same as those for **a**–**e**

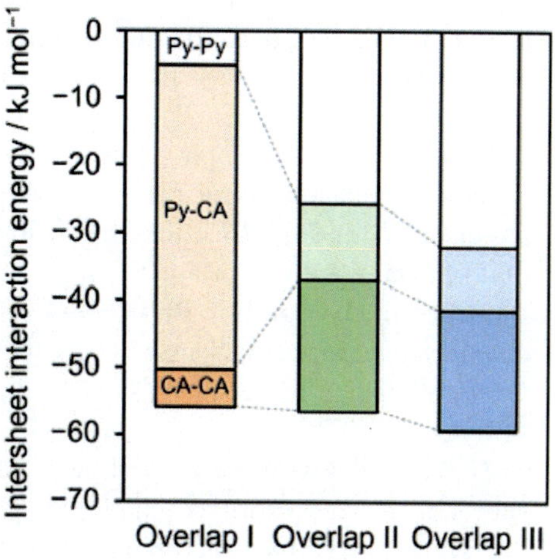

Fig. 4.12 Intersheet interaction energies of Overlap I (left), II (center), and III (right) in **PyCA** at 100 K. For each overlapping pattern, the contribution from the molecular pairs of CA–CA, Py–CA, and Py–Py are shown by dark, light, and no shading areas, respectively. The intersheet interaction energies were calculated for an isolated pair of sheets. The results are given as quantities per mole of **PyCA**

The difference in the energetic contributions of Overlaps I, II, and III prompted us to study the chemical modification of one component molecule for the control of stacking patterns. Given that the heavier halogen has a larger atomic polarizability (F, 0.557; Cl, 2.16; Br, 3.11; I, 5.35 Å^3) [65], it is intuitively expected that replacing the chlorine in CA with a heavier halogen will lead to an increase in the interaction energies of the CA-related terms, that is, Py–CA and CA–CA. Thus, Overlaps I and II, which exhibit a larger contribution from the CA-related terms, become more stable than Overlap III. Therefore, in the following section, we present the effect of halogen substitution on the stacking patterns using both experimental and computational approaches.

4.3.2 *Effect of Halogen Substitution*

4.3.2.1 Synthesis and Crystal Structures of Halogen-Substituted Cocrystals

The combination of anilic acids bearing different halogens [XA (FA, X = F; BA, X = Br; and IA, X = I in Fig. 4.1a)] with Py afforded their corresponding cocrystals (**PyXA**), that is, **PyFA**, **PyBA**, and **PyIA**, with the same composition as **PyCA**. Cocrystallization was performed using an acetonitrile solution containing Py and XA. For the synthesis of each cocrystal, two methods, cooling of the hot solution and evaporation of the solution at room temperature, were conducted, and both afforded the same crystal structure (see Sect. 4.2 for details). This is in contrast to the case observed for **PyCA**, where the cooling and evaporation processes provided metastable High-A and stable High-B phases, respectively. Despite the different vdW radii of the halogens (Fig. 4.13a), all the cocrystals form 2D assemblies, similar to **PyCA** (Fig. 4.13b–e). In addition, in contrast to **PyCA**, the halogen-substituted cocrystals showed no phase transitions in the temperature range studied (100–373 K) [66]. The crystallographic data of the cocrystals are listed in Tables 4.1 and 4.2.

However, the overlapping patterns were indeed different for each cocrystal, as expected. Facial views of the two hydrogen-bonded tapes in two adjacent layers are displayed in Fig. 4.14, together with those of **PyCA**. **PyBA** and **PyIA** adopt a uniform stack with Overlap II, which is isomorphous to Low-B **PyCA**. Here, the formation of Overlap II in **PyBA** and **PyIA** is consistent with the expectations based on our calculations (see Sect. 4.3.1.2). In contrast, **PyFA** adopts a uniform stack with Overlap I, which is the same as one of the overlapping patterns of Low-A observed in **PyCA**.

Notably, **PyFA** and **PyCA** showed fair cleavage properties parallel to the 2D sheets, which can afford extremely thin plates (Fig. 4.15); the fabrication of organic thin plates is an important topic in the field of organic photonics [67, 68].

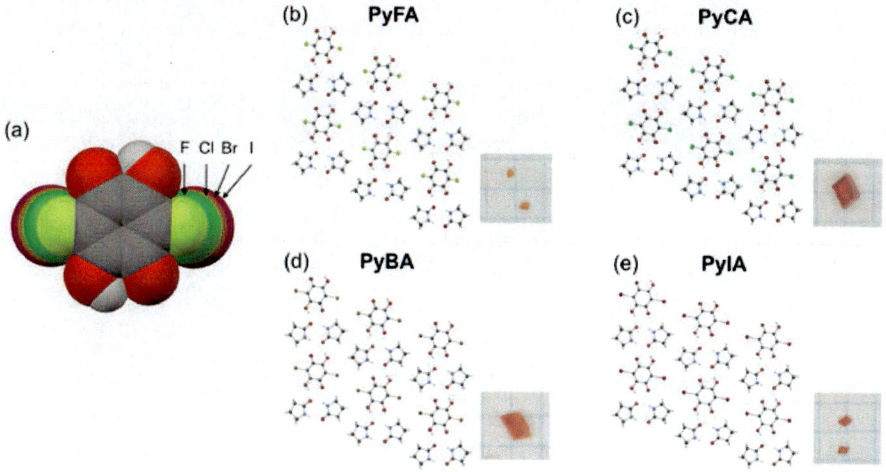

Fig. 4.13 **a** Molecular models of the XAs based on the structure of CA in High-B of PyCA (298 K), wherein the sizes of the spheres correspond to the van der Waals radius of each atom [61]. For FA, BA, and IA, the Cl atoms of CA were replaced with F, Br, and I, respectively. The C–X bond lengths were modified using literature values [77]. Two-dimensional sheets of **b** **PyFA**, **c** **PyCA** (High-B), **d** **PyBA**, and **e** **PyIA** at 298 K. The molecules are drawn with thermal ellipsoids with 50% probability (gray, C; white, H; purple, N; red, O; yellow, F; green, Cl; orange, Br; magenta, I). Inset shows the photograph of each cocrystal on millimeter-marked paper

4.3.2.2 Discussion on the Appearance of Overlap II in PyBA and PyIA

Here, we discuss why the substitution of Cl with the heavier Br or I leads to a change in the stable overlapping pattern from Overlap III to II. As mentioned in Sect. 4.3.1.2, a possible scenario for the change in the stability of the stacking patterns is that substitution with a heavier halogen induces an increase in atomic polarizability, followed by an increase in XA-related intermolecular interactions. In this case, Overlap II, with a larger contribution from XA-related interactions, can become more stable than Overlap III (Fig. 4.12).

To confirm the increase in the XA-related terms upon halogen substitution, we calculated the intersheet interactions of **PyBA** and **PyIA** based on the crystal structures at 100 K (intersheet distance: 3.47 and 3.53 Å, respectively). As shown in Fig. 4.16a, the Py–Py term remains almost unchanged upon halogen substitution (−25.69, −26.32, and −26.27 kJ mol^{-1} for **PyCA**, **PyBA**, and **PyIA**, respectively), whereas the XA–XA term is more significantly influenced (−19.48, −21.33, and −24.71 kJ mol^{-1} for **PyCA**, **PyBA**, and **PyIA**, respectively).

To elucidate the origin of the increase in the XA-related terms, we performed an energy decomposition analysis of the XA–XA terms for **PyCA**, **PyBA**, and **PyIA** based on the symmetry-adapted perturbation theory (SAPT) method [59, 60], which can separate the molecular interactions into electrostatic, induction, dispersion, and exchange terms (see Sect. 4.2 for details). As shown in Fig. 4.16b, the dispersion term (−22.68, −25.47, and −32.38 kJ mol^{-1} for **PyCA**, **PyBA**, and **PyIA**, respectively)

Table 4.2 Cell parameters and crystallographic information for **PyFA** and Low-A of **PyCA**

	PyBA	PyBA	PyIA	PyIA
Formula	C_{14} H_{16} Br_2 N_2 O_6	C_{14} H_{16} Br_2 N_2 O_6	C_{14} H_{16} I_2 N_2 O_6	C_{14} H_{16} I_2 N_2 O_6
Formula weight	468.10	468.10	562.10	562.10
Temperature (K)	298	100	298	100
Wavelength (Å)	0.71075	0.71075	0.71075	0.71075
Crystal system	Triclinic	Triclinic	Triclinic	Triclinic
Space group	$P-1$	$P-1$	$P-1$	$P-1$
Color	Orange	Orange	Orange	Orange
a (Å)	4.84680(10)	4.76730(10)	4.9520(3)	4.86130(9)
b (Å)	8.9572(2)	8.9312(2)	8.9952(5)	8.96389(17)
c (Å)	10.1708(2)	9.9806(2)	10.3424(3)	10.16806(15)
α (°)	103.864(2)	103.676(2)	104.311(4)	104.0253(15)
β (°)	97.8230(10)	96.361(2)	97.740(4)	96.3861(14)
γ (°)	93.7970(10)	93.868(2)	94.279(4)	94.5383(16)
V (Å^3)	422.451(16)	408.427(16)	439.53(4)	424.615(13)
Z	1	1	1	1
Calcd density (g cm^{-3})	1.840	1.903	2.123	2.198
μ (mm^{-1})	4.844	5.011	3.611	3.738
$F(000)$	232	232	268	268
Crystal size (mm^3)	0.40 × 0.19 × 0.11	0.40 × 0.19 × 0.11	0.26 × 0.24 × 0.16	0.26 × 0.24 × 0.16
Total reflections	18,377	17,417	9420	18,070
Unique reflections	2217	2149	2152	2253
Parameters used	110	110	109	109
R_{int}	0.0269	0.0236	0.0146	0.0139
Goodness-of-fit	1.163	1.115	1.127	1.206
R_1 [$I > 2\sigma(I)$]	0.0265	0.0178	0.0201	0.0130
wR_2 (all reflections)	0.0621	0.0420	0.0528	0.0353
max, min $\Delta\rho$ (e Å^{-3})	0.68, −0.43	0.55, −0.38	1.02, −0.84	0.59, −0.68
CCDC deposition nos	2,076,402	2,076,403	2,076,404	2,076,405

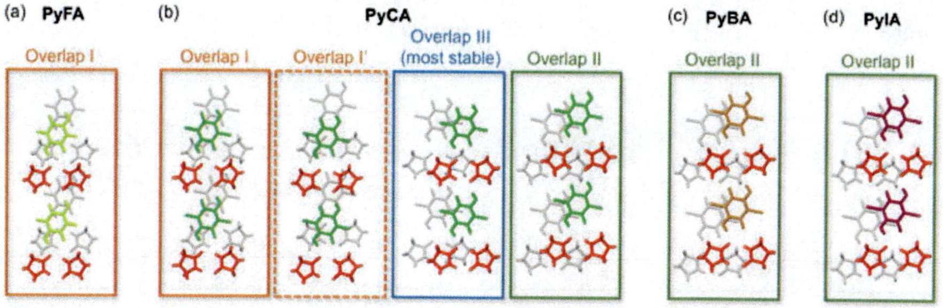

Fig. 4.14 Facial view of two stacked tapes of **a PyFA**, **b PyCA**, **c PyBA**, and **d PyIA**. For **PyCA**, Overlaps I, I′, II, and III were extracted from the crystal structures of Low-A at 100 K, High-A at 298 K, Low-B at 100 K, and High-B at 298 K, respectively. Colored molecules (red, Py; yellow, FA; green, CA; orange, BA; magenta, IA) are placed in front of the gray molecules

was more strongly influenced by halogen substitution than the electrostatic term (−12.94, −14.12, and −16.62 kJ mol^{-1}). This indicates that enhancement of the induced dipole, rather than a change in the intramolecular charge distribution in the unperturbed monomers, is the primary cause for the increase in the XA-related term upon halogen substitution. Notably, this result is consistent with the calculations of the surface electrostatic potentials, showing no significant change in the charge distribution upon halogen substitution (Fig. 4.11a–d).

4.3.2.3 Discussion on the Appearance of Overlap I in PyFA

In contrast to the results for **PyBA** and **PyIA**, substitution with a lighter F is expected to result in a decrease in the polarizability and XA-related terms in **PyFA**. Thus, Overlap III must be the most stable overlapping pattern in **PyFA** because it has the largest intersheet interactions and the smallest ratio of XA-related terms in **PyCA** (Fig. 4.12). However, Overlap I was found to be the most stable overlapping pattern in **PyFA**. To understand the reason, we focused on the conformational disorder of Py. Because the High-A ↔ Low-A transition in **PyCA** is a transition between Low-A, which adopts Overlap I with intersheet vdW contacts without disorder, and High-A, which adopts Overlap I′ without intersheet vdW contacts with disorder, it is apparent that the entropic term associated with the disorder compensates for the enthalpic term associated with the intersheet interactions in **PyCA**. Given that such disorder was also observed in **PyFA** with Overlap I, as evidenced by the elongated thermal ellipsoid ($U_{eq} = 0.1013(14)$ Å^2 at 298 K) of the 4-positioned carbon in Py (Fig. 4.17a, bottom), the entropic contribution may stabilize Overlap I with an expected smaller enthalpic contribution than that of Overlap III in **PyFA**. To verify these considerations, the contribution of the entropic term was roughly estimated. Assuming a disorder of Py between the two conformations (two envelope conformations bending up and down), each Py molecule in motion affords the entropy of $k\ln 2$, where k is the Boltzmann constant. Considering that half of Py molecules is dynamically active in

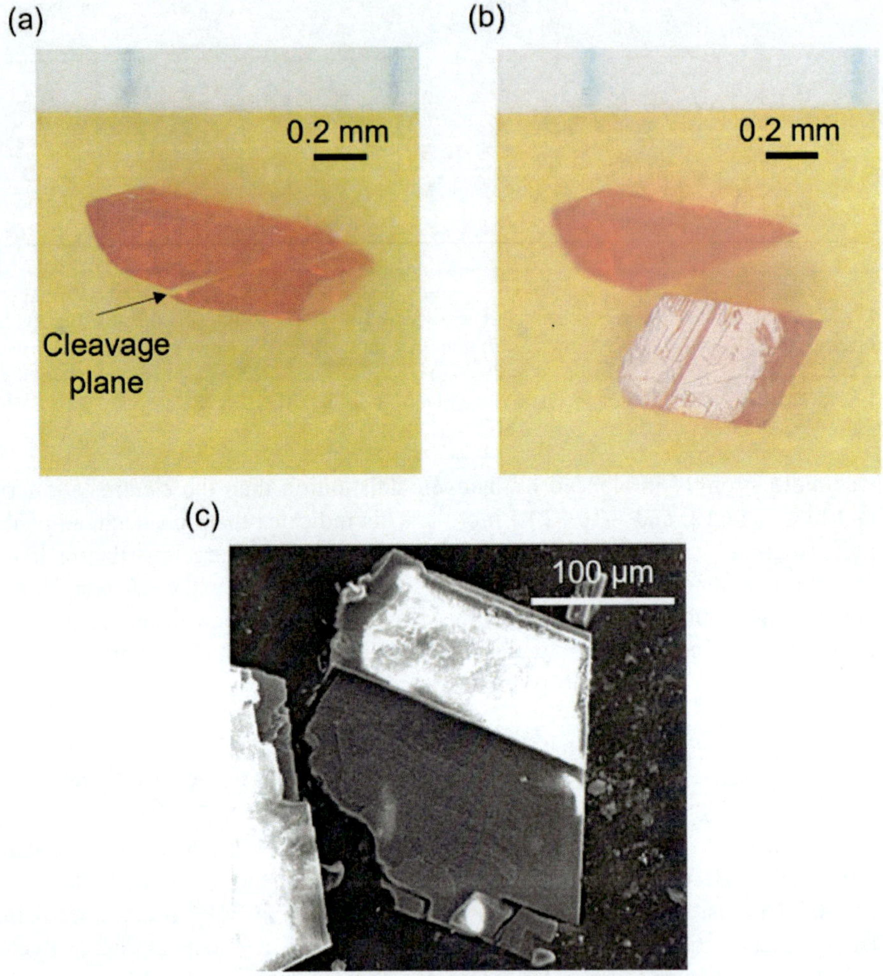

Fig. 4.15 **a**, **b** Mechanical cleavage of a High-B crystal of **PyCA** parallel to the 2D assembly, the orientation of which has been determined via single-crystal XRD experiments after the cleavage. **a** and **b** show the same two pieces obtained by mechanical cleavage of a block-shaped crystal. The orientation of the piece placed at the bottom right is different for the two figures. **c** SEM image of a piece obtained by peeling off a High-B crystal of **PyCA** with a carbon tape

High-A of **PyCA** [= $(Py)_2(CA)$], the entropy resulting from the motion is $N_A k \ln 2 = 5.76\ \mathrm{J\ K^{-1}\ mol^{-1}}$ per mole of **PyCA**, where N_A is Avogadro's number. Notably, the stabilization owing to this entropy reaches a few kJ $\mathrm{mol^{-1}}$ in the temperature range studied (100–300 K), and the value is of the same order of magnitude as the energy difference between Overlaps I and III in **PyCA** (3.51 kJ $\mathrm{mol^{-1}}$). In the case of **PyFA**, all Py molecules show conformational disorder; therefore, the entropic gain is twice the above estimation.

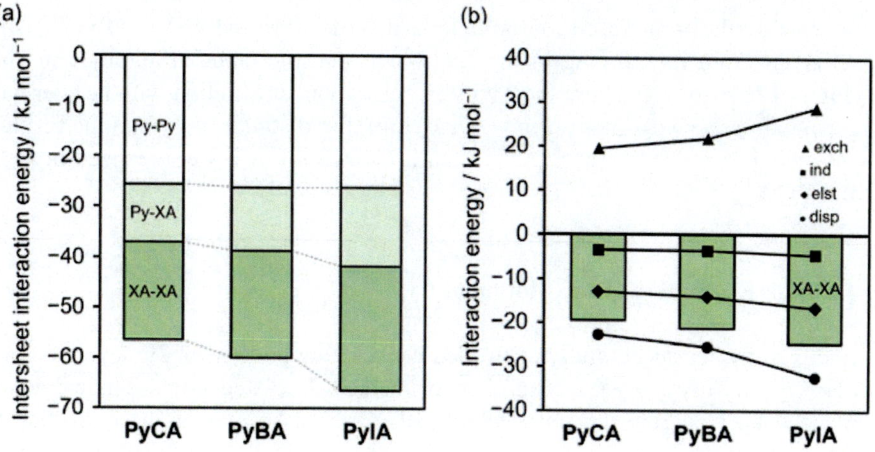

Fig. 4.16 **a** Intersheet interaction energy of Overlap II for **PyCA** (Low-B, left), **PyBA** (center), and **PyIA** (right) at 100 K. For each cocrystal, the dark, light, and no-shading regions show the contribution from the molecular pairs Py–Py, Py–XA, and XA–XA, respectively, where XA = CA, BA, or IA. **b** Partitioned intersheet interaction energies obtained for the XA–XA molecular pairs in Overlap II (exch, exchange; elst, electrostatic; ind, induction; disp, dispersion). The green bar graphs show the total interaction energies, namely the sum of all the components. The intersheet interaction energies were calculated for an isolated pair of sheets. The results are given as quantities per mole of **PyXA**

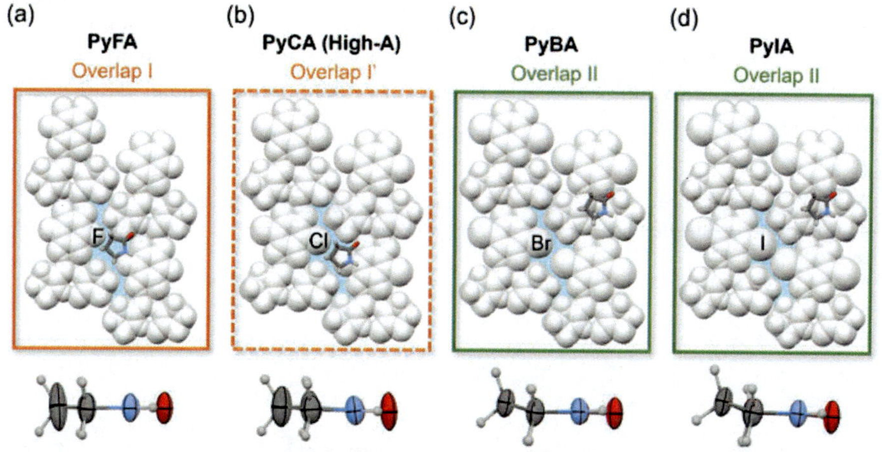

Fig. 4.17 Intrasheet molecular packing viewed perpendicular to the molecular plane (top) and side view of Py (bottom) in **a PyFA**, **b PyCA** (High-A), **c PyBA**, and **d PyIA** at 298 K. In the top panel, a colored capped stick model of a Py molecule (gray, C; white, H; purple, N; red, O) is placed in front of a sheet drawn using a white-space-filling model showing the vdW surface. The pale-blue-colored area shows the cavities formed between the two tapes. In the bottom panel, a Py molecule is drawn with thermal ellipsoids with 50% probability. Atoms are colored in the same manner as Py in the top panel

Notably, the pronounced conformational disorder of Py in **PyFA** and **PyCA** (High-A) arises from its trapping in the cavity between the tapes in an adjacent sheet (Fig. 4.17a,b, top). Because the cavity shrinks upon substitution with heavier halogens with larger vdW radii (Fig. 4.17c,d, top), the disorder of Py and the resulting entropic gain should be negligible, even if Overlaps I or I′ were realized in **PyBA** and **PyIA**.

4.3.2.4 Evaluation of the Motion of Py

In this study, we successfully controlled the stacking patterns of 2D-layered molecular crystals using **PyXA**. The change in the XA-related intersheet interactions induced by changing the polarizability of XA is a key factor in controlling the stacking patterns. Moreover, the structural disorder is also a primary factor determining the stacking patterns. However, the current evidence for the disorder of Py is only the SCXRD results, which does not usually discriminate between static and dynamic disorders in most cases. Hence, we used the nuclear magnetic resonance technique, which is a powerful tool for investigating molecular dynamics because its resonance frequency is comparable to the low-frequency range of molecular motion. We initially conducted ^{1}H NMR (nuclear magnetic resonance) experiments on the polycrystalline **PyCA** sample and found that it has a very long spin–lattice relaxation time (T_1) of approximately 600 s (Fig. 4.18). We then performed ^{35}Cl NQR experiments to reduce the measurement time. NQR spectroscopy, which probes the electric field gradient at the nucleus site, is sensitive to the local environment and might detect the dynamic motion of Py, even though the Cl nuclei are in the CA molecules [69, 70].

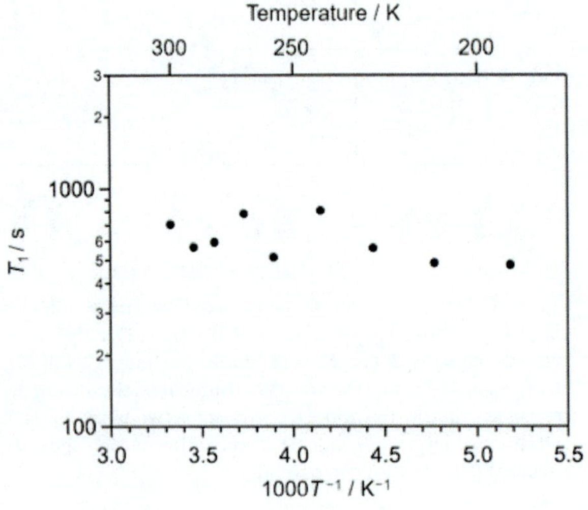

Fig. 4.18 Temperature dependence of spin–lattice relaxation time (T_1) of ^{1}H NMR of **PyCA** (High-A) observed at 59.33–59.42 MHz

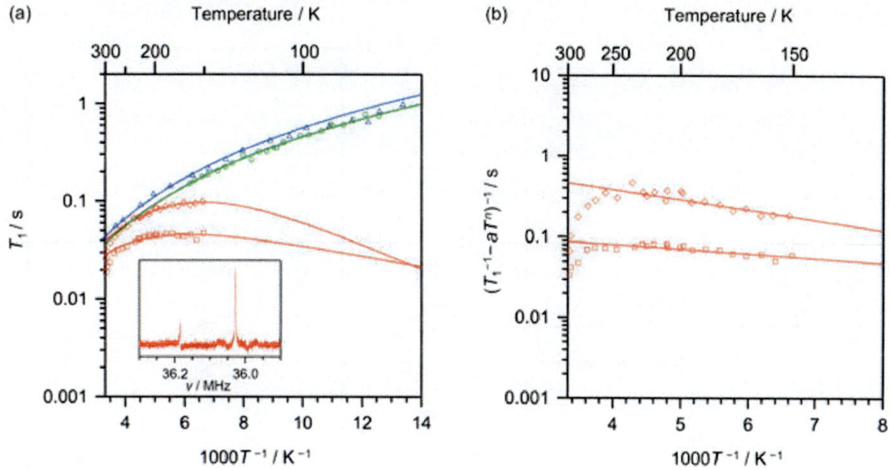

Fig. 4.19 **a** Temperature dependence of T_1 in ^{35}Cl NQR spectra for **PyCA**. Red squares (□) and diamonds (◇) represent the data for High-A. Green circles (○) and blue triangles (△) represent the data for Low-B and High-B, respectively. For solid lines, see text. The inset shows the spectrum at room temperature (292 K). **b** Temperature dependence of $(T_1{}^{-1}-aT^{n})^{-1}$ following Arrhenius-type behavior

We carried out variable-temperature ^{35}Cl NQR measurements (room temperature → 100 K → room temperature) of polycrystalline **PyCA**. At room temperature, two signals ascribed to the crystallographically non-equivalent Cl sites in High-A were observed (Fig. 4.19a, inset). Given that distinct signals were observed in High-A, it is obvious that the disorder of Py observed using XRD is dynamic because static disorder would result in the disappearance of the NQR signals due to the disturbance of the electric field around the resonant Cl.

Upon cooling, a signal for Low-B began to appear at approximately 165 K, whereas the signals for High-A faded (Fig. 4.20). In the subsequent heating process, one signal for High-B emerged at approximately 190 K and became intense, whereas the signal for Low-B gradually disappeared. The signals attributable to Low-A were not detected, probably because the High-A ↔ Low-A transition has a very low probability of occurrence.

The temperature dependences of T_1 for High-A and Low-B during the cooling process are shown in Fig. 4.19, along with the results for High-B. The temperature dependences of Low-B and High-B were well fitted using the following equation, considering the lattice vibration [71, 72].

$$T_1^{-1} = aT^n$$
$$(a = 5.2 \pm 1.2 \times 10^{-5} s^{-1} K^{-n},\ n = 2.31 \pm 0.05\ \text{for Low-}B;$$
$$a = 3.6 \pm 0.9 \times 10^{-5} s^{-1} K^{-n},\ n = 2.35 \pm 0.05\ \text{for High-}B).$$

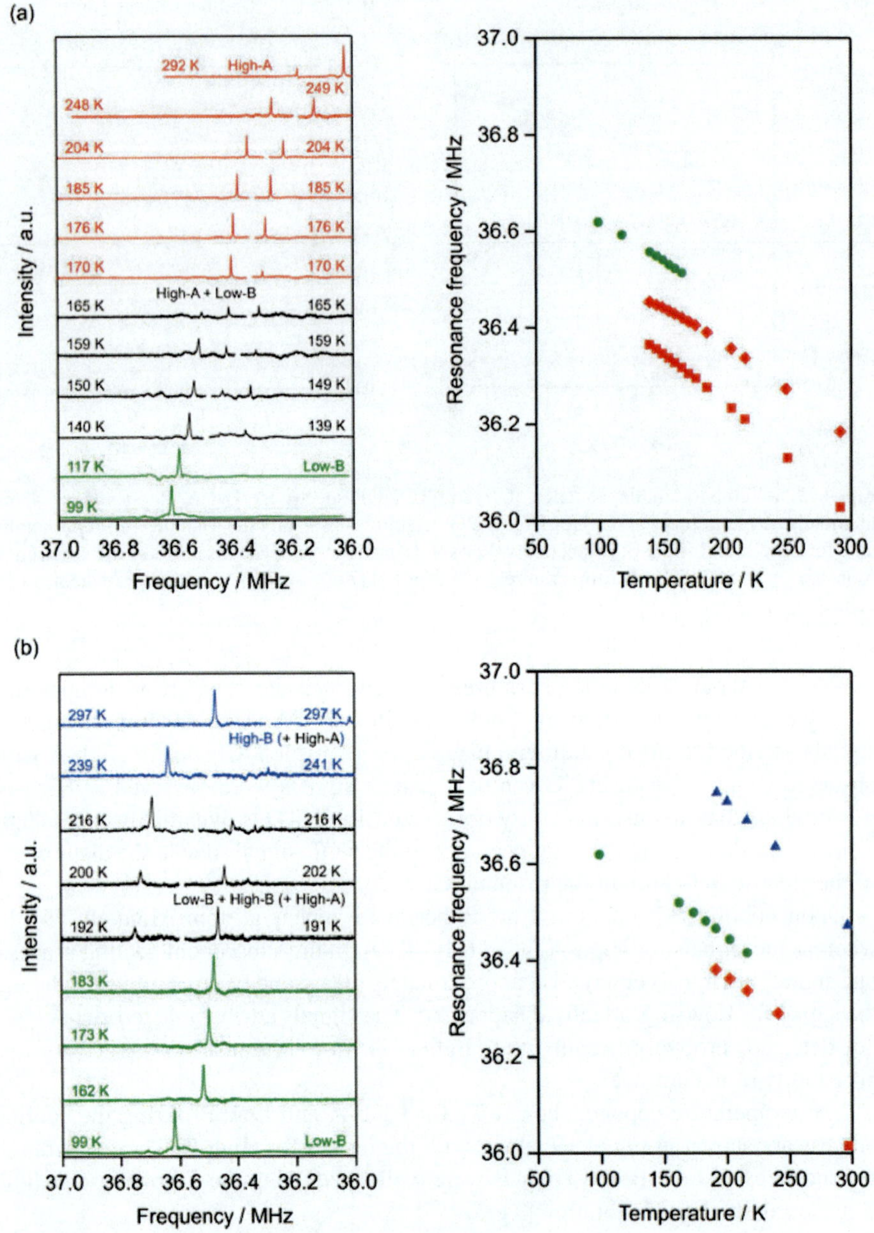

Fig. 4.20 Temperature dependences of spectrum (left) and resonance frequency (right) of ^{35}Cl NQR of **PyCA** upon **a** cooling from 298 to 100 K and **b** subsequent heating from 100 to 298 K

In contrast, the T_1 value for High-A exhibited a downward deviation from the above equation at approximately 270–150 K. The temperature dependence is well explained using the following equation, assuming the coexistence of lattice vibration and a molecular motion following Arrhenius-type relaxation with an activation energy of ~ 2 kJ mol^{-1} [71–73]:

$$
\begin{aligned}
&T_1^{-1} = aT^n + b\exp(E_a/RT) \\
&(a = 3.6 \times 10^{-5}s^{-1}K^{-n},\ n = 2.35,\ b = 8.2 \pm 1.8 \times 10^{-1}s^{-1}, \\
&E_a = 2.4 \pm 0.4\,\mathrm{kJ\,mol^{-1}} \\
&\text{and}\, a = 3.6 \times 10^{-5}s^{-1}K^{-n},\ n = 2.35,\ b = 7.4 \pm 1.0s^{-1}, \\
&E_a = 1.1 \pm 0.3\,\mathrm{kJ\,mol^{-1}}),
\end{aligned}
$$

where R is the gas constant.

Figure 4.19b shows the Arrhenius plot of T_1, wherein the contribution of lattice vibration was subtracted.

This relaxation can be readily attributed to the motion of Py derived from the two envelope conformations, where the decrease in T_1 during the cooling process clearly indicates that the frequency of the motion approaches the NQR frequency (~36 MHz) in the faster frequency region. It is important to note that Low-B and High-B, in which the motion of Py almost freezes, as indicated by the SCXRD measurements, show no sign of such a T_1 drop. The NQR results therefore provide strong evidence that the freezing of motion triggers the transition.

It should be noted that the estimation of the activation energy for the molecular motion of Py contains some uncertainty. We performed fitting for High-A using data below 270 K because there seems to be another relaxation attributable neither to the lattice vibration nor to the molecular motion of Py at higher temperatures (not discussed in detail here). In the temperature range of < 270 K, there is no range where the relaxation originates mainly from the lattice vibration, which results in uncertainty in the estimation of the contribution of the lattice vibration. In this study, we used the fitting result of High-B to subtract the contribution of lattice vibration for High-A. It should be noted that the High-B data cover a wider temperature range than the Low-B data, including the temperature range in which molecular motion was observed for High-A. Although the assumption that the contribution of lattice vibration for High-A is the same as that for Low-B leads to the different activation energies for the motion (see Fig. 4.21), it is evident that our NQR measurements detected the motion of Py in High-A because the T_1 value steadily decreases as the temperature approaches the High-A → Low-B transition temperature (~165 K) at which the motion of Py almost freezes, as indicated by the SCXRD measurements.

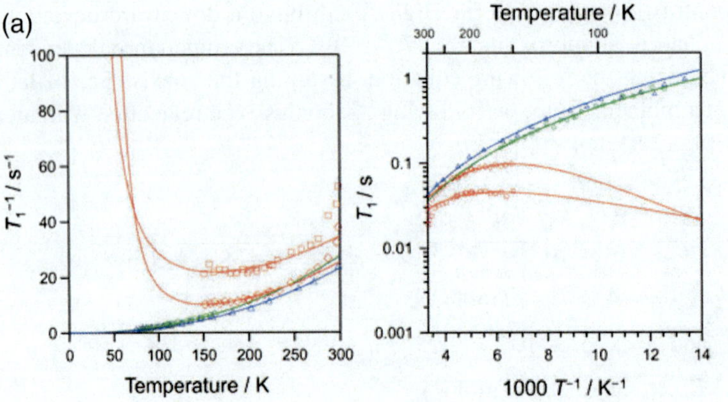

Fitting results	a / s^{-1}K^{-n}	n	b / s^{-1}	E_a / kJ mol^{-1}
Fitting for High-A (□) assuming lattice vibration same as High-B	3.6×10^{-5} (fixed)	2.35 (fixed)	7.4 ± 1.0	1.1 ± 0.3
Fitting for High-A (◇) assuming lattice vibration same as High-B	3.6×10^{-5} (fixed)	2.35 (fixed)	8.2 ± 1.8 $\times 10^{-1}$	2.4 ± 0.4

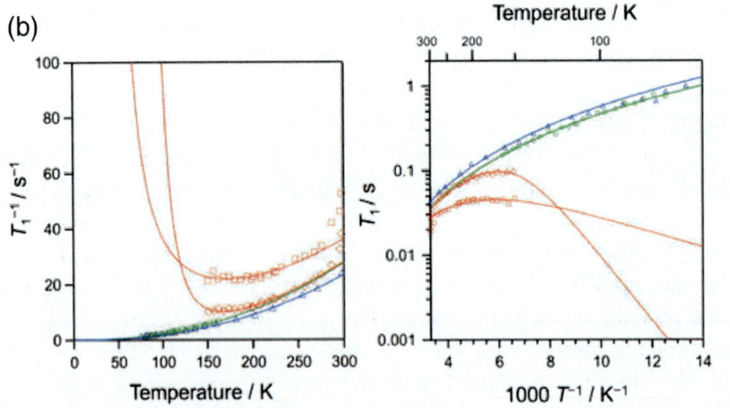

Fitting results	a / s^{-1}K^{-n}	n	b / s^{-1}	E_a / kJ mol^{-1}
Fitting for High-A (□) assuming lattice vibration same as Low-B	5.2×10^{-5} (fixed)	2.31 (fixed)	4.2 ± 0.6	1.7 ± 0.3
Fitting for High-A (◇) assuming lattice vibration same as Low-B	5.2×10^{-5} (fixed)	2.31 (fixed)	1.7 ± 1.1 $\times 10^{-2}$	7.2 ± 0.9

Fig. 4.21 Fitting results for temperature dependence of T_1 of ^{35}Cl NQR spectra of **PyCA**. Left and right figures show the plot of $T_1{}^{-1}$ as a function of temperature (T) and T_1 as a function of T^{-1}, respectively. Red squares (□) and diamonds (◇) represent data of High-A. Green circles (○) and blue triangles (△) represent data of Low-B and High-B, respectively. Blue and green solid lines represent the fitting results for High-B and Low-B, respectively, using $T_1{}^{-1} = aT^n$. Red solid lines represent fitting results for High-A using $T_1{}^{-1} = aT^n + b\exp(E_a/RT)$. For the fitting of High-A, the contributions from the lattice vibration (aT^n) obtained by the fitting of High-B and Low-B were applied in **a** and **b**, respectively. Bottom tables show the fitting parameters obtained for High-A

4.4 Conclusion

In this study, we demonstrated, for the first time, rational control over stacking patterns via chemical modification by investigating the stacking patterns of 2D molecular assemblies in hydrogen-bonded cocrystals from an energy perspective using experimental and computational methods. Hydrogen-bonded crystals have recently attracted attention owing to their potential for using proton dynamics to produce intriguing physical properties such as ferroelectricity and protonic solitons [74–76].

We prepared a series of hydrogen-bonded cocrystals composed of 2-pyrrolidone (Py) and anilic acids bearing different halogens (X = F, Cl, Br, and I). Among all the cocrystals, only **PyCA** exhibited structural phase transitions between the four polymorphs via reversible High-A ↔ Low-A and irreversible High-A → Low-B → High-B transitions. SCXRD and NQR experiments revealed that the intersheet intermolecular interactions and molecular motion associated with the two conformations of Py are key factors in determining the overlapping patterns; note that the contribution of molecular motion is unique to organic systems. In addition, we found that changes in the overlapping patterns caused by halogen substitution originated from changes in (1) the intersheet interactions associated with the atomic polarizability of the halogen and (2) the motion of Py associated with the intrasheet cavity size, depending on the size of the halogen. Notably, the occurrence of structural phase transitions only in **PyCA** can be attributed to the moderate polarizability and cavities of the 2D sheets.

The approach demonstrated here, namely the strong intermolecular interactions (e.g., hydrogen bonds) serve to assemble building blocks, whereas the weak intermolecular interactions (e.g., π–π and lone pair···π interactions) that are tunable through chemical modifications determine their assembling patterns, can be applied to various systems, including other dimensional systems. The trapping or stabilization of enthalpically unfavorable patterns using entropic gain related to molecular flexibilities also has high potential for use in diverse systems. These findings may pave the way for the practical tuning of the physical and chemical properties of a wide range of molecular materials.

This chapter was reproduced from Ref. [78] with permission from John Wiley and Sons.

References

1. Miller JS (ed) (1983) Extended linear chain compounds, vol 1, 2, 3. Plenum Press, New York
2. Grasso V (ed) (1986) Electronic structure and electronic transitions in layered materials. D. Reidel Publishing, Dordrecht
3. Desiraju GR (1995) Angew Chem Int Ed Engl 34:2311–2327
4. Moulton B, Zaworotko MJ (2001) Chem Rev 101:1629–1658
5. Bernstein J (2020) Polymorphism in molecular crystals, 2nd edn. Oxford University Press, New York
6. Israelachvili JN (2011) Intermolecular and surface forces, 3rd edn. Academic Press, New York

7. Nyman J, Day GM (2015) CrystEngComm 17:5154–5165
8. Aakerçy CB, Fasulo M, Schultheiss N, Desper J, Moore C (2007) J Am Chem Soc 129:13772–13773
9. Banerjee R, Mondal R, Howard JAK, Desiraju GR (2006) Cryst Growth Des 6:999–1009
10. Zerkowski JA, MacDonald JC, Seto CT, Wierda DA, Whitesides GM (1994) J Am Chem Soc 116:2382–2391
11. Palacin S, Chin DN, Simanek EE, MacDonald JC, Whitesides GM, McBride MT, Palmore GTR (1997) J Am Chem Soc 119:11807–11816
12. Chin DN, Palmore GTR, Whitesides GM (1999) J Am Chem Soc 121:2115–2122
13. Paine RT, Narula CK (1990) Chem Rev 90:73–91
14. Han GH, Duong DL, Keum DH, Yun SJ, Lee YH (2018) Chem Rev 118:6297–6336
15. Chakraborty G, Park IH, Medishetty R, Vittal JJ (2021) Chem Rev 121:3751–3891
16. Novoselov KS, Geim AK, Morozov SV, Jiang D, Zhang Y, Dubonos SV, Grigorieva IV, Firsov AA (2004) Science 306:666–669
17. Geng K, He T, Liu R, Dalapati S, Tan KT, Li Z, Tao S, Gong Y, Jiang Q, Jiang D (2020) Chem Rev 120:8814–8933
18. Naik I, Rastogi AK (2011) Pramana 76:957–963
19. Cao Y, Fatemi V, Fang S, Watanabe K, Taniguchi T, Kaxiras E, Jarillo-Herrero P (2018) Nature 556:43–50
20. Cao Y, Fatemi V, Demir A, Fang S, Tomarken SL, Luo JY, Sanchez-Yamagishi JD, Watanabe K, Taniguchi T, Kaxiras E, Ashoori RC, Jarillo-Herrero P (2018) Nature 556:80–84
21. Huskić I, Novendra N, Lim DW, Topić F, Titi HM, Pekov IV, Krivovichev SV, Navrotsky A, Kitagawa H, Friščić T (2019) Chem Sci 10:4923–4929
22. Tang WQ, Zhao YJ, Xu M, Xu JY, Meng SS, Yin YD, Zhang QH, Gu L, Liu DH, Gu ZY (2021) Angew Chem Int Ed 60:6920–6925
23. Davey RJ, Maginn SJ, Andrews SJ, Black SN, Buckley AM, Cottier D, Dempsey P, Plowman R, Rout JE, Stanley DR, Taylor A (1994) J Chem Soc Faraday Trans 90:1003–1009
24. Karothu DP, Weston J, Desta IT, Naumov P (2016) J Am Chem Soc 138:13298–13306
25. Sasaki T, Miyamoto Y, Takamizawa S (2020) Cryst Growth Des 20:4779–4782
26. Hunter CA, Sanders JKM (1990) J Am Chem Soc 112:5525–5534
27. Thakuria R, Nath NK, Saha BK (2019) Cryst Growth Des 19:523–528
28. Mooibroek TJ, Gamez P, Reedijk J (2008) CrystEngComm 10:1501–1515
29. Nishio M (2004) CrystEngComm 6:130–158
30. Grabowski SJ (2006) In: Hydrogen bonding-new insights. Springer, Dordrecht
31. Maréchal Y (2007) In: The hydrogen bond and the water molecule: the physics and chemistry of water, aqueous and bio media, Elsevier, Amsterdam
32. Steiner T (2002) Angew Chem Int Ed 41:48–76
33. MacDonald JC, Whitesides GM (1994) Chem Rev 94:2383–2420
34. Williams LJ, Jagadish B, Lyon SR, Kloster RA, Carducci MD, Mash EA (1999) Tetrahedron 55:14281–14300
35. Williams LJ, Jagadish B, Lansdown MG, Carducci MD, Mash EA (1999) Tetrahedron 55:14301–14322
36. Takasawa R, Yoshikawa I, Araki K (2004) Org Biomol Chem 2:1125–1132
37. Sada K, Inoue K, Tanaka T, Tanaka A, Epergyes A, Nagahama S, Matsumoto A, Miyata M (2004) J Am Chem Soc 126:1764–1771
38. Gong B, Zheng C, Skrzypczak-Jankun E, Yan Y, Zhang J (1998) J Am Chem Soc 120:11194–11195
39. Gong B, Zheng C, Zeng H, Zhu J (1999) J Am Chem Soc 121:9766–9767
40. Zerkowski JA, Seto CT, Wierda DA, Whitesides GM (1990) J Am Chem Soc 112:9025–9026
41. Zerkowski JA, Seto CT, Whitesides GM (1992) J Am Chem Soc 114:5473–5475
42. Zerkowski JA, Whitesides GM (1994) J Am Chem Soc 116:4298–4304
43. Zerkowski JA, Mathias JP, Whitesides GM (1994) J Am Chem Soc 116:4305–4315
44. Russell VA, Etter MC, Ward MD (1994) J Am Chem Soc 116:1941–1952
45. Russell VA, Evans CC, Li W, Ward MD (1997) Science 276:575–579

46. Ermer O (1988) J Am Chem Soc 110:3747–3754
47. Gotoh K, Ishida H (2011) Acta Crystallogr Sect C 67:o500–o504
48. Donoshita M, Hayashi M, Ikeda R, Yoshida Y, Morikawa S, Sugimoto K, Kitagawa H (2018) Chem Commun 54:8571–8574
49. Akutagawa T, Koshinaka H, Sato D, Takeda S, Noro S, Takahashi H, Kumai R, Tokura Y, Nakamura T (2009) Nat Mater 8:342–347
50. Hayami S, Murata K, Urakami D, Kojima Y, Akita M, Inoue K (2008) Chem Commun 6510–6512
51. Su X, Zuo F, Schlueter JA, Kelly ME, Williams JM (1998) Phys Rev B 57:R14056–R14059
52. Torrey HA, Hunter WH (1912) J Am Chem Soc 34:702–716
53. Murase R, Abrahams BF, D'Alessandro DM, Davies CG, Hudson TA, Jameson GNL, Moubaraki B, Murray KS, Robson R, Sutton AL (2017) Inorg Chem 56:9025–9035
54. Sheldrick GM (2014) SHELXT Version 2014/5. Acta Cryst A70:C1437
55. Sheldrick GM (2014) SHELXL Version 2014/3. Acta Cryst A64:112–122
56. CrystalStructure (2018) Version 4.3, Rigaku Corporation, Tokyo, Japan
57. Gaussian 09, Revision A.02, Frisch MJ, Trucks GW, Schlegel HB, Scuseria GE, Robb MA, Cheeseman JR, Scalmani G, Barone V, Mennucci B, Petersson GA, Nakatsuji H, Caricato M, Li X, Hratchian HP, Izmaylov AF, Bloino J, Zheng G, Sonnenberg JL, Hada M, Ehara M, Toyota K, Fukuda R, Hasegawa J, Ishida M, Nakajima T, Honda Y, Kitao O, Nakai H, Vreven T, Montgomery JA, Peralta JE, Ogliaro F, Bearpark M, Heyd JJ, Brothers E, Kudin KN, Staroverov VN, Kobayashi R, Normand J, Raghavachari K, Rendell A, Burant JC, Iyengar SS, Tomasi J, Cossi M, Rega N, Millam JM, Klene M, Knox JE, Cross JB, Bakken V, Adamo C, Jaramillo J, Gomperts R, Stratmann RE, Yazyev O, Austin AJ, Cammi R, Pomelli C, Ochterski JW, Martin RL, Morokuma K, Zakrzewski VG, Voth GA, Salvador P, Dannenberg JJ, Dapprich S, Daniels AD, Farkas Ö, Foresman JB, Ortiz JV, Cioslowski J, Fox DJ (2009) Gaussian 09, Gaussian Inc., Wallingford, CT
58. Psi4 v1.3.2, Parrish RM, Burns LA, Smith DGA, Simmonett AC, DePrince III AE, Hohenstein EG, Bozkaya U, Sokolov AY, Di Remigio R, Richard RM, Gonthier JF, James AM, McAlexander HR, Kumar A, Saitow M, Wang X, Pritchard BP, Verma P, Schaefer III HF, Patkowski K, King RA, Valeev EF, Evangelista FA, Turney JM, Crawford TD, Sherrill CD (2017) J Chem Theory Comput 13:3185–3197
59. Jeziorski B, Moszynski R, Szalewicz K (1994) Chem Rev 94:1887–1930
60. Parker TM, Burns LA, Parrish RM, Ryno AG, Sherrill CD (2014) J Chem Phys 140:094106
61. Bondi A (1964) J Phys Chem 68:441–451
62. Wang H, Wang W, Jin WJ (2016) Chem Rev 116:5072–5104
63. Politzer P, Murray JS (2018) J Comput Chem 39:464–471
64. Gavezzotti A (2002) J Phys Chem B 106:4145–4154
65. Miller TM (2019) Polarizabilities of atoms and molecules. In: Rumble JR (ed) CRC handbook of chemistry and physics, 100th ed. CRC Press, Boca Raton, FL, pp 10–96–10–97
66. In the temperature region above 298 K, we used samples sealed in glass capillaries
67. Zhuo M-P, Tao Y-C, Wang X-D, Wu Y, Chen S, Liao L-S, Jiang L (2018) Angew Chem Int Ed 57:11300–11304
68. Liu Y, Hu H, Xu L, Qiu B, Liang J, Ding F, Wang K, Chu M, Zhang W, Ma M, Chen B, Yang X, Zhao YS (2020) Angew Chem Int Ed 59:4456–4463
69. Das TP, Hahn EL (1958) Nuclear quadrupole resonance spectroscopy. In: Solid state physics (Suppl. 1), Academic Press, New York
70. Rigamonti A (1984) Adv Phys 33:115–191
71. Tobu Y, Ikeda R, Nihei TA, Gotoh K, Ishida H, Asaji T (2012) Phys Chem Chem Phys 14:12347–12354
72. Nogaj B (1988) Bull Chem Soc Jpn 61:549–556
73. Woessner DE, Gutowsky HS (1963) J Chem Phys 39:440–456
74. Horiuchi S, Ishibashi S (2020) J Phys Soc Jpn 89:051009
75. Davydov AS (1991) In: Solitons in molecular systems, 2nd ed. Kluwer Academic Publishers, Dordrecht, (English translation)

76. Takasu I, Sugawara T, Mochida T (2004) J Phys Chem B 108:18495–18499
77. Allen FH, Watson DG, Brammer A, Orpen AG, Taylor R (1999) Typical interatomic distances: organic compounds. In: Prince E (ed) International tables for crystallography volume C: Mathematical, physical and chemical tables, 2nd ed. Kluwer Academic Publishers, Dordrecht, pp 782–803
78. Donoshita M, Yoshida Y, Hayashi M, Ikeda R, Tanaka S, Yamamura Y, Saito K, Kawaguchi S, Sugimoto K, Kitagawa H (2021) Angew Chem Int Ed 60:22839–22848

Chapter 5
General Conclusion

Abstract The findings reported and described in each chapter are summarized in this chapter.

This Ph.D. thesis described the design of crystal structures using hydrogen bonds for molecular-layered cocrystals and proton–electron mixed conductors.

Chapter 1 provided the background information for the research involved in this study. The importance of the structure–property relationship and, hence, the design of crystal structures (i.e., crystal engineering) in molecular crystals was emphasized. Hydrogen bonding was highlighted as the most promising intermolecular interaction for designing crystal structures. Subsequently, the scope for advancement in the design of crystal structures based on hydrogen bonds was noted. In this context, two specific directions were identified: exploration of new functionalities and higher-order control of molecular packing; this thesis addressed both directions in Chaps. 2 and 3 and 4, respectively.

In Chap. 2, the design of crystal structures using hydrogen bonds was demonstrated as a rational method to construct molecular electron-conducting wires encapsulated in a proton-conducting matrix. Hydrogen-bonded networks designed at the molecular level served not only as proton-conduction pathways, but also as a container for electron donor molecules to align them in an appropriate one-dimensional manner.

In Chap. 3, the author demonstrated the use of hydrogen-bonded networks to achieve anisotropic changes in crystal shape. This study focused on a cocrystal composed of 2-pyrrolidone (Py) and chloranilic acid (CA) (2:1) with a two-dimensional sheet motif. The mechanism of the irreversible two-step single-crystal-to-single-crystal transformation induced by the translation of two-dimensional molecular sheets was discussed in terms of changes in the intersheet interactions and conformational disorder of Py.

In Chap. 4, the author demonstrated and described rational control over the stacking patterns of two-dimensional sheets principally maintained by hydrogen bonds. Based on calculations pertaining to the intersheet intermolecular interactions of the cocrystal described in Chap. 3, the stacking patterns of the two-dimensional

M. Donoshita, *Design of Crystal Structures Using Hydrogen Bonds on Molecular-Layered Cocrystals and Proton–Electron Mixed Conductor*, Recognizing Outstanding Ph.D. Research, https://doi.org/10.1007/978-981-99-7062-9_5

sheets were controlled using halogen-substituted cocrystals. Based on experimental and computational studies, the mechanism was discussed in terms of the balance between the enthalpy related to the intersheet intermolecular interactions and the entropy related to the molecular motional degrees of freedom of Py.

The design of crystal structures, or crystal engineering, is closely related to materials science and the specific physical and chemical properties of materials. In addition, crystal engineering is associated with several other research fields, such as host–guest chemistry, supramolecular chemistry, and biological chemistry, where intermolecular interactions, including hydrogen bonds, play pivotal roles. The author hopes that the results reported in this thesis will contribute greatly to the development of not only crystal engineering and associated fields of research but also the whole of science.

Curriculum Vitae

Masaki Donoshita

Institute for Materials Chemistry and Engineering,

Kyushu University, 744 Motooka, Nishi-ku, Fukuoka 819–0395, Japan.

+ 81–92-802–6452

donoshita@ms.ifoc.kyushu-u.ac.jp.

Education

April 2019–March 2022:

Ph.D. in Chemistry, Kyoto University

Supervisor: Professor Hiroshi Kitagawa.

April 2017–March 2019:

M.Sc. in Chemistry, Kyoto University

Supervisor: Professor Hiroshi Kitagawa.

April 2013–March 2017:

B.Sc. in Chemistry, Kyoto University

Supervisor: Professor Hiroshi Kitagawa.

Employment

April 2022–present, Assistant Professor at Institute for Materials Chemistry and Engineering, Kyushu University.

April 2019–March 2022, Japan Society of Promotion of Science Research Fellow at Graduate School of Science, Kyoto University.

Research Interest

Molecular assembly.

Physical property.

Molecular dynamics.

M. Donoshita, *Design of Crystal Structures Using Hydrogen Bonds on Molecular-Layered Cocrystals and Proton–Electron Mixed Conductor*, Recognizing Outstanding Ph.D. Research, https://doi.org/10.1007/978-981-99-7062-9

276

F

oK

Springer Theses

Recognizing Outstanding Ph.D. Research

Aims and Scope

The series "Springer Theses" brings together a selection of the very best Ph.D. theses from around the world and across the physical sciences. Nominated and endorsed by two recognized specialists, each published volume has been selected for its scientific excellence and the high impact of its contents for the pertinent field of research. For greater accessibility to non-specialists, the published versions include an extended introduction, as well as a foreword by the student's supervisor explaining the special relevance of the work for the field. As a whole, the series will provide a valuable resource both for newcomers to the research fields described, and for other scientists seeking detailed background information on special questions. Finally, it provides an accredited documentation of the valuable contributions made by today's younger generation of scientists.

Theses may be nominated for publication in this series by heads of department at internationally leading universities or institutes and should fulfill all of the following criteria

- They must be written in good English.
- The topic should fall within the confines of Chemistry, Physics, Earth Sciences, Engineering and related interdisciplinary fields such as Materials, Nanoscience, Chemical Engineering, Complex Systems and Biophysics.
- The work reported in the thesis must represent a significant scientific advance.
- If the thesis includes previously published material, permission to reproduce this must be gained from the respective copyright holder (a maximum 30% of the thesis should be a verbatim reproduction from the author's previous publications).
- They must have been examined and passed during the 12 months prior to nomination.
- Each thesis should include a foreword by the supervisor outlining the significance of its content.
- The theses should have a clearly defined structure including an introduction accessible to new PhD students and scientists not expert in the relevant field.

Indexed by zbMATH.

Mona Fuhrländer

Design Methods for Reducing Failure Probabilities with Examples from Electrical Engineering

Doctoral Thesis accepted by
Technische Universität Darmstadt, Germany

Springer

Author
Dr. Mona Fuhrländer
Computational Electromagnetics
Technische Universität Darmstadt
Darmstadt, Germany

Supervisors
Prof. Dr. Sebastian Schöps
Computational Electromagnetics
Technische Universität Darmstadt
Darmstadt, Germany

Prof. Dr. Winnifried Wollner
Department of Mathematics
Universität Hamburg
Hamburg, Germany

Apl. Prof. Dr. Ing. habil. Helmut Gräb
School of Computation, Information
and Technology
Technische Universität München (TUM)
München, Germany

ISSN 2190-5053 ISSN 2190-5061 (electronic)
Springer Theses
ISBN 978-3-031-37018-2 ISBN 978-3-031-37019-9 (eBook)
https://doi.org/10.1007/978-3-031-37019-9

This Springer imprint is published by the registered company Springer Nature Switzerland AG
The registered company address is: Gewerbestrasse 11, 6330 Cham, Switzerland

To my father

Supervisor's Foreword

The excellent Ph.D. thesis of Mona Fuhrländer deals with the calculation and optimization of the yield of electrotechnical products. The yield measures the probability of a product to fulfill its design specifications, i.e., to not fail. Its improvement is a classic topic of computer aided engineering that has been investigated for example in the analogue circuit community. However, it's nowadays also addressed in many other communities, e.g., in mathematics, stochastics, but especially also in reliability engineering, which is close to civil and mechanical engineering. This interdisciplinarity is also very well-reflected in the thesis which eventually combines machine learning methods with error control and robust optimization for product design. The topic, although established as explained, is gaining further interest due to the increasing focus on sustainability in manufacturing. In conclusion, the new and efficient simulation and optimization methods developed by Mona Fuhrländer allow to design new products that fulfill their specification even considering manufacturing imperfections while minimizing their environmental footprint.

The thesis is structured as follows: after a motivational chapter, industry-relevant model problems are explained in Chap. 2, i.e., their physics and corresponding finite element method are introduced. Two problems are exemplified, a low-frequency and a high-frequency problem from electrical engineering. At the end of each section, the relevant quantities of interest (QoIs) are described, laying the foundation for the subsequent chapters. Chapter 3 introduces the stochastic foundations necessary for (sampling-based) quantification and robust optimization methods. There are corresponding sub-chapters on stochastics, optimization and uncertainty quantification. Chapter 4 is devoted to methods for efficient yield estimation. The topic is formally introduced and an overview of existing methods is given. It introduces a new hybrid method based on Monte Carlo, stochastic collocation (related to polynomial interpolation) and adjoint error indicators, which allows adaptive model refinement. Then, the stochastic collocation method is replaced by Gaussian Process Regression (GPR) because, among other advantages, it allows more flexible options for model updates and parallelization strategies. Chapter 5 deals with (robust) optimization. In the preparation for gradient-based optimization methods, derivatives of the yield function are calculated and various single-objective and multi-objective optimization problems are formulated. Some of these formulations are new, but most of them are variants of problems from the literature, which have rarely been presented in such a systematic way. An adaptive Newton Monte Carlo method is proposed for yield optimization, which adaptively controls the number of Monte Carlo samples. Finally, the Hermite least-squares method and the well-known Bound constrained Optimization BY Quadratic Approximation (BOBYQA) method are combined for optimization with mixed gradient information. In Chap. 6, all the proposed methods are demonstrated with examples, two more realistic ones (a waveguide and an electric motor) as well as a long list of academic benchmarks are discussed. The thesis is concluded with a suitable conclusion and an interesting outlook in Chap. 7.

Darmstadt, Germany
April 2023

Prof. Dr. rer. nat. Sebastian Schöps

Parts of this thesis have been published in the following articles:

M. Fuhrländer and S. Schöps, *Hermite least squares optimization: a modification of BOBYQA for optimization with limited derivative information*, Optimization and Engineering, 2023.

M. Huber, M. Fuhrländer, and S. Schöps, *Multi-objective yield optimization for electrical machines using Gaussian processes to learn faulty designs*, IEEE Transactions on Industry Applications, vol. 59, no. 2, pp. 1340–1350, 2023.

M. Fuhrländer and S. Schöps, *Efficient yield optimization with limited gradient information*, in *Progress in Industrial Mathematics at ECMI 2021*, vol. 39, pp. 203–209, Springer, 2022.

M. Fuhrländer and S. Schöps, *Yield optimization using hybrid Gaussian process regression and a genetic multi-objective approach*, Advances in Radio Science, vol. 19, no. B, pp. 41–48, 2021.

M. Fuhrländer and S. Schöps, *A blackbox yield estimation workflow with Gaussian process regression applied to the design of electromagnetic devices*, Journal of Mathematics in Industry, vol. 10, no. 1, pp. 1–17, 2020.

M. Fuhrländer, N. Georg, U. Römer, and S. Schöps, *Yield optimization based on adaptive Newton-Monte Carlo and polynomial surrogates*, International Journal for Uncertainty Quantification, vol. 10, no. 4, pp. 351–373, 2020.

Acknowledgements

First of all, I would like to thank my supervisor Sebastian Schöps for the opportunity to work on this topic, for his guidance and support, and for his outstanding supervision. I highly appreciate the extraordinary working atmosphere he created and the many opportunities he gave me to improve my scientific qualifications as well as my soft skills.

Second, I thank Winnifried Wollner for the fruitful discussions on optimization and Helmut Gräb for raising my interest in the field of yield estimation. I thank both for co-examining this work.

Further, I like to thank various people for their collaboration and support: Niklas Georg and Ulrich Römer for the collaboration on adjoint error indicators for yield estimation, Morten Huber for his contribution to the investigation of the PMSM, Julien Bect for the discussions on Gaussian process regression, Frank Mosler for supervising my master thesis, Nick Mackinnon for proofreading this work, Carina Schuster, Heike Koch and Achim Wagner for their help and assistance with organizational issues and Christian Schmitt for solving all IT problems.

I would also like to thank all the current and former colleagues from the CEM group, the TEMF institute and the Graduate School CE. I am very grateful for our scientific and non-scientific discussions, our virtual and non-virtual tea times, our KWTs and retreats, game nights and GSC events, cake meetings and salad days, and in general for the great company over the past years.

Finally, I like to express my deep thanks to my husband, my parents and my family for supporting me and believing in me at all times.

This work is supported by the Graduate School CE within the Centre for Computational Engineering at Technische Universität Darmstadt.

Darmstadt, Germany
December 2022

Mona Fuhrländer

Contents

Acronyms

CST	CST Studio Suite® 2021
DFO	Derivative free optimization
DoF	Degree of freedom
FE	Finite element
FEM	Finite element method
GP	Gaussian process
GPR	Gaussian process regression
MC	Monte Carlo
MOO	Multi-objective optimization
MVP	Magnetic vector potential
PDE	Partial differential equation
PEC	Perfect electric conductor
PFS	Performance feature specifications
PM	Permanent magnet
PMSM	Permanent magnet synchronous machine
QoI	Quantity of interest
RMSE	Root mean square error
SC	Stochastic collocation
SOO	Single-objective optimization
SQP	Sequential quadratic programming
TE	Transverse electric
UQ	Uncertainty quantification

Symbols

Electromagnetic Field Quantities

$\mathbf{E}$	Electric field strength
$\mathbf{D}$	Electric flux density
$\mathbf{J}$	Electric current density
$\mathbf{H}$	Magnetic field strength
$\mathbf{B}$	Magnetic flux density
$\mathbf{M}_{\mathrm{pm}}$	Magnetization of a permanent magnet
$\mathbf{A}$	Magnetic vector potential
ϱ	Electric charge density
ϵ	Permittivity
μ	Permeability
ν	Reluctivity

Physical Quantities

S	S-parameter of the waveguide
ω	Angular frequency
τ	Torque of an electrical machine
C	Size of a permanent magnet
ϕ_j	Magnetization direction of the j-th permanent magnet
B_j	Magnitude of the magnetic field induced by the j-th permanent magnet

Finite Element Method

$\mathfrak{D}$	Differential operator of the model problem
$\mathfrak{f}$	Right-hand side of the primal model problem
$\mathfrak{q}$	Right-hand side of the adjoint model problem
$\mathfrak{u}$	Solution of the primal model problem
$\mathfrak{z}$	Solution of the adjoint model problem
$\mathbf{K}$	System matrix of the discretized finite element system
$\mathbf{f}$	Right-hand side of the discretized primal finite element system
$\mathbf{q}$	Right-hand side of the discretized dual finite element system
$\mathbf{u}$	Vector of degrees of freedom of the primal finite element system
$\mathbf{z}$	Vector of degrees of freedom of the dual finite element system
V	Solution space
h	Mesh size of the finite element discretization
N_h	Number of degrees of freedom
$\mathbf{w}_j$	Basis functions
$\mathbf{v}_i$	Test functions
$\varepsilon_{\mathrm{fe},h}$	Finite element error
$\varepsilon^{\mathrm{c}}_{\mathrm{fe},h}$	Computable formulation of the finite element error

Probability Theory

$\mathbb{P}$	Probability
$\mathbb{E}$	Expectation value
$\mathbb{V}$	Variance
Cov	Covariance
$\sum$	Covariance matrix
σ	Standard deviation
$\bar{\xi}$	Mean value of a random variable ξ
φ	Probability density function

Uncertainty Quantification

Q	Quantity of interest
$\widetilde{Q}$	Approximation of Q
$\mathbf{d}$	Deterministic design parameter
$\boldsymbol{\xi}$	Uncertain (design) parameter
r	Range parameter
T_{r}	Range parameter interval
T_{d}	Discretized range parameter interval

$\mathbf{c}$	Performance feature specification threshold
Y	Yield
N_{MC}	Number of Monte Carlo sample points
$\boldsymbol{\xi}^{(j)}$	j-th Monte Carlo sample point
N_{B}	Batch size for parallelized hybrid approaches
s	Safety factor in hybrid approaches
C_{H}	Hybrid sorting criterion in GPR-Hybrid approach
C_{EGL}	Echard, Gayton, Lemaire (EGL) sorting criterion in GPR-Hybrid approach
$\mathcal{T}$	Training data set
$k/\mathbf{k}/\mathcal{K}$	Kernel function/vector/matrix of a Gaussian process
$m/\mathbf{m}$	Mean function/vector of a Gaussian process
$\varepsilon_{\mathrm{MC}}$	Monte Carlo estimation error
$\varepsilon_{\mathrm{MC},Y}$	Yield estimation error
$\varepsilon_{\mathrm{sc}}$	Stochastic collocation error
$\varepsilon_{\mathrm{GPR}}$	Gaussian process regression error
$\mathcal{I}^{\varepsilon}$	Error interval in hybrid approaches
#HF	Computational costs measured by number of high-fidelity function evaluations
$\#\mathrm{HF}_{\mathrm{L}}$	Computational costs considering different levels of high-fidelity function evaluations
$\#\mathrm{HF}_{\mathrm{eff}}$	Effective computational costs considering parallel high-fidelity function evaluations

Hermite-Type Optimization

$\mathbb{X}$	Solution space in optimization
$\tilde{m}$	Quadratic model of objective function
$c/\mathbf{g}/\mathbf{H}$	Scalar/vectorial/matrix coefficients of $\tilde{m}$
Λ	Poisedness constant
l^i	Lagrange basis polynomial
$\mathbf{M}$	Interpolation or least squares system matrix
$\mathbf{b}$	Interpolation or least squares right-hand side
p_1	Number of interpolation points in BOBYQA and Hermite-type approaches
q_1	Number of basis polynomials in BOBYQA and Hermite-type approaches
y^j	j-th interpolation point in BOBYQA and Hermite-type approaches
n_{kd}	Number of known derivatives in Hermite-type approaches
$\mathcal{D}/\mathcal{I}_{\mathcal{D}}$	Set/index set of available first order derivatives in Hermite-type approaches
$\mathcal{D}_2/\mathcal{I}_{\mathcal{D}_2}$	Set/index set of available second order derivatives in Hermite-type approaches

Mathematical Notations

$\mathbb{R}$	Set of real numbers
$\mathbb{C}$	Set of complex numbers
$\mathbb{N}$	Set of natural numbers
i	Imaginary unit
∇	Gradient
$\nabla \cdot$	Div-operator
$\nabla \times$	Curl-operator
D	Computational domain
$\mathbf{n}$	Unit outward normal vector
$\mathbf{e}^j$	j-th unit vector
$\mathbf{x}_{\text{lb}}$ / $\mathbf{x}_{\text{ub}}$	Lower/upper bound (elementwise) of a vector $\mathbf{x}$
$\Phi / \boldsymbol{\Phi}$	Set/vector of polynomial basis functions
$\mathbf{A}^\dagger$	Adjoint of an operator or a matrix $\mathbf{A}$

Chapter 1
Introduction

This introductory chapter describes the motivation behind the research topic and the research goals. First, the reader is introduced to the field of yield estimation and optimization and a brief overview of related work is provided. Then, the research goals and the contribution of this work are detailed. Finally, we outline the structure of this thesis.

1.1 Motivation and Background

Electrotechnical devices, such as antennas, electrical machines or waveguides, are designed in order to achieve high performance requirements. Technical development allows the design of devices closer to their physical limits. However, there are often manufacturing imperfections, natural material variations or unpredictable environmental influences. This may lead to deviations in the geometry or material parameters, or boundary conditions, which may lead to deviations in operation. This can cause violations of performance requirements which implies that the device is useless due to malfunction. The aim of measuring and reducing this risk is a necessity for efficient and sustainable production; thereby resources, time and money can be saved. Hence, the quantification of uncertainty and optimization under uncertainty becomes increasingly popular. The estimation and optimization methods discussed and proposed in this work are motivated by the design of electromagnetic devices, but are not limited to this application. The workflows are equally applicable for the design of mechanical components, for example.

The *yield* is the fraction of realizations in a manufacturing process under uncertainties that fulfills all performance requirements. We follow the definition of [1] and interpret it as the counterpart of the *failure probability* (i.e., yield = 1− failure probability). Thus, it may be interpreted as the *probability of success* of the manufacturing process. However, the term failure probability is also often, e.g. in [2],

M. Fuhrländer, *Design Methods for Reducing Failure Probabilities with Examples from Electrical Engineering*, Springer Theses,
https://doi.org/10.1007/978-3-031-37019-9_1

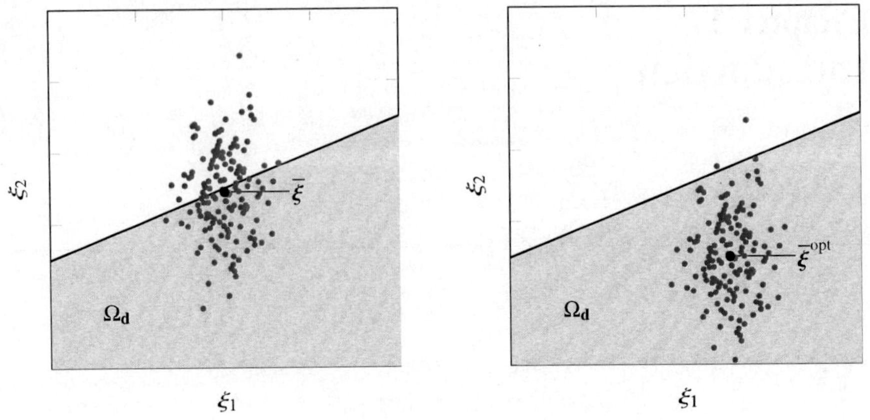

Fig. 1.1 Illustration of yield estimation with a sampling based method. Example with one performance requirement and two uncertain parameters ξ_1 and ξ_2. Left: original design, right: optimized design

used to refer to failures in lifetime management of a product. To avoid confusion, we formulate our results in terms of the yield which is particularly common in chip design, i.e., we investigate the *maximization* of the yield instead of the *minimization* of the failure probability.

In practice, the yield cannot be calculated directly, but must be estimated. Commonly, sampling methods are employed. Therefore, a large sample set of (virtual) designs is generated and the performance requirements are checked. The quotient of sample points fulfilling the requirements and all considered sample points forms the yield estimate, cf. MC analysis. In Fig. 1.1 the sampling method for yield estimation is illustrated on an example with one performance requirement and two uncertain parameters ξ_1 and ξ_2. The gray shaded area indicates the *safe domain*, i.e., the set of design parameter combinations for which all performance requirements are fulfilled. The white area represents the failure domain. On the left we see a possible initial design, on the right an optimized design.

When designing complex electrotechnical devices, the performance requirements are typically restrictions involving the partial differential equations (PDEs) describing the electromagnetic field, i.e., Maxwell's equations. For solving Maxwell's equations, numerical methods are employed, e.g. the FEM. Even though advances in research and computational capacity make solving PDEs increasingly efficient, it remains a computationally expensive task. Hence, evaluating the PDEs for each sample point in a large sample set quickly becomes computationally prohibitive. This motivates the aim for efficient yield estimation and optimization techniques.

In practice, often the simultaneous optimization of reliability (i.e., yield) and other performance indicators (e.g. size, costs,...) is requested. In MOO a trade-off between the different objective functions is sought. This usually requires many evaluations of the objective functions, including the yield estimate. Without efficient yield estimation techniques, MOO for yield optimization is not realizable for industrial problems.

In order to use gradient based optimization methods for yield optimization, all partial derivatives of the yield estimate must be available. Depending on the type of uncertainties encountered in the manufacturing process, it occurs that the partial derivatives with respect to some optimization variables are available, but with respect to other optimization variables they are not. Finite differences approximations to enable gradient based solvers would require expensive extra evaluations of the yield estimate; DFO solvers would not take advantage of the gradient information available. Both are disadvantageous for the overall efficiency of the optimization process. This motivates the development of an optimization method suited to this kind of mixed gradient information (without approximating the missing gradients).

1.2 Related Work

Our definition of the yield originates from [3]. The same term is used in [4–6]. The classic MC analysis is described in [7, Chap. 5] and applied for yield estimation for example in [3]. Since this approach is computationally prohibitive in case of involved performance requirements, there is much research on increasing the efficiency of sampling based techniques. One way is to reduce the size of the design sample set. To avoid degradation in accuracy, methods like importance sampling [8] and subset simulation [9–11] are proposed. Another way is to reduce the computational effort of evaluating one sample point. In multilevel MC methods this is achieved by considering different accuracy levels of the underlying functions [12], e.g. by varying the accuracy in the simulation model of the PDEs. In surrogate based approaches, the underlying function is approximated by a function that can be cheaply evaluated. Then, the MC analysis is conducted on the resulting surrogate model [13]. In the literature, surrogate models based on linear regression [14], SC [15], GPR [16] and neural networks [17] are investigated. GPR for yield or failure probability estimation is for example discussed in [18, 19], SC for approximating PDEs under uncertainty for example in [20, 21]. Hybrid approaches, evaluating some sample points on the surrogate model and a small subset on the original high fidelity model are proposed in [1, 22–25].

Yield optimization using worst case analysis is discussed in [3, 26], using Bayesian optimization for example in [4, 27]. In Bayesian optimization, the GP approximates the objective function itself, not the underlying PDEs. MOO is employed in the context of uncertainty optimization in [27, 28]. Applied to electrical machines, it is investigated in [26, 29, 30].

Optimization with mixed gradient information is usually approached with one of the following strategies: Firstly, by using derivative approximations, for example finite differences for first order derivatives, and Broyden-Fletcher-Goldfarb-Shanno (BFGS) updates for second order derivatives [31], and secondly, by using DFO methods without exploiting available derivative information [32]. Commonly used DFO methods are swarm algorithms [33, 34] or the PDFO framework [35], including the BOBYQA method [36]. In [37, 38], modifications of the original BOBYQA method are proposed, in [32, Chap. 11] a version with provable convergence is discussed.

1.3 Contribution

In this work efficient yield estimation and optimization methods are investigated. The industrial application of MOO under uncertainty is discussed and a new optimization method for mixed gradient information is proposed. The research contributions can be summarized as follows:

1. Development of a hybrid yield estimation approach based on SC and adjoint error indicators (SC-Hybrid). Testing on a dielectrical waveguide shows that the computational effort can be drastically reduced compared to classic MC methods. High estimation accuracy is ensured by construction, since all error sources are controlled, i.e., finite element (FE) error, SC error and MC error, cf. [39].
2. Development of a non-intrusive hybrid yield estimation approach based on GPR (GPR-Hybrid). GP updates, parallelization and sample sorting strategies are investigated, cf. [40].
3. Demonstration that the efficiency of the GPR-Hybrid approach allows yield estimation and (multi-objective) yield optimization in real world applications such as the design of waveguides, cf. [40, 41], or electrical machines, cf. [42]. This enables the design of (electrotechnical) components for highly reliable manufacturing processes, under consideration of all kinds of uncertainties. The rejections due to malfunction can be significantly reduced by product design involving reliability (yield) optimization.
4. Development of an adaptive Newton-MC method for yield optimization with full gradient information, cf. [39, 41]. Numerical tests show high efficiency, due to adaptive sample size increase and combination with the above mentioned hybrid approaches for yield estimation.
5. Modification of the adaptive Newton-MC method, cf. [43], and application of the SQP method, cf. [44], for optimization with mixed gradient information using approximations of first and second order derivatives.
6. Development of Hermite-type optimization methods for mixed gradient information (without approximations of the derivatives). Proof of global convergence, performance study based on a test set of nonlinear optimization problems and demonstration of practical applicability for yield optimization on the example of a waveguide, cf. [44].

1.4 Outline

The thesis is structured as follows. In Chap. 2 the fundamental concepts of modeling electromagnetic phenomena are introduced. We start with the formulation of Maxwell's equations in Sect. 2.1 and the material relations in Sect. 2.2. In this work, the PDEs originating from Maxwell's equations and describing electromagnetic field quantities are solved numerically with the FEM. Hence, in Sect. 2.3, the method is briefly covered. Two benchmark problems will be used to investigate the performance of the proposed methods: a rectangular waveguide and a permanent magnet synchronous machine (PMSM). The corresponding problem formulations, boundary conditions and quantities of interest (QoIs) are introduced in Sects. 2.4 and 2.5, respectively. The mathematical concepts discussed in Chap. 3 and the yield estimation and optimization approaches proposed in Chaps. 4 and 5 are independent of the application to electromagnetic phenomena and can be easily adapted for example to the design of mechanical devices. The reader who is not interested in this application area may skip Chap. 2 and start with Chap. 3.

In Chap. 3 we summarize mathematical foundations which are relevant for the robust design methods in the subsequent chapters. Section 3.1 is dedicated to the basics of stochastic and probability theory, Sect. 3.2 recalls the fundamentals of optimization and briefly describes some well-established optimization methods, and Sect. 3.3 provides an overview of techniques commonly used in uncertainty quantification.

Chapters 4 and 5 contain the core contributions of this thesis. Chapter 4 is dedicated to efficient yield estimation. After defining the yield in Sect. 4.1 and providing a review of existing yield estimation methods in Sect. 4.2, the new hybrid approaches are proposed. In Sect. 4.3 the SC-Hybrid approach is introduced, which is a method based on SC and adjoint error indicators and promotes model refinement. The GPR-Hybrid approach is proposed in Sect. 4.4. Model updates and parallelization strategies for this non-intrusive GPR based hybrid method are discussed.

Chapter 5 addresses optimization. In preparation for gradient based optimization solvers, first and second order derivatives of the yield function are calculated in Sect. 5.1. In Sect. 5.2 single- and multi-objective yield optimization problems are formulated. In Sect. 5.3 the adaptive Newton-MC method for yield optimization is proposed. The Hermite least squares and Hermite BOBYQA (Bound constrained Optimization BY Quadratic Approximation) methods for optimization with mixed gradient information are introduced in Sect. 5.4.

A numerical evaluation of all proposed methods is provided in Chap. 6. Therefore, the waveguide and PMSM model problems are specified in Sect. 6.1. Then, the yield estimation methods, the adaptive Newton-MC yield optimization and the MOO approaches are tested on the model problems in Sects. 6.2, 6.3 and 6.4, respectively. In Sect. 6.5 the Hermite-type approaches are investigated, first on a large test set for benchmarking, then on the waveguide problem. Finally, the thesis is concluded in Chap. 7.

References

1. Tyagi A, Jonsson X, Beelen T, Schilders W (2018) Hybrid importance sampling Monte Carlo approach for yield estimation in circuit design. J Math Ind 8. https://doi.org/10.1186/s13362-018-0053-4
2. Rahimpour S, Tarzamni H, Kurdkandi NV, Husev O, Vinnikov D, Tahami F (2022) An overview of lifetime management of power electronic converters. IEEE Access 10:109688–109711. https://doi.org/10.1109/ACCESS.2022.3214320
3. Gräb HE (2007) Analog design centering and sizing. Springer (2007)
4. Wang M, Lv W, Yang F, Yan C, Cai W, Zhou D, Zeng X (2017) Efficient yield optimization for analog and SRAM circuits via Gaussian process regression and adaptive yield estimation. IEEE Trans Comput Aided Des Integr Circuits Syst 37(10):1929–1942
5. Ciccazzo A, Pillo GD, Latorre V (2014) Yield optimization in electronic circuits design. In: European consortium for mathematics in industry. Springer, pp 445–452
6. Shavezipur M, Ponnambalam K, Khajepour A, Hashemi S (2008) Fabrication uncertainties and yield optimization in MEMS tunable capacitors. Sens Actuat A 147(2):613–622
7. Hammersley JM, Handscomb DC (1964) Monte Carlo methods. Methuen & Co Ltd
8. Gallimard L (2019) Adaptive reduced basis strategy for rare-event simulations. Int J Numer Meth Eng 1:1–20. https://doi.org/10.1002/nme.6135
9. Au SK, Beck JL (2001) Estimation of small failure probabilities in high dimensions by subset simulation. Probab Eng Mech 16:263–277. https://doi.org/10.1016/S0266-8920(01)00019-4
10. Bect J, Li L, Vazquez E (2017) Bayesian subset simulation. SIAM/ASA J Uncertain Quantific 5(1):762–786. https://doi.org/10.1137/16m1078276
11. Kouassi A, Bourinet J, Lalléchére S, Bonnet P, Fogli M (2016) Reliability and sensitivity analysis of transmission lines in a probabilistic EMC context. IEEE Trans Electromagn Compatib 58(2):561–572. https://doi.org/10.1109/TEMC.2016.2520205
12. Giles MB (2015) Multilevel Monte Carlo methods. Acta Numer 24. https://doi.org/10.1017/S09624929
13. Bogoclu C, Roos D (2016) A benchmark of contemporary metamodeling algorithms. https://doi.org/10.7712/100016.2039.7645
14. Rao CR, Toutenburg H (1999) Linear models: least squares and alternatives, 2 edn. Springer (1999)
15. Babuška I, Nobile F, Tempone R (2007) A stochastic collocation method for elliptic partial differential equations with random input data. SIAM J Numer Anal 45(3):1005–1034. https://doi.org/10.1137/100786356
16. Rasmussen CE, Williams CK (2006) Gaussian processes for machine learning. The MIT Press
17. Goodfellow I, Bengio Y, Courville A (2016) Deep learning. The MIT Press
18. Xiao S, Oladyshkin S, Nowak W (2020) Reliability analysis with stratified importance sampling based on adaptive kriging. Reliab Eng Syst Safety 197
19. Zhang J, Taflanidis AA (2019) Accelerating MCMC via kriging-based adaptive independent proposals and delayed rejection. Comput Methods Appl Mech Eng 355:1124–1147
20. Kouri DP, Heinkenschloss M, Ridzal D, van Bloemen Waanders BG (2013) A trust-region algorithm with adaptive stochastic collocation for PDE optimization under uncertainty. SIAM J Sci Comput 35(4):A1847–A1879
21. Loukrezis D, Römer U, De Gersem H (2019) Assessing the performance of Leja and Clenshaw-Curtis collocation for computational electromagnetics with random input data. Int J Uncertain Quantific 9(1)
22. Li J, Xiu D (2010) Evaluation of failure probability via surrogate models. J Comput Phys 229(23):8966–8980. https://doi.org/10.1016/j.jcp.2010.08.022
23. Li J, Li J, Xiu D (2011) An efficient surrogate-based method for computing rare failure probability. J Comput Phys 230(24):8683–8697
24. Bect J, Ginsbourger D, Li L, Picheny V, Vazquez E (2012) Sequential design of computer experiments for the estimation of a probability of failure. Stat Comput 22(3):773–793

25. Butler T, Wildey T (2018) Utilizing adjoint-based error estimates for surrogate models to accurately predict probabilities of events. Int J Uncertain Quantif 8(2):143–159. https://doi.org/10.1615/Int.J.UncertaintyQuantification.2018020911
26. Bontinck Z, Lass O, Schöps S, De Gersem H, Ulbrich S, Rain O (2018) Robust optimisation formulations for the design of an electric machine. IET Sci, Meas Technol 12(8):939–948
27. Feliot P, Bect J, Vazquez E (2017) A bayesian approach to constrained single- and multi-objective optimization. J Global Optim 67(1):97–133
28. Taguchi G, Chowdhury S, Wu Y (2005) Taguchi's quality engineering handbook. Wiley. https://doi.org/10.1002/9780470258354
29. Di Barba P (2010) Multiobjective shape design in electricity and magnetism. Lecture notes in electrical engineering. Springer (2010)
30. Lei G, Bramerdorfer G, Liu C, Guo Y, Zhu J (2020) Robust design optimization of electrical machines: A comparative study and space reduction strategy. IEEE Trans Energy Convers 36(1):300–313
31. Ulbrich M, Ulbrich S (2012) Nichtlineare Optimierung. Birkhäuser
32. Conn AR, Scheinberg K, Vicente LN (2009) Introduction to derivative-free optimization. SIAM
33. Audet C, Hare W (2017) Derivative-free and blackbox optimization. Springer. https://doi.org/10.1007/978-3-319-68913-5
34. Srinivas N, Deb K (1994) Multiobjective optimization using nondominated sorting in genetic algorithms. Evol Comput 2(3):221–248. https://doi.org/10.1162/evco.1994.2.3.221
35. Ragonneau TM, Zhang Z (2022) PDFO - Powell's derivative-free optimization solvers. https://www.pdfo.net/docs.html. Accessed 25 June 2022
36. Powell MJ (2009) The BOBYQA algorithm for bound constrained optimization without derivatives. Cambridge NA Report NA2009/06, vol 26. University of Cambridge, Cambridge, pp 26–46 (2009)
37. Cartis, C, Fiala, J, Marteau, B, Roberts, L (2019) Improving the flexibility and robustness of model-based derivative-free optimization solvers. ACM Trans Math Softw 45(3). https://doi.org/10.1145/3338517
38. Cartis C, Roberts L, Sheridan-Methven O (2021) Escaping local minima with local derivative-free methods: a numerical investigation. Optimization 1–31. https://doi.org/10.1080/02331934.2021.1883015
39. Fuhrländer M, Georg N, Römer U (2020) Schöps S (2020) Yield optimization based on adaptive Newton-Monte Carlo and polynomial surrogates. Int J Uncertain Quantific 10(4):351–373. https://doi.org/10.1615/Int.J.UncertaintyQuantification.2020033344
40. Fuhrländer M, Schöps S (2020) A blackbox yield estimation workflow with Gaussian process regression applied to the design of electromagnetic devices. J Math Ind 10(1):1–17
41. Fuhrländer M, Schöps S (2021) Yield optimization using hybrid Gaussian process regression and a genetic multi-objective approach. Adv Radio Sci 19(B):41–48
42. Huber M, Fuhrländer M, Schöps S (2013) Multi-objective yield optimization for electrical machines using Gaussian processes to learn faulty designs. IEEE Trans Ind Appl 59(2):1340–1350
43. Fuhrländer M, Schöps S (2022) Efficient yield optimization with limited gradient information. In: Progress in industrial mathematics at ECMI 2021, vol 39. Springer, pp 203–209
44. Fuhrländer M, Schöps S (2023) Hermite least squares optimization: a modification of BOBYQA for optimization with limited derivative information. Optim Eng. https://doi.org/10.1007/s11081-023-09795-y

Chapter 2
Modeling

The yield estimation and the optimization methods discussed in Chaps. 3–5 are independent of electromagnetic phenomena. The concepts can be applied to any kind of design processes which are affected by uncertainties. The reader who is not interested in the application to electrotechnical devices can skip this chapter, in which an introduction to electromagnetism is provided.

Before any estimation or optimization technique can be applied, a computer model of the physical object is required. This model is based on mathematical equations. When dealing with spatially resolved problems, these equations are typically partial differential equations (PDEs). In this chapter we will firstly recall Maxwell's equations, the fundamental equations describing electromagnetic phenomena. We briefly summarize the FEM for solving the resulting system of PDEs. In this work, two benchmark problems are discussed, a rectangular waveguide and a permanent magnet synchronous machine (PMSM). Both problems are introduced. For the waveguide problem, we formulate Maxwell's equations in the time harmonic case, and for the PMSM problem in the magnetostatic case. From this, we derive the curl-curl equations for each problem, discuss suitable boundary conditions, define corresponding quantities of interest (QoIs) and build their finite element (FE) approximations.

2.1 Maxwell's Equations

Macroscopic electromagnetic fields are described by Maxwell's equations [1–4]. Let $\mathbf{E}$ denote the *electric field strength*, $\mathbf{D}$ the *electric flux density*, $\mathbf{J}$ the *electric current density*, $\mathbf{H}$ the *magnetic field strength* and $\mathbf{B}$ the *magnetic flux density*. All these quantities are vector fields $\mathbb{R} \times \mathbb{R}^3 \to \mathbb{R}^3$ depending on time $t \in \mathbb{R}$ and space $\mathbf{r} = (x, y, z) \in \mathbb{R}^3$. Let ϱ denote the electric charge density, which is a scalar field $\mathbb{R} \times \mathbb{R}^3 \to \mathbb{R}$ also depending on time and space. Further, let $V \subseteq \mathbb{R}^3$ denote a volume with boundary surface ∂V and $S \subseteq \mathbb{R}^2$ a surface with boundary contour ∂S.

M. Fuhrländer, *Design Methods for Reducing Failure Probabilities with Examples from Electrical Engineering*, Springer Theses,
https://doi.org/10.1007/978-3-031-37019-9_2

Following [3, Sect. 1.1.1], Maxwell's equations in rest, i.e., with V and S independent of t, in *integral form* are given by

$$\int_{\partial S} \mathbf{E} \cdot \mathrm{d}\mathbf{s} = -\int_S \frac{\partial}{\partial t}\mathbf{B} \cdot \mathrm{d}\mathbf{S}, \tag{2.1}$$

$$\int_{\partial S} \mathbf{H} \cdot \mathrm{d}\mathbf{s} = \int_S \left(\frac{\partial}{\partial t}\mathbf{D} + \mathbf{J}\right) \cdot \mathrm{d}\mathbf{S}, \tag{2.2}$$

$$\int_{\partial V} \mathbf{D} \cdot \mathrm{d}\mathbf{S} = \int_V \varrho \, \mathrm{d}V, \tag{2.3}$$

$$\int_{\partial V} \mathbf{B} \cdot \mathrm{d}\mathbf{S} = 0. \tag{2.4}$$

Let $\mathbf{n}$ denote the unit outward normal vector to the surface S and $\boldsymbol{\tau}$ the unit tangent vector to the curve ∂S. Then, we can write $\mathrm{d}\mathbf{S} \equiv \mathbf{n}\,\mathrm{d}S$ for the oriented infinitesimal surface element. Further, $\mathrm{d}\mathbf{s} \equiv \mathbf{n}\,\mathrm{d}s$ is an oriented infinitesimal curve element and $\mathrm{d}V$ an infinitesimal volume element.

Equation (2.1) is called *Faraday's law of induction* and states that a change over time in the magnetic flux induces a circulation of the electric field. For example, if a magnet is moved through a wire loop, an electric current in the wire is induced. Equation (2.2) is called *Maxwell-Ampère's circuit law*. It states that electric current or a change in the electric flux, i.e., displacement current, within a closed loop induces a circulating magnetic field around that loop. The Eqs. (2.3) and (2.4) are called the *(electric) Gauss law* and the *magnetic Gauss law*, respectively. The electric Gauss law states that the total electric flux through the closed surface of a volume is proportional to the electric charge within this volume. This implies that the electric field lines start in positive charges and end in negative charges. On the other hand, the magnetic Gauss law states that the total magnetic flux through the closed surface of a volume is equal to zero. This implies that there is no magnetic analogue to electric charges, i.e., there are no magnetic monopoles, but only pairs of north and south pole in nature.

For a smooth vector field $\mathbf{F}$ the Stokes theorem

$$\int_{\partial S} \mathbf{F} \cdot \mathrm{d}\mathbf{s} = \int_S \nabla \times \mathbf{F} \cdot \mathrm{d}\mathbf{S} \tag{2.5}$$

and the Gauss theorem

$$\int_{\partial V} \mathbf{F} \cdot \mathrm{d}\mathbf{S} = \int_V \nabla \cdot \mathbf{F} \cdot \mathrm{d}V \tag{2.6}$$

hold. Here, $\nabla\cdot$ denotes the div-operator and $\nabla\times$ the curl-operator. Applying (2.5) and (2.6) allows transformation of Maxwell's equations (2.1)–(2.4) into their *differential form*, see e.g. [3, Sect. 1.1.2]. They read

$$\nabla \times \mathbf{E} = -\frac{\partial}{\partial t}\mathbf{B}, \tag{2.7}$$

$$\nabla \times \mathbf{H} = \frac{\partial}{\partial t}\mathbf{D} + \mathbf{J}, \tag{2.8}$$

$$\nabla \cdot \mathbf{D} = \varrho, \tag{2.9}$$

$$\nabla \cdot \mathbf{B} = 0. \tag{2.10}$$

In general, Maxwell's equations can be defined on the whole space domain $\mathbb{R}^3$ and time domain $\mathbb{R}^{\geq 0}$, see e.g. [5, Chap. 1]. However, in the remainder of this work we only consider the domains $D \subset \mathbb{R}^3$ and $T \subset \mathbb{R}^{\geq 0}$ for space and time, respectively. For simplicity, the computational domain D is assumed to be bounded, simply connected and to have a Lipschitz continuous boundary.

In $\mathbb{R}^3$ the de Rham sequence describes the relation between function spaces based on kernels and images of the differential operators ∇ (grad), $\nabla\cdot$ (div) and $\nabla\times$ (curl). It can be shown that the electromagnetic fields can be identified with the function spaces in the de Rham sequence [5, Sects. 3.7, 5.1]. To ensure correct approximations of the field quantities when solving Maxwell's equations with discretization, it is important to find finite dimensional spaces following the same structure.

2.2 Material Relations

Maxwell's equations are not sufficient to describe electromagnetic fields. They are complemented by equations expressing the properties of the medium in which the wave propagates, the so-called *material relations* or *constitutive relations* [3, Sect. 1.1.3]. For brevity of notation we assume that the material law behaves linearly, i.e., independent of the field strength, isotropically, i.e., the material properties are direction-independent, only spatially inhomogeneously, i.e., dependent on the space, and independent of time and frequency. However, these assumptions serve only to simplify the notation and have no effect on the methods proposed in this thesis. Let $\epsilon \equiv \epsilon(\mathbf{r})$ denote the electric permittivity and $\mu \equiv \mu(\mathbf{r})$ the magnetic permeability. With the assumptions mentioned above, both quantities are scalar fields. They are each composed of the vacuum permittivity ϵ_0 and the relative permittivity ϵ_r, respectively vacuum and relative permeability μ_0 and μ_r. We can write $\epsilon = \epsilon_0\epsilon_\mathrm{r}$ and $\mu = \mu_0\mu_\mathrm{r}$. We obtain the following material relations

$$\mathbf{D} = \epsilon\mathbf{E}, \tag{2.11}$$

$$\mathbf{B} = \mu\mathbf{H}. \tag{2.12}$$

In case of anisotropic materials, the scalar material constants would be replaced by tensor fields.

2.3 Finite Element Method

In order to calculate the field quantities introduced in the previous sections, Maxwell's equations (2.7)–(2.10) and the material relations (2.11)–(2.12) are coupled and we obtain a PDE. In Sects. 2.4 and 2.5 we will derive the PDEs for two problems that will be further investigated in this work. However, this section is dedicated to solving the resulting PDEs in general. In practice, numerical methods are usually applied to solve Maxwell's equations. In this work we focus on the FEM, which is a well established method to solve PDEs in the frequency domain [5–7]. Hence, in this section we will briefly recall the basic procedure and the required function spaces.

Function Spaces

Let D be a simply connected and bounded computational domain and let $\mathbf{a}(\mathbf{r}) \equiv \mathbf{a}$ and $\mathbf{b}(\mathbf{r}) \equiv \mathbf{b}$ be three-dimensional vector functions. We define their inner product by

$$(\mathbf{a}, \mathbf{b})_D = \int_D \mathbf{a} \cdot \overline{\mathbf{b}} \, \mathrm{d}\mathbf{r}, \tag{2.13}$$

where $\overline{\mathbf{b}}$ is the complex conjugate. Let

$$\left(L^2(D)\right)^3 := \{\mathbf{a} : (\mathbf{a}, \mathbf{a})_D < \infty\} \tag{2.14}$$

denote the complex vector function space of square integrable three dimensional vector functions on D. Then, the solution space is defined as the Hilbert space

$$H(\text{curl}, D) := \left\{\mathbf{a} \in \left(L^2(D)\right)^3 : (\nabla \times \mathbf{a}, \nabla \times \mathbf{a})_D < \infty\right\}, \tag{2.15}$$

cf. [5], where $\nabla \times \mathbf{a}$ is the weak curl of $\mathbf{a}$, i.e., for all infinitely differentiable functions $\mathbf{g}$ with compact support holds

$$\int_D (\nabla \times \mathbf{a}) \cdot \mathbf{g} \, \mathrm{d}\mathbf{r} = \int_D \mathbf{a} \cdot (\nabla \times \mathbf{g}) \, \mathrm{d}\mathbf{r}. \tag{2.16}$$

Note that the existence of partial derivatives is assumed in the weak sense.

In preparation for the use of adjoint error indicators, we define the adjoint $\mathbf{A}^\dagger$ of an operator $\mathbf{A}$ by the property [8, Sect. 7.6]

$$(\mathbf{A}\mathbf{a}, \mathbf{b})_D = (\mathbf{a}, \mathbf{A}^\dagger \mathbf{b})_D. \tag{2.17}$$

An operator $\mathbf{A}$ is called self-adjoint, if

$$\mathbf{A} = \mathbf{A}^{\dagger} \tag{2.18}$$

in (2.17) holds [8, Sect. 7.7].

The Method

We consider a model problem originating from Maxwell's equations

$$\mathfrak{D}\mathfrak{u}(\mathbf{x}) = \mathfrak{f} \text{ in } D, \tag{2.19}$$

where $\mathfrak{D}$ denotes a differential operator, $\mathfrak{f}$ a forcing term and $\mathfrak{u}(\mathbf{x})$ the continuous solution. The computational domain is divided into a finite number of elements. These elements are of simple geometric form, e.g. triangles in 2D or tetrahedrons in 3D. This discretization is realized by constructing a mesh of the original object, i.e., its computational domain. A coarse triangular mesh on the computational domain $[0, 1] \times [0, 1]$ is illustrated in Fig. 2.1. The mesh size parameter in this example is $h = 1/\sqrt{2}$.

We introduce a subspace V of $H(\text{curl}, D)$ and a finite dimensional function space $V_h \subset V$. Let $\{\mathbf{w}_j\}_{j=1}^{N_h} \subset V_h$ be basis functions with $\mathbf{w}_j : D \to \mathbb{R}^3$. In Fig. 2.2, for example, first and second order basis functions for a one dimensional problem are illustrated. The solution of (2.19) can be approximated as a linear combination of the basis functions, i.e.,

$$\mathfrak{u}(\mathbf{x}) \approx \mathfrak{u}_{\mathrm{h}}(\mathbf{x}) = \sum_{j=1}^{N_h} u_j \mathbf{w}_j(\mathbf{x}), \tag{2.20}$$

where N_h denotes the number of degrees of freedom (DoFs). Next, we introduce some test functions $\mathbf{v}_i$ living in V. Note that the test and basis functions are prob-

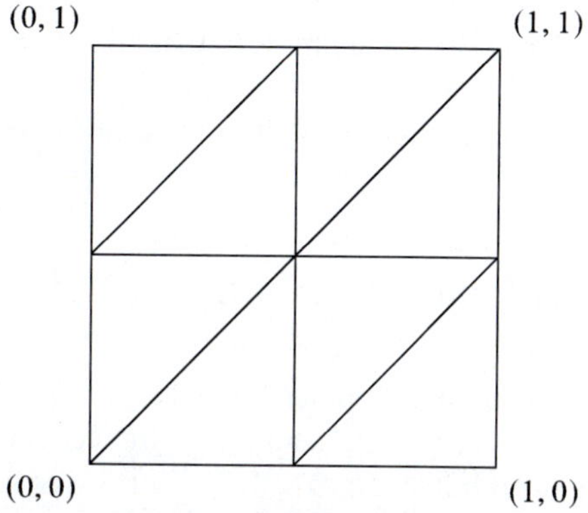

Fig. 2.1 Illustration of a coarse triangular discretization of the computational domain $[0, 1] \times [0, 1]$, with mesh size $h = 1/\sqrt{2}$

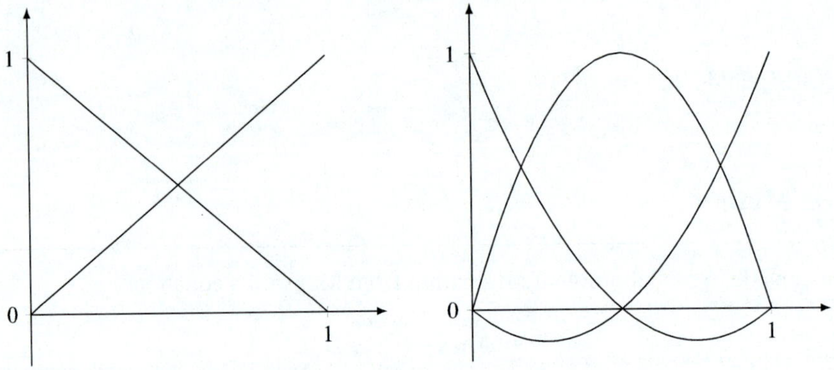

Fig. 2.2 Illustration of first and second order basis functions for a 1D problem

lem dependent and will be discussed in more detail when the specific problems are introduced, cf. Sects. 2.4 and 2.5. We insert the approximation (2.20) into the problem (2.19), multiply with the test functions and integrate over the computational domain. We obtain

$$\sum_{j=1}^{N_h} u_j \int_D (\mathfrak{D}\mathbf{w}_j(\mathbf{x}))\mathbf{v}_i(\mathbf{x})\,\mathrm{d}D = \int_D \mathfrak{f}\,\mathbf{v}_i(\mathbf{x})\,\mathrm{d}D \quad \forall i = 1, \ldots, N_h. \tag{2.21}$$

This can be rewritten into matrix notation

$$\mathbf{Ku} = \mathbf{f} \Leftrightarrow \mathbf{u} = \mathbf{K}^{-1}\mathbf{f}. \tag{2.22}$$

where $\mathbf{K}$ and $\mathbf{f}$ can be computed and then the system can be solved for the unknowns $u_j, j = 1, \ldots, N_h$. In the following, basis functions and test functions are chosen to be equal (Ritz-Galerkin), i.e., $\mathbf{w}_i = \mathbf{v}_i, i = 1, \ldots, N_h$.

Other numerical methods for solving Maxwell's equations are for example the boundary element method (BEM) [9, 10], the finite integration technique (FIT) [11], the discontinuous Galerkin method [12] or the finite difference method [13, 14]. However, the numerical method applied for solving the QoI makes no conceptional difference for the estimation and optimization techniques discussed in this work. Hence, these methods are not further discussed.

Adjoint Error Indicator for FEM

There are different approaches to measure the FEM approximation error, e.g. residual based error estimators [15, 16], recovery based error indicators [17], or adjoint based error indicators [18, 19]. While an error estimator provides a strict upper bound, the error indicator is a measure for the locally expected size of the error. Indicators are commonly used for determining the region where an approximation model should be improved, e.g., by increasing the number of finite elements. An overview over different error estimation approaches for FEM is provided in [20]. In this work we will focus on adjoint based error indicators.

We formulate the continuous adjoint equation to (2.19) and obtain

$$\mathfrak{D}^{\dagger}\mathfrak{z}(\mathbf{x}) = \mathfrak{q} \text{ in } D. \tag{2.23}$$

Let $Q(\mathfrak{u}(\mathbf{x})) \equiv Q(\mathbf{x})$ denote a QoI depending on the solution of (2.19), and $Q_h(\mathfrak{u}_h(\mathbf{x})) \equiv Q_h(\mathbf{x})$ the QoI depending on the approximation, respectively. Following [18, 19], an adjoint error indicator is introduced in order to quantify the FE error. The FE error indicator is derived by

$$\begin{aligned}
\varepsilon_{\text{fe},h}(\mathbf{x}) &= |Q(\mathbf{x}) - Q_h(\mathbf{x})| \\
&= \left|\left(\mathfrak{q}, \mathfrak{u}(\mathbf{x})\right)_D - \left(\mathfrak{q}, \mathfrak{u}_h(\mathbf{x})\right)_D\right| \\
&= \left|\left(\mathfrak{q}, \mathfrak{u}(\mathbf{x}) - \mathfrak{u}_h(\mathbf{x})\right)_D\right| \\
&\overset{(2.23)}{=} \left|\left(\mathfrak{D}^{\dagger}\mathfrak{z}(\mathbf{x}), \mathfrak{u}(\mathbf{x}) - \mathfrak{u}_h(\mathbf{x})\right)_D\right| \\
&\overset{(2.17)}{=} \left|\left(\mathfrak{z}(\mathbf{x}), \mathfrak{D}(\mathfrak{u}(\mathbf{x}) - \mathfrak{u}_h(\mathbf{x}))\right)_D\right| \\
&\overset{(2.19)}{=} \left|\left(\mathfrak{z}(\mathbf{x}), \mathfrak{f} - \mathfrak{D}\mathfrak{u}_h(\mathbf{x})\right)_D\right|.
\end{aligned} \tag{2.24}$$

A computable expression can only be obtained by employing a FE approximation of the adjoint solution $\mathfrak{z}$. However, since $\mathfrak{z}_h$ is orthogonal to the residual implying that the corresponding error estimate is zero, we cannot simply choose $\mathfrak{z}_h$. An adjoint solution with higher accuracy is required. Therefore we employ the FE approximation of $\mathfrak{z}$ with an increased number of finite elements, denoted by $\mathfrak{z}_{h/2}$. Please note that the solution spaces must be nested, i.e., $V_{h/2} \subset V_h$, to ensure higher accuracy of the solution $\mathfrak{z}_{h/2}$. The computable formulation of (2.24) is then given by

$$\varepsilon^{\text{c}}_{\text{fe},h}(\mathbf{x}) = \left|\left(\mathfrak{z}_{h/2}(\mathbf{x}), \mathfrak{f} - \mathfrak{D}\mathfrak{u}_h(\mathbf{x})\right)_D\right|. \tag{2.25}$$

However, this expression requires solving the adjoint equation (2.23) with higher accuracy, implying more computing effort. Later we will use an approximation of this error.

2.4 The Waveguide Problem

One benchmark problem we discuss in this work is a rectangular waveguide. Therefore, we reformulate Maxwell's equations for the time harmonic case. This is not only relevant for the specific waveguide problem, but generally for high frequency electromagnetic wave phenomena, e.g. in the context of scattering problems, antennas, microwave or optical waveguides, or resonant cavities [21, Sect. 2.1].

Maxwell's Equations for Time Harmonic Problems

In a time harmonic problem the field quantities are sinusoidal over time, this makes the use of complex numbers possible. The time dependence of the data and the fields is proportional to $e^{i\omega t}$, where $\omega \in \Omega \subset \mathbb{R}^+$ is the angular frequency and i the imaginary unit. This allows us to use the electric field phasor $\mathbf{E}(\mathbf{r})$ and write

$$\mathbf{E}(\mathbf{r}, t) = \mathrm{Re}(\mathbf{E}(\mathbf{r})e^{i\omega t}). \tag{2.26}$$

for the electric field strength. In general, the field $\mathbf{E}(\mathbf{r})$ is complex and space dependent, i.e., $\mathbf{E} : D \to \mathbb{C}^3$. The same holds for all other field quantities appearing in (2.7)–(2.10). Further, the derivative with respect to the time $\partial/\partial t$ becomes $i\omega$. Then, the Maxwell's equations for time harmonic problems are given by

$$\nabla \times \mathbf{E} \quad = \quad -i\omega\mathbf{B} \overset{(2.12)}{=} -i\omega\mu\mathbf{H}, \tag{2.27}$$

$$\nabla \times \mathbf{H} \quad = \quad i\omega\mathbf{D} + \mathbf{J} \overset{(2.11)}{=} i\omega\epsilon\mathbf{E} + \mathbf{J}, \tag{2.28}$$

$$\nabla \cdot (\epsilon\mathbf{E}) \overset{(2.11)}{=} \nabla \cdot \mathbf{D} \quad = \quad \varrho, \tag{2.29}$$

$$\nabla \cdot (\mu\mathbf{H}) \overset{(2.12)}{=} \nabla \cdot \mathbf{B} \quad = \quad 0. \tag{2.30}$$

For a detailed mathematical discussion we refer to [2, Sect. 6.9] and [5, Sect. 1.2].

Strong Formulation

We aim to derive the E-field formulation of Maxwell's equations. Starting with (2.27), multiplying by μ^{-1} from the left and applying $\nabla\times$ yields

$$\nabla \times (\mu^{-1}\nabla \times \mathbf{E}) \quad = \quad -i\omega(\nabla \times \mathbf{H}) \tag{2.31}$$

$$\overset{(2.28)}{=} \quad -i\omega(i\omega\epsilon\mathbf{E} + \mathbf{J}) \tag{2.32}$$

$$= \quad \omega^2\epsilon\mathbf{E} - i\omega\mathbf{J}. \tag{2.33}$$

In the following we assume the absence of charges and currents, i.e., $\varrho = 0$ and $\mathbf{J} = 0$. Using $\mu = \mu_0\mu_r$ and multiplying with μ_0, we obtain the so-called curl-curl equation describing the electric wave in the computational domain D by

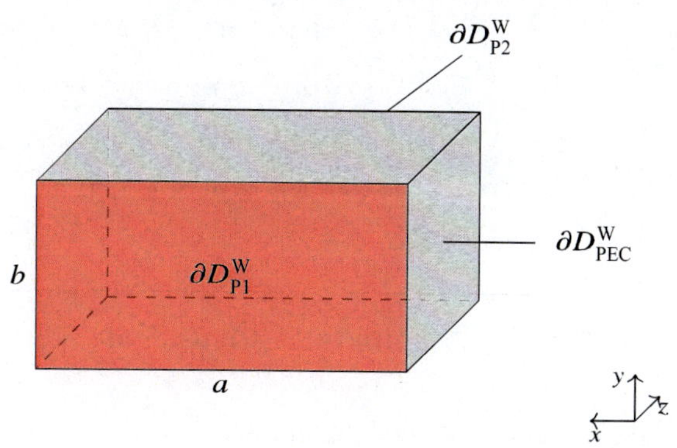

Fig. 2.3 Rectangular waveguide with two ports $\partial D^{\mathrm{W}}_{\mathrm{P1}}$ and $\partial D^{\mathrm{W}}_{\mathrm{P2}}$, PEC boundary conditions on the wave- guide walls $\partial D^{\mathrm{W}}_{\mathrm{PEC}}$, width a and height b

$$\nabla \times (\mu_{\mathrm{r}}^{-1} \nabla \times \mathbf{E}) - \omega^2 \mu_0 \epsilon \mathbf{E} = 0. \tag{2.34}$$

Let us consider a rectangular waveguide model on the computational domain D^{W}, with width a and height b, as illustrated in Fig. 2.3. The boundary of the computational domain D^{W} is split into three parts, i.e., $\partial D^{\mathrm{W}} = \partial D^{\mathrm{W}}_{\mathrm{P1}} \cup \partial D^{\mathrm{W}}_{\mathrm{P2}} \cup \partial D^{\mathrm{W}}_{\mathrm{PEC}}$. The boundaries $\partial D^{\mathrm{W}}_{\mathrm{P1}}$ and $\partial D^{\mathrm{W}}_{\mathrm{P2}}$ refer to the two ports (in red). At the walls of the waveguide (in gray) we assume perfect electric conductor (PEC) boundary conditions, which corresponds to Dirichlet boundary conditions, i.e.,

$$\mathbf{n} \times \mathbf{E} = 0 \quad \text{on } \partial D^{\mathrm{W}}_{\mathrm{PEC}}, \tag{2.35}$$

where $\mathbf{n}$ denotes the outer unit normal vector. We assume that the waveguide is excited at $\partial D^{\mathrm{W}}_{\mathrm{P1}}$ by an incident transverse electric (TE) TE_{10} wave

$$\mathbf{E}^{\mathrm{inc}} = E_0 \mathbf{E}^{\mathrm{TE}}_{10} \mathrm{e}^{-ik_{z10}z} \quad \text{with} \quad \mathbf{E}^{\mathrm{TE}}_{10} := \sin\left(\frac{\pi x}{a}\right) \mathbf{e}_y, \tag{2.36}$$

where E_0 denotes the amplitude of the incident wave, $k_{z10} = \sqrt{\omega^2 \mu_0 \epsilon_0 - \left(\frac{\pi}{a}\right)^2}$ the propagation constant and $\mathbf{e}_y$ the unit vector in y-direction. Following [6, Sect. 8.5], we derive the boundary conditions for the ports by

$$\mathbf{n} \times (\nabla \times \mathbf{E}) - ik_{z10}(\mathbf{n} \times \mathbf{E}) \times \mathbf{n} = -2ik_{z10}\mathbf{E}^{\mathrm{inc}} \qquad \text{on } \partial D^{\mathrm{W}}_{\mathrm{P1}}, \tag{2.37}$$

$$\mathbf{n} \times (\nabla \times \mathbf{E}) - ik_{z10}(\mathbf{n} \times \mathbf{E}) \times \mathbf{n} = 0 \qquad \text{on } \partial D^{\mathrm{W}}_{\mathrm{P2}}. \tag{2.38}$$

We summarize (2.34), (2.35), (2.37) and (2.38), and obtain the strong formulation of the waveguide problem as

$$
\begin{aligned}
\nabla \times (\mu_r^{-1} \nabla \times \mathbf{E}) - \omega^2 \mu_0 \epsilon \mathbf{E} &= 0 && \text{in } D^{\mathrm{W}}, \\
\mathbf{n} \times \mathbf{E} &= 0 && \text{on } \partial D^{\mathrm{W}}_{\mathrm{PEC}}, \\
\mathbf{n} \times (\nabla \times \mathbf{E}) - i k_{z10} (\mathbf{n} \times \mathbf{E}) \times \mathbf{n} &= -2 i k_{z10} \mathbf{E}^{\mathrm{inc}} && \text{on } \partial D^{\mathrm{W}}_{\mathrm{P1}}, \\
\mathbf{n} \times (\nabla \times \mathbf{E}) - i k_{z10} (\mathbf{n} \times \mathbf{E}) \times \mathbf{n} &= 0 && \text{on } \partial D^{\mathrm{W}}_{\mathrm{P2}}.
\end{aligned}
\tag{2.39}
$$

Scattering Parameter Calculation

We introduce the fundamental scattering parameter (S-parameter) of the TE_{10} mode on the first port $\partial D^{\mathrm{W}}_{\mathrm{P1}}$. Once (2.39) is solved, i.e., $\mathbf{E}$ is known, the S-parameter can be calculated in a post-processing by

$$
S \equiv S^{\mathrm{TE}}_{11} := \frac{2}{E_0 a b} \left(\mathbf{E} - \mathbf{E}^{\mathrm{inc}}, \mathbf{E}^{\mathrm{TE}}_{10} \right)_{\partial D^{\mathrm{W}}_{\mathrm{P1}}}. \tag{2.40}
$$

Without loss of generality we assume that the port $\partial D^{\mathrm{W}}_{\mathrm{P1}}$ is located at $z = 0$. As the electric field strength, the S-parameter depends on the frequency ω, i.e., $S \equiv S_\omega$. Note that the S-parameter is an affine-linear functional of $\mathbf{E}$.

For the example of the waveguide, the S-parameter (2.40) is considered as QoI. Hence, we define the parametrized solution operator Q^{W} as

$$
Q^{\mathrm{W}}(\mathbf{p}) := S, \tag{2.41}
$$

where $\mathbf{p}$ denotes the parametrization containing for example geometry and material information, boundary conditions and frequencies. Since the S-parameter depends on the solution $\mathbf{E}$ of (2.39), cf. (2.40), and the solution $\mathbf{E}$ depends on the parametrization $\mathbf{p}$, the operator Q^{W} involves solving a PDE and could be written as $Q^{\mathrm{W}}(\mathbf{p}) = \widehat{Q}^{\mathrm{W}}(\mathbf{E}(\mathbf{p}))$.

Weak Formulation

In order to solve the waveguide problem numerically with FEM, the strong formulation (2.39) needs to be transformed into the weak formulation which will be discretized later. We follow our work in [22].

First, for brevity of notation, we introduce the trace operators [5, Eqs. (3.45–3.46)]

$$
\pi_{\mathrm{t}}[\mathbf{u}] := \mathbf{n} \times \mathbf{u}|_{\partial D^{\mathrm{W}}}, \tag{2.42}
$$

$$
\pi_{\mathrm{T}}[\mathbf{u}] := \left(\mathbf{n} \times \mathbf{u}|_{\partial D^{\mathrm{W}}} \right) \times \mathbf{n}. \tag{2.43}
$$

In order to derive the weak formulation of (2.39) we build the inner products of (2.34) with the test function $\mathbf{E}' \in V^{\mathrm{W}}$, where $V^{\mathrm{W}} \subset H(\mathrm{curl}, D^{\mathrm{W}})$ will be defined later, and integrate by parts [5, Theorem 3.31]. We obtain

$$\left(\mu_{\mathrm{r}}^{-1}\nabla \times \mathbf{E}, \nabla \times \mathbf{E}'\right)_{D^{\mathrm{W}}} - \omega^2\mu_0\left(\epsilon\,\mathbf{E}, \mathbf{E}'\right)_{D^{\mathrm{W}}} + \left(\pi_{\mathrm{t}}[\mu_r^{-1}\nabla \times \mathbf{E}], \pi_{\mathrm{T}}[\mathbf{E}']\right)_{\partial D^{\mathrm{W}}} = 0. \tag{2.44}$$

We insert the boundary conditions. The boundary integral in (2.44) vanishes on $\partial D^{\mathrm{W}}_{\mathrm{PEC}}$ due to the application of PEC boundary conditions (2.35), also for $\mathbf{E}'$. On $\partial D^{\mathrm{W}}_{\mathrm{P1}} \cup \partial D^{\mathrm{W}}_{\mathrm{P2}}$ the boundary conditions (2.37)–(2.38) are employed. We obtain the weak formulation of the waveguide problem: find $\mathbf{E} \in V^{\mathrm{W}}$ s.t.

$$\begin{aligned}\left(\mu_r^{-1}\nabla \times \mathbf{E}, \nabla \times \mathbf{E}'\right)_{D^{\mathrm{W}}} - \omega^2\mu_0\left(\epsilon\,\mathbf{E}, \mathbf{E}'\right)_{D^{\mathrm{W}}} + ik_{z10}\left(\pi_{\mathrm{T}}[\mathbf{E}], \pi_{\mathrm{T}}[\mathbf{E}']\right)_{D^{\mathrm{W}}_{\mathrm{P1}}\cup D^{\mathrm{W}}_{\mathrm{P2}}} \\ = 2ik_{z10}\left(\mathbf{E}^{\mathrm{inc}}, \pi_{\mathrm{T}}[\mathbf{E}']\right)_{D^{\mathrm{W}}_{\mathrm{P1}}} \qquad \forall \mathbf{E}' \in V^{\mathrm{W}}.\end{aligned} \tag{2.45}$$

Accounting for the PEC boundary condition (2.35) and seeking a well defined boundary integral in (2.45), the space V^{W} is chosen as [21, Eq. (2.47)]

$$\begin{aligned}V^{\mathrm{W}} := \Big\{\mathbf{E} \in H(\mathrm{curl}, D^{\mathrm{W}}) : \pi_{\mathrm{T}}[\mathbf{E}]\big|_{\partial D^{\mathrm{W}}_{\mathrm{P1}}} \in \left(L^2(\partial D^{\mathrm{W}}_{\mathrm{P1}})\right)^3 \wedge \pi_{\mathrm{T}}[\mathbf{E}]\big|_{\partial D^{\mathrm{W}}_{\mathrm{P2}}} \in \left(L^2(\partial D^{\mathrm{W}}_{\mathrm{P2}})\right)^3 \\ \wedge\ \pi_{\mathrm{t}}[\mathbf{E}]\big|_{\partial D^{\mathrm{W}}_{\mathrm{PEC}}} = 0\Big\}.\end{aligned} \tag{2.46}$$

Discretization

The last step before the waveguide problem can be solved with FEM is its discretization. We introduce a finite dimensional function space $V_h^{\mathrm{W}} \subset V^{\mathrm{W}}$, for possible choices we refer e.g. to [5]. Independent of the choice of V_h^{W}, the approximated electric field strength $\mathbf{E}_h$ can be written as

$$\mathbf{E} \approx \mathbf{E}_h \equiv \mathbf{E}_h(\mathbf{r}) = \sum_{j=1}^{N_h^{\mathrm{W}}} e_j \mathbf{w}_j^{\mathrm{W}}(\mathbf{r}) \equiv \sum_{j=1}^{N_h^{\mathrm{W}}} e_j \mathbf{w}_j^{\mathrm{W}}, \tag{2.47}$$

where N_h^{W} is the number of DoFs, $\{\mathbf{w}_j^{\mathrm{W}}\}_{j=1}^{N_h^{\mathrm{W}}} \subset V_h^{\mathrm{W}}$ are basis functions with $\mathbf{w}_j^{\mathrm{W}} : D^{\mathrm{W}} \to \mathbb{R}^3$ and $e_j \in \mathbb{C}$, $j = 1, \ldots, N_h^{\mathrm{W}}$ are the DoFs. In this work, we use first-kind, second order Nédélec basis functions defined on a tetrahedral mesh of the computational domain D^{W}. By that it is ensured that the finite element spaces have the same relationship as the continuous spaces, i.e., they are linked by the discrete de Rham sequence [5, Sect. 5.1]. The fact that the Nédélec basis functions ensure a conforming approximation allows to apply the Cea's Lemma [5, Lemma 2.37], which is the discrete analogue of the Lax-Milgram Lemma and states that a unique solution exists [5, Sect. 2.3]. For details on Nédélec basis functions and their curl-conforming property we refer to the original work by Nédélec [23] or discussions in [5, 24].

The discrete solution $\mathbf{e} = (e_1, \ldots, e_{N_h^{\mathrm{W}}})^{\top} \in \mathbb{C}^{N_h^{\mathrm{W}}}$ is computed by solving the linear system

$$\underbrace{\left(\mathbf{A} - \omega^2 \mathbf{M}^{\epsilon} + ik_{z10}\mathbf{M}^{\mathrm{port}}\right)}_{=:\mathbf{K}^{\mathrm{W}}(\omega)} \mathbf{e} = \mathbf{f}^{\mathrm{W}}, \tag{2.48}$$

where $\mathbf{K}^{\mathrm{W}}(\omega) \in \mathbb{C}^{N_h^{\mathrm{W}} \times N_h^{\mathrm{W}}}$ denotes the system matrix and $\mathbf{f}^{\mathrm{W}} \in \mathbb{C}^{N_h^{\mathrm{W}}}$ the discretized right-hand side. In (2.48), the stiffness matrix $\mathbf{A}$, the mass-matrix $\mathbf{M}^{\epsilon}$, the matrix $\mathbf{M}^{\mathrm{port}}$ and the right-hand side $\mathbf{f}^{\mathrm{W}}$ are given by

$$\mathbf{A}_{ij} = (\mu_{\mathrm{r}}^{-1} \nabla \times \mathbf{w}_j^{\mathrm{W}}, \nabla \times \mathbf{w}_i^{\mathrm{W}})_{D^{\mathrm{W}}}, \qquad \mathbf{M}_{ij}^{\epsilon} = \mu_0 (\epsilon \mathbf{w}_j^{\mathrm{W}}, \mathbf{w}_i^{\mathrm{W}})_D, \tag{2.49}$$

$$\mathbf{M}_{ij}^{\mathrm{port}} = \left(\pi_{\mathrm{T}}[\mathbf{w}_j^{\mathrm{W}}], \pi_{\mathrm{T}}[\mathbf{w}_i^{\mathrm{W}}]\right)_{\partial D_{\mathrm{P1}}^{\mathrm{W}} \cup \partial D_{\mathrm{P2}}^{\mathrm{W}}}, \qquad \mathbf{f}_j^{\mathrm{W}} = 2ik_{z10}\left(\mathbf{E}^{\mathrm{inc}}, \pi_{\mathrm{T}}[\mathbf{w}_j^{\mathrm{W}}]\right)_{\partial D_{\mathrm{P1}}^{\mathrm{W}}}. \tag{2.50}$$

Then, the discrete QoI is given by the discrete counterpart of the S-parameter (2.40), i.e.,

$$Q_h^{\mathrm{W}}(\mathbf{p}) := S_h \equiv S_{10,h}^{\mathrm{TE}} = \mathbf{j}_{10}^{\mathrm{TE}}\left(\mathbf{e} - \mathbf{e}^{\mathrm{inc}}\right) \in \mathbb{C}, \tag{2.51}$$

where $\mathbf{e}^{\mathrm{inc}}$ is the projection of (2.36) on the basis (2.47) and

$$\left(\mathbf{j}_{10}^{\mathrm{TE}}\right)_j = \frac{2}{E_0 ab} (\mathbf{w}_j^{\mathrm{W}}, \mathbf{E}_{10}^{\mathrm{TE}})_{\partial D_{\mathrm{P1}}^{\mathrm{W}}}. \tag{2.52}$$

Later we aim to interpolate S_h, so we assume S_h to be sufficiently smooth with respect to the parametrization $\mathbf{p}$. Similar assumptions and numerical studies to justify these have also been made in [25] for example.

2.5 The Permanent Magnet Synchronous Machine Problem

The second application discussed in this work is a PMSM. Therefore, we introduce the magnetostatic formulation of Maxwell's equations, following [26].

Maxwell's Equations in the Magnetostatic Case

In the *magnetoquasistatic* case it is assumed that the fields are slowly oscillating and that the electric energy is low compared to the magnetic energy [27]. The displacement current density can be neglected compared to the current density, i.e., $\partial \mathbf{D}/\partial t = 0$. Thus, Maxwell's equations in the magnetoquasistatic case are formulated as the general Maxwell's equations (2.7)–(2.10), with the difference that (2.8) is replaced by

$$\nabla \times \mathbf{H} = \mathbf{J} = \mathbf{J}_c + \mathbf{J}_s, \tag{2.53}$$

where $\mathbf{J}_c$ denotes the conduction current density and $\mathbf{J}_s$ the source current density [28, Eq. (2.15)]. In the *magnetostatic* case it is assumed that $\mathbf{J}_c = 0$, i.e., that there are no eddy currents, and that there are no changes over time, which implies that also $\partial \mathbf{B}/\partial t = 0$. Hence, the electric and magnetic fields can be decoupled. From the Maxwell's equations (2.7)–(2.10) and the material relations (2.11)–(2.12) remain [28, Sect. 2.4]

$$\nabla \times \mathbf{H} = \mathbf{J}_s \tag{2.54}$$

$$\nabla \cdot \mathbf{B} = 0 \tag{2.55}$$

$$\mathbf{B} = \mu \mathbf{H}. \tag{2.56}$$

Since the PMSM contains permanent magnets (PMs), there are materials which induce a constant magnetic field without requiring electric currents. We introduce $\mathbf{M}_{\mathrm{pm}} \equiv \mathbf{M}_{\mathrm{pm}}(\mathbf{r})$ as the magnetization of a permanent magnet (PM). The material relation (2.56) is replaced by

$$\mathbf{B} = \mu(\mathbf{H} + \mathbf{M}_{\mathrm{pm}}). \tag{2.57}$$

From (2.55), we introduce the magnetic vector potential (MVP) $\mathbf{A} \equiv \mathbf{A}(t, \mathbf{r})$ such that

$$\mathbf{B} = \nabla \times \mathbf{A} \tag{2.58}$$

is fulfilled and replace (2.55) with (2.58).

Strong Formulation

We combine (2.54), (2.57) and (2.58) to obtain the magnetostatic approximation of Maxwell's equations. Starting with (2.58), multiplying by μ^{-1} from the left and applying $\nabla\times$ yields

$$\nabla \times (\mu^{-1} \nabla \times \mathbf{A}) = \nabla \times (\mu^{-1}\mathbf{B}) \tag{2.59}$$

$$\overset{(2.57)}{=} \nabla \times (\mathbf{H} + \mathbf{M}_{\mathrm{pm}}) \tag{2.60}$$

$$\overset{(2.54)}{=} \mathbf{J}_s + \nabla \times \mathbf{M}_{\mathrm{pm}}. \tag{2.61}$$

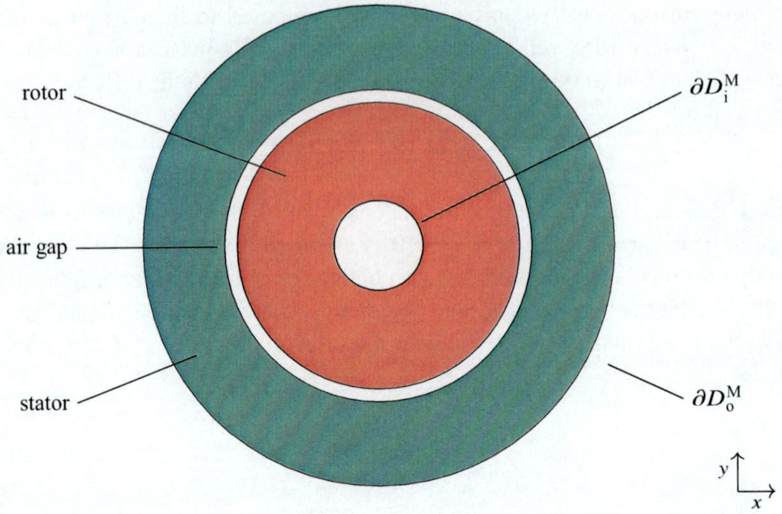

Fig. 2.4 Cross-sectional view of a simplified PMSM with stator and rotor, on the computational domain D^{M} with the outer boundary $\partial D_{\mathrm{o}}^{\mathrm{M}}$ and the inner boundary $\partial D_{\mathrm{i}}^{\mathrm{M}}$

We introduce $\nu := \mu^{-1}$ as the magnetic reluctivity. Then, the curl-curl equation describing the phenomena of the PMSM on the computational domain D^{M} is given by

$$\nabla \times (\nu \nabla \times \mathbf{A}) = \mathbf{J}_{\mathrm{s}} + \nabla \times \mathbf{M}_{\mathrm{pm}}. \tag{2.62}$$

Let us consider a PMSM on the computational domain D^{M} as illustrated in Fig. 2.4. We see a cross-sectional view of the PMSM model, consisting of a rotor, including the PMs, and a stator. For the inner and the outer boundary $\partial D_{\mathrm{i}}^{\mathrm{M}}$ and $\partial D_{\mathrm{o}}^{\mathrm{M}}$, respectively, we assume homogeneous Dirichlet boundary conditions, i.e.,

$$\mathbf{n} \times \mathbf{A} = 0 \ \text{ on } \partial D^{\mathrm{M}} = \partial D_{\mathrm{i}}^{\mathrm{M}} \cup \partial D_{\mathrm{o}}^{\mathrm{M}}. \tag{2.63}$$

The Eqs. (2.62) and (2.63) form the strong formulation of the PMSM problem, which will be investigated in this work.

In practice, Eq. (2.62) is often solved as a 2D problem, see Sect. 6.1. Then, Eq. (2.62) transforms into the Poisson equation. Together with the boundary condition (2.63) and the assumption that D^{M} has a Lipschitz continuous boundary (see Sect. 2.1), using the Lax-Milgram Lemma, it can be shown that there exists a unique weak solution of the 2D problem [5, Theorem 3.14].

Torque Calculation

Based on the MVP $\mathbf{A}$ different quantities such as efficiency, total harmonic distortion and the torque can be computed. In this work we focus on the torque. In [29, Chap. 6] different methods on the calculation of the torque are discussed. Following the Maxwell stress method [30], [31, Sect. 1.5], [32, Eq. (121)] the torque is given by

$$\tau = \int_{S_r} \mathbf{t} \times (\boldsymbol{\sigma}_{\mathrm{M}} \cdot \mathbf{n})\, \mathrm{d}S_r, \tag{2.64}$$

where S_r denotes the surface enclosing the rotor, $\mathbf{t}$ the vector connecting the origin of the machine with the surface S_r, $\mathbf{n}$ the unit normal vector with respect to the surface S_r and $\boldsymbol{\sigma}_{\mathrm{M}}$ the Maxwell stress tensor defined by

$$\boldsymbol{\sigma}_{\mathrm{M},ij} = \nu_0 \left(B_i B_j - \frac{1}{2} |\mathbf{B}|^2 \delta_{ij} \right) \tag{2.65}$$

with Kronecker delta δ_{ij}, $i, j = 1, 2, 3$. Using the relation $\mathbf{B} = \nabla \times \mathbf{A}$, cf. (2.58), we see that the calculation of the torque is a post-processing step once the MVP is obtained by FEM for example. In practice, Arkkio's formulation [32, Eq. (124)] is often used because it is considered to be numerically more stable.

In the PMSM example the torque is considered as QoI. Analogously to (2.41), we define

$$Q^{\mathrm{M}}(\mathbf{p}) := \tau \tag{2.66}$$

with parametrization $\mathbf{p}$ containing all problem specific information. Since the torque depends on the solution $\mathbf{A}$ of (2.62)–(2.63), and the solution $\mathbf{A}$ depends on the parametrization $\mathbf{p}$, the operator Q^{M} involves solving a PDE and could be written as $Q^{\mathrm{M}}(\mathbf{p}) = \widehat{Q}^{\mathrm{M}}(\mathbf{A}(\mathbf{p}))$.

Weak Formulation

Analogously to the waveguide problem, we multiply (2.62) with test functions $\mathbf{A}'$ and integrate by parts. We obtain

$$\begin{aligned} &\left(\nu \nabla \times \mathbf{A}, \nabla \times \mathbf{A}'\right)_{D^{\mathrm{M}}} + \left(\mathbf{n} \times (\nu \nabla \times \mathbf{A}), \mathbf{A}'\right)_{\partial D^{\mathrm{M}}} \\ &\quad = \left(\mathbf{J}_{\mathrm{s}}, \mathbf{A}'\right)_{D^{\mathrm{M}}} + \left(\mathbf{M}_{\mathrm{pm}}, \nabla \times \mathbf{A}'\right)_{D^{\mathrm{M}}} + \left(\mathbf{n} \times \mathbf{M}_{\mathrm{pm}}, \mathbf{A}'\right)_{\partial D^{\mathrm{M}}}. \end{aligned} \tag{2.67}$$

We insert the boundary condition (2.63) into (2.67), i.e., we define the solution space

$$V^{\mathrm{M}} := \left\{ \mathbf{A} \in H(\mathrm{curl}, D^{\mathrm{M}}) : \mathbf{n} \times \mathbf{A}\big|_{\partial D^{\mathrm{M}}} = 0 \right\} \tag{2.68}$$

and choose the test functions such that $\mathbf{A}' \in V^{\mathrm{M}}$. The boundary integral vanishes and the weak formulation reads: find $\mathbf{A} \in V^{\mathrm{M}}$ s.t.

$$\left(\nu \nabla \times \mathbf{A}, \nabla \times \mathbf{A}'\right)_{D^{\mathrm{M}}} = \left(\mathbf{J}_{\mathrm{s}}, \mathbf{A}'\right)_{D^{\mathrm{M}}} + \left(\mathbf{M}_{\mathrm{pm}}, \nabla \times \mathbf{A}'\right)_{D^{\mathrm{M}}}. \tag{2.69}$$

Discretization

The PMSM shall be solved by FEM. Analogously to (2.47) in the waveguide problem we introduce a finite dimensional function space $V_h^{\mathrm{M}} \subset V^{\mathrm{M}}$ and obtain the approximated MVP by

$$\mathbf{A} \approx \mathbf{A}_h \equiv \mathbf{A}_h(\mathbf{r}) = \sum_{j=1}^{N_h^{\mathrm{M}}} a_j \mathbf{w}_j^{\mathrm{M}}(\mathbf{r}) \equiv \sum_{j=1}^{N_h^{\mathrm{M}}} a_j \mathbf{w}_j^{\mathrm{M}}, \tag{2.70}$$

where N_h^{M} is the number of DoFs, $\{\mathbf{w}_j^{\mathrm{M}}\}_{j=1}^{N_h^{\mathrm{M}}} \subset V_h^{\mathrm{M}}$ are basis functions with $\mathbf{w}_j^{\mathrm{M}} : D^{\mathrm{M}} \to \mathbb{R}^3$ and $a_j \in \mathbb{R}$, $j = 1, \ldots, N_h^{\mathrm{M}}$ are the DoFs. The discrete solution $\mathbf{a} = (a_1, \ldots, a_{N_h^{\mathrm{M}}})^\top \in \mathbb{R}^{N_h^{\mathrm{M}}}$ can be computed by solving the linear system

$$\mathbf{K}^{\mathrm{M}}\mathbf{a} = \mathbf{f}^{\mathrm{M}}, \tag{2.71}$$

where the system matrix $\mathbf{K}^{\mathrm{M}} \in \mathbb{R}^{N_h^{\mathrm{M}} \times N_h^{\mathrm{M}}}$ and the discretized right-hand side $\mathbf{f}^{\mathrm{M}} \in \mathbb{R}^{N_h^{\mathrm{M}}}$ are given by

$$\mathbf{K}_{ij}^{\mathrm{M}} = \left(\nu \nabla \times \mathbf{w}_j^{\mathrm{M}}, \nabla \times \mathbf{w}_i^{\mathrm{M}}\right)_{D^{\mathrm{M}}} \qquad \mathbf{f}_j^{\mathrm{M}} = \left(\mathbf{J}_{\mathrm{s}}, \mathbf{w}_j^{\mathrm{M}}\right)_{D^{\mathrm{M}}} + \left(\mathbf{M}_{\mathrm{pm}}, \nabla \times \mathbf{w}_j^{\mathrm{M}}\right)_{D^{\mathrm{M}}}. \tag{2.72}$$

As for the waveguide, first-kind, second order Nédélec basis functions can be used.

To account for the rotating movement of the rotor inside the stator, typically two models are employed, one for the stator and one for the rotor. In combination with FEM, there exist several methods to realize that, see for example [33] for an overview. In the Lagrange multiplier method for example, the two models are coupled by a coupling term, consisting of Lagrange multipliers [34]. In the locked step approach the time step is chosen such that the rotor shift corresponds to a multiple of the mesh size in the air gap. By that, conformal meshes are ensured [35]. In the moving band technique the meshes of rotor and stator are fixed, but a third mesh in the air gap is generated and only this is deformed or re-meshed if required [36].

2.6 Summary

In this chapter the concepts of modeling electromagnetic phenomena were discussed. Therefore, we introduced Maxwell's equations as the fundamental PDEs and described the basic procedure of solving them with FEM. As an example to investigate high frequency electromagnetic wave phenomena we introduced a rectangular waveguide. We reformulated Maxwell's equations for the time harmonic case and, using the material relations, derived the curl-curl equation describing the electric wave within the waveguide. Boundary conditions for the ports and the walls were formulated and inserted into the strong formulation. The scattering parameter was defined as QoI. As a low frequency example we introduced a PMSM. We simplified Maxwell's equations for the magnetostatic case, set Dirichlet boundary conditions and derived the curl-curl wave equation in terms of the MVP. The torque of the machine was introduced as QoI.

In the beginning of this chapter we mentioned that the yield estimation and optimization methods introduced in the following chapters are independent of the considered model problem. Hence, we define the QoI more generally. Let

$$Q(\mathbf{p}) \quad \text{on} \quad D(\mathbf{p}) \equiv D \tag{2.73}$$

denote a parametrized solution operator with parametrization $\mathbf{p}$ containing for example geometry and material information or boundary conditions. The computational domain D also depends on the parametrization $\mathbf{p}$. The QoI can be the S-parameter (2.40) or the torque (2.64), but it can also be any other operator. The methods proposed in the following are especially relevant if solving the QoI is computationally expensive. Typically, the QoI involves solving PDEs numerically, although this is not a requirement for the proposed methods. However, it motivates the term PDE constrained optimization.

In the next chapter fundamental mathematical definitions and methods for robust design are summarized.

References

1. Maxwell JC (1864) A dynamical theory of electromagnetic field. R Soc Trans 155:526–597
2. Jackson JD (1998) Classical electrodynamics, 3 edn. Wiley, New York. https://doi.org/10.1017/CBO9780511760396
3. Assous F, Ciarlet P, Labrunie S (2018) Mathematical foundations of computational electromagnetism. Springer. https://doi.org/10.1007/978-3-319-70842-3
4. Meschede D (2006) Gerthsen Physik. Springer
5. Monk P (2003) Finite element methods for Maxwell's equations. Oxford University Press
6. Jin JM (2014) The finite element method in electromagnetics, 3rd edn. Wiley
7. Zienkiewicz OC, Taylor RL, Zhu JZ (2005) The finite element method: its basis and fundamentals. Elsevier Science
8. Ovchinnikov S (2018) Functional analysis: an introductory course. Springer

9. Kurz S, Rain O, Rjasanow S (2007) Fast boundary element methods in computational electromagnetism. Springer, pp 249–279. https://doi.org/10.1007/978-3-540-47533-0
10. Sauter SA, Schwab C (2011) Boundary element methods. Springer
11. Weiland T (1977) A discretization method for the solution of Maxwell's equations for six-component fields. Int J Electron Commun 31:116–120
12. Hesthaven JS, Warburton T (2008) Nodal discontinuous Galerkin methods: algorithms, analysis, and applications. Texts in applied mathematics. Springer (2008)
13. Taflove A (1995) Computational electrodynamics: the finite-difference time-domain-method. Artech House
14. Weiland T (1996) Time domain electromagnetic field computation with finite difference methods. Int J Numer Model Electron Networks Devices Fields 9(4):295–319
15. Babuška I, Rheinboldt WC (1978) A-posteriori error estimates for the finite element method. Int J Numer Meth Eng 12(10):1597–1615
16. Ainsworth M, Oden JT (1993) A unified approach to a posteriori error estimation using element residual methods. Numer Math 65(1):23–50
17. Zienkiewicz OC, Zhu JZ (1987) A simple error estimator and adaptive procedure for practical engineering analysis. Int J Numer Meth Eng 24(2):337–357
18. Eriksson K, Estep D, Hansbo P, Johnson C (1995) Introduction to adaptive methods for differential equations. Acta Numer 4:105–158
19. Zienkiewicz OC, Zhu JZ (1992) The superconvergent patch recovery and a posteriori error estimates. Part 1: the recovery technique. Int J Numer Methods Eng 33(7):1331–1364
20. Ainsworth M, Oden JT (2000) A posteriori error estimation in finite element analysis. Wiley, New York
21. Georg N (2021) Surrogate modeling and uncertainty quantification for radio frequency and optical applications. PhD thesis, Technische Universität Darmstadt
22. Fuhrländer M, Georg N, Römer U, Schöps S (2020) Yield optimization based on adaptive Newton-Monte Carlo and polynomial surrogates. Int J Uncertain Quantif 10(4):351–373. https://doi.org/10.1615/Int.J.UncertaintyQuantification.2020033344
23. Nédélec JC (1980) Mixed finite elements in R3. J Numer Math 35(3):315–341. https://doi.org/10.1007/BF01396415
24. Hiptmair R (2002) Finite elements in computational electromagnetism. Acta Numer 11:237–339. https://doi.org/10.1017/S0962492902000041
25. Georg N, Römer U (2020) Conformally mapped polynomial chaos expansions for Maxwell's source problem with random input data. Int J Numer Model Electron Netw Devices Fields 33(6):e2776
26. Steinmetz T, Kurz S, Clemens M (2011) Domains of validity of quasistatic and quasistationary field approximations. COMPEL - Int J Comput Math Electr Electron Eng 30(4):1237–1247. https://doi.org/10.1108/03321641111133154
27. Bontinck, Z (2018) Simulation and robust optimization for electric devices with uncertainties. PhD thesis, Technische Universität Darmstadt
28. Cortes Garcia, I (2021) Mathematical analysis and simulation of field models in accelerator circuits. PhD thesis, Technische Universität Darmstadt
29. Salon SJ (1995) Finite element analysis of electrical machines, vol 101. Kluwer Academic Publishers Boston
30. Kulchytska-Ruchka I (2021) Parallel-in-time simulation of electromagnetic energy converters. PhD thesis, Technische Universität Darmstadt
31. Pyrhonen J, Jokinen T, Hrabovcova V (2013) Design of rotating electrical machines. Wiley
32. Arkkio, A (1987) Analysis of induction motors based on the numerical solution of the magnetic field and circuit equations. PhD thesis, Helsinki University of Technology
33. Wallinger GJ (2019) A method to couple non-conforming finite element meshes. PhD thesis, Graz University of Technology

34. Lai H, Rodger D, Leonard P (1992) Coupling meshes in 3D problems involving movements. IEEE Trans Magn 28(2):1732–1734
35. Preston T, Reece A, Sangha P (1988) Induction motor analysis by time-stepping techniques. IEEE Trans Magn 24(1):471–474
36. Davat B, Ren Z, Lajoie-Mazenc M (1985) The movement in field modeling. IEEE Trans Magn 21(6):2296–2298

Chapter 3
Mathematical Foundations of Robust Design

After describing the fundamental concepts of modeling electromagnetic phenomena in the previous chapter, this chapter is dedicated to collecting mathematical definitions and methods, which are relevant for robust design optimization. The chapter is divided into three sections. Firstly, we recall basic definitions from the field of probability theory. This is required, since for example manufacturing uncertainties are modeled as random variables. Secondly, we summarize the foundations of optimization and provide a brief review of existing optimization methods. This is a preparation for the yield optimization methods proposed in Chap. 5. Thirdly, we introduce three approximation methods commonly used in the field of uncertainty quantification, namely MC analysis, SC and GPR. The yield estimation methods proposed in Chap. 4 build on these concepts.

As discussed in the previous chapter, we consider the solution operator $Q(\mathbf{p})$ on the computational domain $D \equiv D(\mathbf{p})$ as QoI. The parametrization $\mathbf{p}$ includes all problem specific information, e.g. material and geometry parameters, boundary conditions, etc.. The operator $Q(\mathbf{p})$ can be the S-parameter from Sect. 2.4 or the torque from Sect. 2.5, or it can be any other operator. Usually we assume that solving $Q(\mathbf{p})$ is computationally expensive, e.g. it involves solving a PDE numerically.

3.1 Fundamentals of Probability Theory

Manufacturing imperfections or unpredictable environmental influences cause statistical uncertainties in the material or geometry properties. These uncertainties are represented by random variables. Above we introduced the QoI as solution operator Q depending on the parametrization $\mathbf{p}$. In the following, we assume that the operator is affected by uncertainties, e.g. geometry deviations due to inaccuracies in the manufacturing process, natural material variations or unknown boundary conditions. Hence, for $\mathbf{p}$ we distinguish between deterministic parameters $\mathbf{d}$, which are not affected by uncertainties, and uncertain parameters $\boldsymbol{\xi}$, which are affected by

M. Fuhrländer, *Design Methods for Reducing Failure Probabilities with Examples from Electrical Engineering*, Springer Theses,
https://doi.org/10.1007/978-3-031-37019-9_3

uncertainties. For the QoI we write $Q(\boldsymbol{\xi}, \mathbf{d})$. In order to quantify the impact of these uncertainties on the QoI, we require statistic quantities like expectation and standard deviation. Therefore, we briefly introduce random variables and their properties formally. We follow [1, 2].

3.1.1 Basic Definitions in Probability Theory

Probability Space

The outcome of a random experiment is denoted by ω. The set of all possible outcomes is called the *sample space* and denoted by Ω. A subset A of Ω is called an *event*. If the set Ω is not finite or countable infinite, the definition of a σ-field is required. Let A and B be events in Ω and let $\mathcal{A}$ be a collection of subsets of Ω fulfilling

i. $\emptyset \in \mathcal{A}$ and $\Omega \in \mathcal{A}$,
ii. if $A \in \mathcal{A}$, then the complement $A^c \in \mathcal{A}$ as well,
iii. if $A, B \in \mathcal{A}$, then $A \cup B \in \mathcal{A}$, $A \cap B \in \mathcal{A}$ and $A \setminus B \in \mathcal{A}$.

Then, $\mathcal{A}$ defines a *σ-field* on Ω [1, Definition 1.1.4]. For $\Omega \neq \emptyset$, the tuple $(\Omega, \mathcal{A})$ is called a *measurable space* [2, Definition 5.1]. The *probability* on $(\Omega, \mathcal{A})$ is a mapping $\mathbb{P} : \mathcal{A} \to \mathbb{R}$, fulfilling [1, Definition 1.1.8]

i. $0 \leq \mathbb{P}[A] \leq 1$ for each $A \in \mathcal{A}$,
ii. $\mathbb{P}[\Omega] = 1$,
iii. for all pairwise disjoint events $A_1, A_2, \ldots \in \mathcal{A}$ hold $\mathbb{P}[\cup_{n=1}^{\infty} A_n] = \sum_{n=1}^{\infty} \mathbb{P}[A_n]$ (σ-additivity).

Then, the triple $(\Omega, \mathcal{A}, \mathbb{P})$ is called a *probability space*. Further, let $\varphi : \mathbb{R} \to \mathbb{R}$ be a function with

i. $\varphi(\xi) \geq 0$ for all $\xi \in \mathbb{R}$,
ii. $\int_{\mathbb{R}} \varphi(\xi) \, \mathrm{d}\xi = 1$.

Then φ is called a *density function* [2, Definition 3.1].

Bayes' Rule

We consider a probability space $(\Omega, \mathcal{A}, \mathbb{P})$ and two events $A, B \in \mathcal{A}$. These events are *independent*, if $\mathbb{P}[A \cap B] = \mathbb{P}[A]\mathbb{P}[B]$ [1, Definition 1.2.1]. The *conditional probability* of the event A given the event B is defined by

$$\mathbb{P}[A|B] = \frac{\mathbb{P}[A \cap B]}{\mathbb{P}[B]}, \tag{3.1}$$

for $\mathbb{P}[B] > 0$ [1, Definition 1.2.6]. For independent events $\mathbb{P}[A|B] = \mathbb{P}[A]$ holds. If we are interested in $\mathbb{P}[A|B]$ and assume $\mathbb{P}[B|A]$ to be known, then *Bayes' rule of retrodiction* [1, Theorem 1.2.8] can be used

$$\mathbb{P}[A|B] = \frac{\mathbb{P}[B|A]\mathbb{P}[A]}{\mathbb{P}[B]}. \tag{3.2}$$

In Sect. 3.3.3 we will introduce GPR, which is based on the concept of conditional probability and Bayes' rule.

Random Variables

Let $(\Omega, \mathcal{A}, \mathbb{P})$ be a probability space and $(\Omega_\Xi, \mathcal{A}_\Xi)$ a measurable space. A mapping $\Xi : (\Omega, \mathcal{A}) \to (\Omega_\Xi, \mathcal{A}_\Xi)$ is called a *random variable* [2, Definition 5.8]. Let a real random variable Ξ and a density function φ be given. If

$$\mathbb{P}[\Xi \in A] = \int_A \varphi(\xi)\, \mathrm{d}\xi \tag{3.3}$$

holds for all events $A \in \mathcal{A}$, the function φ is called the *probability density function* of the random variable Ξ [2, Definition 5.8]. In Chap. 6 we will mainly discuss uniform, Gaussian or truncated Gaussian distributed random variables. In preparation for this, we introduce these common probability density functions. For a *uniformly* distributed random variable we write $\Xi \sim \mathcal{U}(a, b)$ and the probability density function is defined by

$$\varphi_{\mathcal{U}(a,b)}(\xi) = \begin{cases} \frac{1}{b-a}, & \text{if } a \leq \xi \leq b \\ 0, & \text{else.} \end{cases} \tag{3.4}$$

For a *Gaussian* distributed random variable we write $\Xi \sim \mathcal{N}(m, \sigma^2)$ and the probability density function is defined by

$$\varphi_{\mathcal{N}(m,\sigma^2)}(\xi) = \frac{1}{\sqrt{2\pi}\,\sigma} \mathrm{e}^{-\frac{(\xi-m)^2}{2\sigma^2}}. \tag{3.5}$$

The Gaussian distribution has infinite support, i.e., the probability that the deviation from the mean value becomes arbitrarily large is greater than zero. When modeling uncertainties, such a behavior is typically not wanted. Therefore, we introduce the *truncated Gaussian* distributed random variable, which is a rescaled Gaussian distributed random variable with zero tails. We write $\Xi \sim \mathcal{N}_\mathcal{T}(m, \sigma^2, a, b)$ and the probability density function is defined by

$$\varphi_{\mathcal{N}_\mathcal{T}(m,\sigma^2,a,b)}(\xi) = \begin{cases} \frac{1}{\sigma} \frac{\varphi_\mathcal{N}\left(\frac{\xi-m}{\sigma}\right)}{\psi_\mathcal{N}\left(\frac{b-m}{\sigma}\right) - \psi_\mathcal{N}\left(\frac{a-m}{\sigma}\right)}, & \text{if } a \leq \xi \leq b \\ 0, & \text{else,} \end{cases} \tag{3.6}$$

where $\varphi_\mathcal{N}(\cdot)$ is the probability density function of the standard normal distribution, i.e.,

$$\varphi_\mathcal{N}(\theta) \equiv \varphi_{\mathcal{N}(0,1)}(\theta) \tag{3.7}$$

and $\psi_\mathcal{N}(\cdot)$ its cumulative distribution function, i.e.,

$$\psi_{\mathcal{N}}(\theta) = \frac{1}{2}\left(1 + \mathrm{erf}\left(\frac{\theta}{\sqrt{2}}\right)\right), \tag{3.8}$$

with error function

$$\mathrm{erf}(\zeta) = \frac{2}{\sqrt{\pi}} \int_0^{\zeta} \mathrm{e}^{-t^2}\, \mathrm{d}t. \tag{3.9}$$

Let Θ define a second random variable on $(\Omega, \mathcal{A}, \mathbb{P})$, with $(\Omega_\Theta, \mathcal{A}_\Theta)$ and $\Theta : (\Omega, \mathcal{A}) \to (\Omega_\Theta, \mathcal{A}_\Theta)$. Then, the random variables Ξ and Θ are independent, if

$$\mathbb{P}[\Xi \in A, \Theta \in B] = \mathbb{P}[\Xi \in A]\mathbb{P}[\Theta \in B] \tag{3.10}$$

for all $A \in \mathcal{A}_\Xi$ and $B \in \mathcal{A}_\Theta$ [2, Definition 5.14].

Statistical Moments

The *expectation value* of the random variable is defined by [2, Definition 5.17]

$$\mathbb{E}[\Xi] = \int_{\mathbb{R}} \xi\, \varphi(\xi)\, \mathrm{d}\xi. \tag{3.11}$$

The k-th moment of the random variable is given by

$$\mathbb{M}^k[\Xi] \equiv \mathbb{E}[\Xi^k] = \int_{\mathbb{R}} \xi^k\, \varphi(\xi)\, \mathrm{d}\xi. \tag{3.12}$$

Centering the second moment at $\mathbb{E}[\Xi]$ yields the *variance* [2, Definition 5.24]

$$\mathbb{V}[\Xi] = \mathbb{E}\left[|\Xi - \mathbb{E}[\Xi]|^2\right] = \int_{\mathbb{R}} |\xi - \mathbb{E}[\Xi]|^2\, \varphi(\xi)\, \mathrm{d}\xi. \tag{3.13}$$

The standard deviation of Ξ is defined by the square root of the variance, i.e.,

$$\mathrm{Std}[\Xi] = \sqrt{\mathbb{V}[\Xi]}. \tag{3.14}$$

The same holds in case of multivariate random variables. However, for multivariate random variables, additionally the covariance is introduced. Let $\Xi = (\Xi_1, \ldots, \Xi_n)^\top$ denote a multivariate random variable and

$$\mathrm{Cov}[\Xi_i, \Xi_j] = \mathbb{E}\Big[|\Xi_i - \mathbb{E}[\Xi_i]||\Xi_j - \mathbb{E}[\Xi_j]|\Big], \qquad \forall i, j = 1, \ldots, n \tag{3.15}$$

the covariance of Ξ_i and Ξ_j. In case of $i = j$ the covariance coincides with the variance. The covariance matrix is then defined by

$$\boldsymbol{\Sigma} = \mathrm{Cov}[\Xi_i, \Xi_j]_{\{i,j=1,\dots,n\}} = \begin{pmatrix} \mathbb{V}[\Xi_1] & \mathrm{Cov}[\Xi_1, \Xi_2] & \dots & \mathrm{Cov}[\Xi_1, \Xi_n] \\ \mathrm{Cov}[\Xi_2, \Xi_1] & \mathbb{V}[\Xi_2] & & \mathrm{Cov}[\Xi_2, \Xi_n] \\ \vdots & & \ddots & \vdots \\ \mathrm{Cov}[\Xi_n, \Xi_1] & \dots & \mathrm{Cov}[\Xi_n, \Xi_{n-1}] & \mathbb{V}[\Xi_n] \end{pmatrix}. \tag{3.16}$$

Law of Large Numbers

The Monte Carlo analysis introduced in Sect. 3.3.1 is based on the law of large numbers. The weak law of large numbers says, that the arithmetic mean of n realizations of a random variable converges in probability to the expectation value if n tends to infinity [2, Theorem 5.10], i.e., for each $\varepsilon > 0$

$$\lim_{n\to\infty} \mathbb{P}\left[\left|\frac{1}{n}\sum_{i=1}^{n} \Xi^{(i)} - \mathbb{E}[\Xi]\right| > \varepsilon\right] = 0. \tag{3.17}$$

3.1.2 *Stochastic Processes*

In Sect. 3.3.3 we introduce GPR as a method to approximate a function by a GP. Therefore, based on [3, Chap. 1], we will define stochastic processes in general and Gaussian processes (GPs) in particular. While a random variable represents one uncertain event at a given time or space, a stochastic process represents a path of random events over time or space. Again we consider a probability space $(\Omega, \mathcal{A}, \mathbb{P})$. A *stochastic process* is a family of random variables $\{Q_\xi\}_{\xi\in\mathcal{X}}$ in this space, indexed by the discrete or continuous set $\mathcal{X}$ describing the evolution over time or space. A stochastic process can be interpreted in two ways. First, for a fixed ξ, Q_ξ can be seen as random variable. Second, for an event $\omega \in \Omega$, we could write $Q_\xi(\omega) = p_\omega(\xi)$, and then $p_\omega(\xi)$ is a function of ξ and a realization or sample function of the stochastic process, i.e., a path representing the evolution of the random process over time or space. The stochastic process is called a *Gaussian process*, if for every finite subset of indices $\{\xi_1, \dots, \xi_n\} \subseteq \mathcal{X}$ the multivariate random variable $Q_\xi = \left(Q_{\xi_1}, \dots, Q_{\xi_n}\right)^\top$ is Gaussian distributed. A GP can be completely defined by its mean function $m(\xi)$ and covariance or kernel function $k(\xi, \xi')$ [4, Sect. 2.2]

$$m(\xi) = \mathbb{E}[Q_\xi] \tag{3.18}$$

$$k(\xi, \xi') = \mathrm{Cov}[Q_\xi, Q_{\xi'}] = \mathbb{E}\left[\left(Q_\xi - m(\xi)\right)\left(Q_{\xi'} - m(\xi')\right)\right]. \tag{3.19}$$

We write

$$\{Q_\xi\}_{\xi\in\mathcal{X}} \sim \mathcal{GP}\left(m(\xi), k(\xi, \xi')\right). \tag{3.20}$$

The squared exponential (or radial basis function (RBF)) kernel

$$k(\xi,\xi') = \mathrm{e}^{-\frac{1}{2}|\xi-\xi'|^2} \tag{3.21}$$

is a common choice in GPR.

3.2 Fundamentals of Optimization

The aim of this work is the optimization under consideration of uncertainties. Therefore we will introduce fundamental definitions and summarize existing optimization methods. We will introduce an objective function f. In yield optimization, cf. Chap. 5, this f will be the yield depending on the QoI Q.

3.2.1 *Basic Definitions in Optimization*

Optimization Problem

Let $f : \mathbb{R}^n \to \mathbb{R}$, $g : \mathbb{R}^n \to \mathbb{R}^{m_1}$ and $h : \mathbb{R}^n \to \mathbb{R}^{m_2}$ be continuously differentiable functions. We formulate a general, single-objective, nonlinear optimization problem by [5, Chap. 15]

$$\min_{\mathbf{x}\in\mathbb{R}^n} f(\mathbf{x}) \tag{3.22}$$

$$\text{s.t.} g(\mathbf{x}) \leq \mathbf{0} \tag{3.23}$$

$$h(\mathbf{x}) = \mathbf{0}. \tag{3.24}$$

The *objective function* f is to be minimized over the *optimization variable* $\mathbf{x}$ (3.22), such that the *inequality constraints* (3.23) and the *equality constraints* (3.24) are fulfilled. Please note that the inequality in (3.23) is to be understood component-wise. The minimization problem can be reformulated into a maximization problem by replacing (3.22) with

$$\max_{\mathbf{x}\in\mathbb{R}^n} -f(\mathbf{x}). \tag{3.25}$$

An optimization problem is *bound constrained*, if there are no equality constraints and the inequality constraints can be written in the simplified form of lower and upper bounds $\mathbf{x}_{\text{lb}}, \mathbf{x}_{\text{ub}} \in \mathbb{R}^n$, i.e.,

$$\min_{\mathbf{x}\in\mathbb{R}^n} f(\mathbf{x}) \tag{3.26}$$

$$\text{s.t.} \mathbf{x}_{\text{lb}} \leq \mathbf{x} \leq \mathbf{x}_{\text{ub}}. \tag{3.27}$$

An optimization problem is called *unconstrained*, if there are no constraints at all.

Optimal Solutions and Convexity

Following [5, Definition 1.1], the vector $\mathbf{x} \in \mathbb{R}^n$ is a *feasible solution* of the optimization problem (3.22)–(3.24) if it fulfills the constraints (3.23)–(3.24). The set of feasible solutions is called the *feasible set* and denoted by $\mathbb{X} \subseteq \mathbb{R}^n$. A feasible solution $\mathbf{x}^{\text{opt}}$ is a *local minimum*, if there exists a neighborhood around $\mathbf{x}^{\text{opt}}$ with width $\varepsilon > 0$ such that

$$f(\mathbf{x}^{\text{opt}}) \leq f(\mathbf{x}), \qquad \forall \mathbf{x} \in \mathbb{X} \text{ with } ||\mathbf{x} - \mathbf{x}^{\text{opt}}|| < \varepsilon. \tag{3.28}$$

A feasible solution $\mathbf{x}^{\text{opt}}$ is a *global minimum*, if

$$f(\mathbf{x}^{\text{opt}}) \leq f(\mathbf{x}), \qquad \forall \mathbf{x} \in \mathbb{X}. \tag{3.29}$$

The set $\mathbb{X}$ is *convex*, if for all $\mathbf{x}_1, \mathbf{x}_2 \in \mathbb{X}$ and $\lambda \in [0, 1]$ such that

$$(1 - \lambda)\mathbf{x}_1 + \lambda \mathbf{x}_2 \in \mathbb{X}. \tag{3.30}$$

Let $\mathbb{X}$ be a convex set. If the function $f : \mathbb{X} \to \mathbb{R}$ fulfills

$$f((1 - \lambda)\mathbf{x}_1 + \lambda \mathbf{x}_2) \leq (1 - \lambda) f(\mathbf{x}_1) + \lambda f(\mathbf{x}_2), \tag{3.31}$$

then f is called a *convex function* [5, Definition 6.1, 6.2]. The local minimum of a convex function is always a global minimum.

Multi-objective Optimization

In case of several objective functions, a MOO problem is formulated by

$$\min_{\mathbf{x} \in \mathbb{X}} f_i(\mathbf{x}) \;\; i = 1, \ldots, k \tag{3.32}$$

writing the simplified notation $\mathbf{x} \in \mathbb{X}$ instead of (3.23)–(3.24). While in SOO the aim is to find the best solution with respect to one objective function, in MOO the aim is to find the best solution with respect to k objective functions. In general, the best solution for one objective function is not the best solution for another objective function. Therefore, the concept of pareto-optimality is introduced [6, Definition 2.1]. A feasible solution $\mathbf{x}^{\text{opt}}$ is called *pareto-optimal*, if there is no other feasible solution $\mathbf{x}$ such that

$$\begin{aligned} &f_i(\mathbf{x}) \leq f_i(\mathbf{x}^{\text{opt}}) \text{ for } i = 1, \ldots, k \text{ and} \\ &f_i(\mathbf{x}) < f_i(\mathbf{x}^{\text{opt}}) \text{ for some } i \in \{1, \ldots, k\}. \end{aligned} \tag{3.33}$$

This definition implies, that a pareto-optimal solution is a solution, such that it is not possible to further minimize the objective function value of one objective function

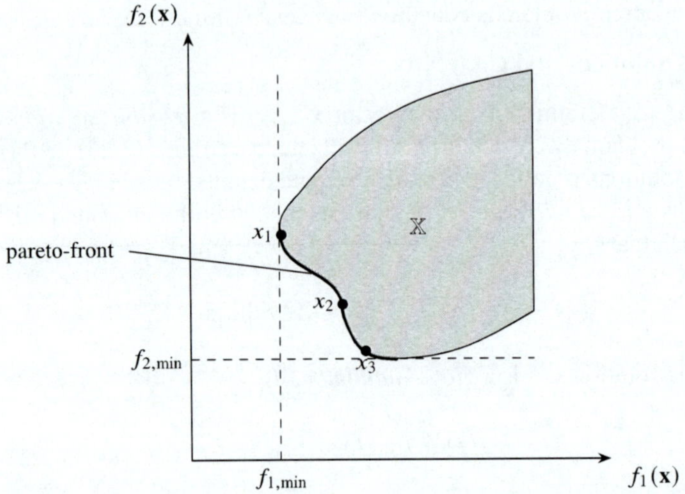

Fig. 3.1 Visualization of the non-convex pareto-front for a bi-objective minimization problem

without increasing the objective function value of another objective function. Usually, the pareto-optimal solution is not one single point, but a set of solutions, the so-called *pareto-front*. In Fig. 3.1 the non-convex pareto-front for a biobjective minimization problem is visualized.

3.2.2 *Review of Existing Optimization Methods*

There are many methods for optimization, see for example [5] for gradient based techniques, [7] for DFO or [8] for heuristic approaches. The methods we discuss in the following are particularly common in computational engineering and are chosen to best prepare for the novel methods proposed in Chap. 5 and applied in Chap. 6. We start with the gradient based Newton method, then we introduce Powell's derivative-free BOBYQA method. We discuss scalarization methods for solving MOO problems by transforming them into SOO problems, before we come to genetic algorithms, solving MOO problems directly.

Newton Method

The Newton method is best known from numerical analysis to find the root of a function. For unconstrained optimization problems, the first order optimality condition is $\nabla f(\mathbf{x}) = \mathbf{0}$ [5, Theorem 5.1]. Hence, the Newton method can be applied for finding the root of the gradient of the objective function. This requires that the objective function is twice continuously differentiable. The basic idea is to iteratively solve the Newton equation

$$\nabla^2 f(\mathbf{x})\mathbf{s} = -\nabla f(\mathbf{x}) \tag{3.34}$$

at a given $\mathbf{x}$ and update the solution $\mathbf{x} = \mathbf{x} + \mathbf{s}$. However, this procedure does not guarantee global convergence, which is the convergence from an arbitrary starting point to a stationary point, i.e., a point fulfilling $\nabla f(\mathbf{x}) = \mathbf{0}$. Therefore, it is necessary to control the search direction and the step size [5, Theorem 10.10]. For the search direction we check if a generalized angular condition is fulfilled, i.e., if for the current $\mathbf{x}$ and the determined search direction $\mathbf{s}$ holds

$$-\nabla f(\mathbf{x})^\top \mathbf{s} \geq \min\left(\alpha_1, \alpha_2 ||\mathbf{s}||^r\right) ||\mathbf{s}||^2, \tag{3.35}$$

where $\alpha_1, \alpha_2, r > 0$. Otherwise, the gradient step is used as search direction, i.e., $\mathbf{s} = -\nabla f(\mathbf{x})$. For the step size the Armijo rule is applied, for two parameters $\beta \in (0, 1)$ and $\gamma \in (0, 1)$ the largest number $\sigma \in \{\beta^0, \beta^1, \dots\}$ is chosen, such that

$$f(\mathbf{x} + \sigma\mathbf{s}) - f(\mathbf{x}) \leq \sigma\gamma\nabla f(\mathbf{x})^\top \mathbf{s} \tag{3.36}$$

holds. Please note that $\gamma \in (0, 1/2)$ is needed to assert fast local convergence. The resulting globalized Newton method from [5, Algorithm 10.9] is summarized in a flowchart, see Fig. 3.2.

Bound Constrained Optimization BY Quadratic Approximation (BOBYQA)

A DFO method for bound constrained problems has been proposed by Powell in [9]. Similar to SQP [5, Chap. 19], in each iteration the original problem is approximated quadratically around the current iterate solution $\mathbf{x}^{(k)}$, the quadratic subproblem is solved in a trust region and $\mathbf{x}^{(k)}$ is updated with the solution of the subproblem. While in SQP gradient information, i.e., second order Taylor expansion, is used for building the quadratic approximation, in BOBYQA a quadratic interpolation problem is solved. The description of the main ideas follows [10] and our work in [11]. Let the set of interpolation points be defined by

$$\mathcal{T} = \{(\mathbf{y}^0, f(\mathbf{y}^0)), \dots, (\mathbf{y}^p, f(\mathbf{y}^p))\}. \tag{3.37}$$

In each iteration a local quadratic approximation $\widetilde{m}^{(k)}(\mathbf{x})$ of $f(\mathbf{x})$ is built, i.e.,

$$f(\mathbf{x}) \approx \widetilde{m}^{(k)}(\mathbf{x}) = c^{(k)} + \mathbf{g}^{(k)^\top}(\mathbf{x} - \mathbf{x}^{(k)}) + \frac{1}{2}(\mathbf{x} - \mathbf{x}^{(k)})^\top \mathbf{H}^{(k)}(\mathbf{x} - \mathbf{x}^{(k)}), \tag{3.38}$$

fulfilling the interpolation conditions

$$f(\mathbf{y}^j) = \widetilde{m}^{(k)}(\mathbf{y}^j) \qquad \forall \mathbf{y}^j \in \mathcal{T}. \tag{3.39}$$

In case of $|\mathcal{T}| = (n + 1)(n + 2)/2$, the interpolation problem can be solved uniquely. Otherwise, i.e., $|\mathcal{T}| < (n + 1)(n + 2)/2$, the interpolation problem is underdetermined and the remaining degrees of freedom are set by minimizing the change in the Hessian matrix $\mathbf{H}$ of the quadratic model (3.38). For this, the matrix Frobenius norm is used, i.e.,

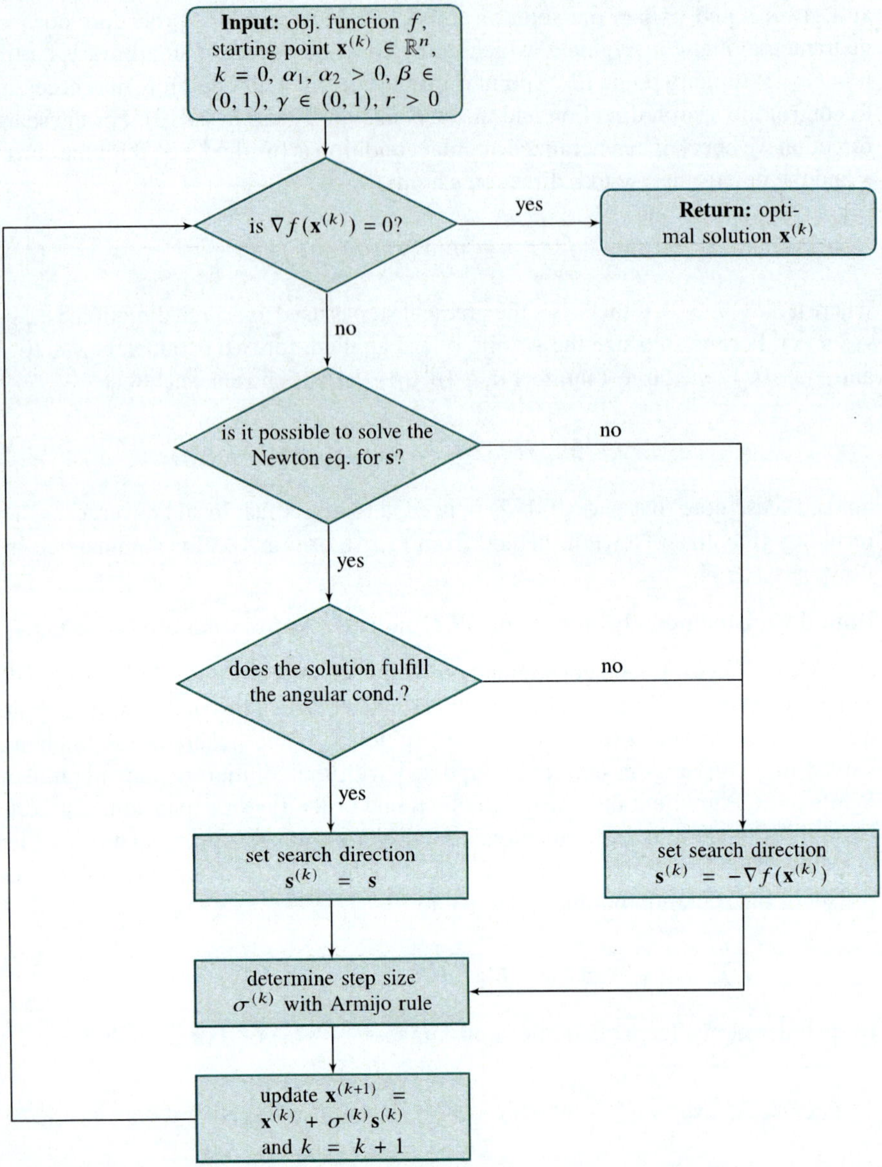

Fig. 3.2 Flowchart for the globalized Newton method, content based on [5, Algorithm 10.9]

$$\min_{c^{(k)},\mathbf{g}^{(k)},\mathbf{H}^{(k)}} ||\mathbf{H}^{(k)} - \mathbf{H}^{(k-1)}||_F^2 \text{ s.t. (3.39) holds,} \tag{3.40}$$

where typically $\mathbf{H}^{(-1)} = \mathbf{0}^{n\times n}$. Having built the quadratic model, the trust region subproblem

$$\begin{aligned} &\min_{\mathbf{x}\in\mathbb{R}^n} \widetilde{m}^{(k)}(\mathbf{x}) \\ &\text{s.t. } ||\mathbf{x} - \mathbf{x}^{(k)}||_2 \le \Delta^{(k)} \\ &\quad \mathbf{x}_{\text{lb}} \le \mathbf{x} \le \mathbf{x}_{\text{ub}} \end{aligned} \tag{3.41}$$

is solved. In (3.41), $\Delta^{(k)} > 0$ denotes the trust region radius in the k-th iteration. Once the optimal solution of the subproblem $\mathbf{x}^{\text{opt}}$ is obtained, it is checked if the decrease in the objective function is sufficient. We compare the ratio between actual and expected decrease, i.e.,

$$r^{(k)} = \frac{\text{actual decrease}}{\text{expected decrease}} = \frac{f(\mathbf{x}^{(k)}) - f(\mathbf{x}^{\text{opt}})}{\widetilde{m}^{(k)}(\mathbf{x}^{(k)}) - \widetilde{m}^{(k)}(\mathbf{x}^{\text{opt}})}. \tag{3.42}$$

If $r^{(k)}$ is considered to be large enough, the solution is accepted, i.e., $\mathbf{x}^{(k+1)} = \mathbf{x}^{\text{opt}}$, and the trust region radius is increased, i.e., $\Delta^{(k+1)} > \Delta^{(k)}$. Otherwise, the solution is not accepted, i.e., $\mathbf{x}^{(k+1)} = \mathbf{x}^{(k)}$, and the trust region radius is decreased, i.e., $\Delta^{(k+1)} < \Delta^{(k)}$. The interpolation set is updated and a new iteration begins. The basic procedure is illustrated in Fig. 3.3.

The convergence of model based optimization methods depends on the accuracy of the approximation. While in gradient based methods, typically Taylor expansion error bounds are considered to show the decreasing error between the model $\widetilde{m}^{(k)}(\mathbf{x})$ and the function $f(\mathbf{x})$ and their derivatives, in DFO other measures are required. Let d denote the polynomial degree of $\widetilde{m}^{(k)}(\mathbf{x})$, i.e., in BOBYQA $d = 2$, G a constant only depending on the function and $\Lambda > 0$ the poisedness constant. In [12] an error bound is then given by

$$||\nabla\widetilde{m}^{(k)}(\mathbf{x}) - \nabla f(\mathbf{x})|| \le \frac{1}{(d+1)!} G\Lambda \sum_{i=0}^{p} ||\mathbf{y}^i - \mathbf{x}||^{d+1}. \tag{3.43}$$

According to (3.43) the poisedness constant Λ is crucial for the accuracy of the quadratic model. It is a measure of how well balanced the interpolation points are. Or, in other words, how linear independent the resulting rows of the interpolation problem are. Figure 3.4 shows the example of an interpolation set with three points. Left, all points lie almost on one line. Hence, the two dimensional space is covered poorly, the poisedness constant would be large. Right, we see a set of well balanced interpolation points.

The following definitions show formally, how the poisedness constant Λ relates to the quality of the interpolation set [7, Definitions 3.1, 3.6, 5.3].

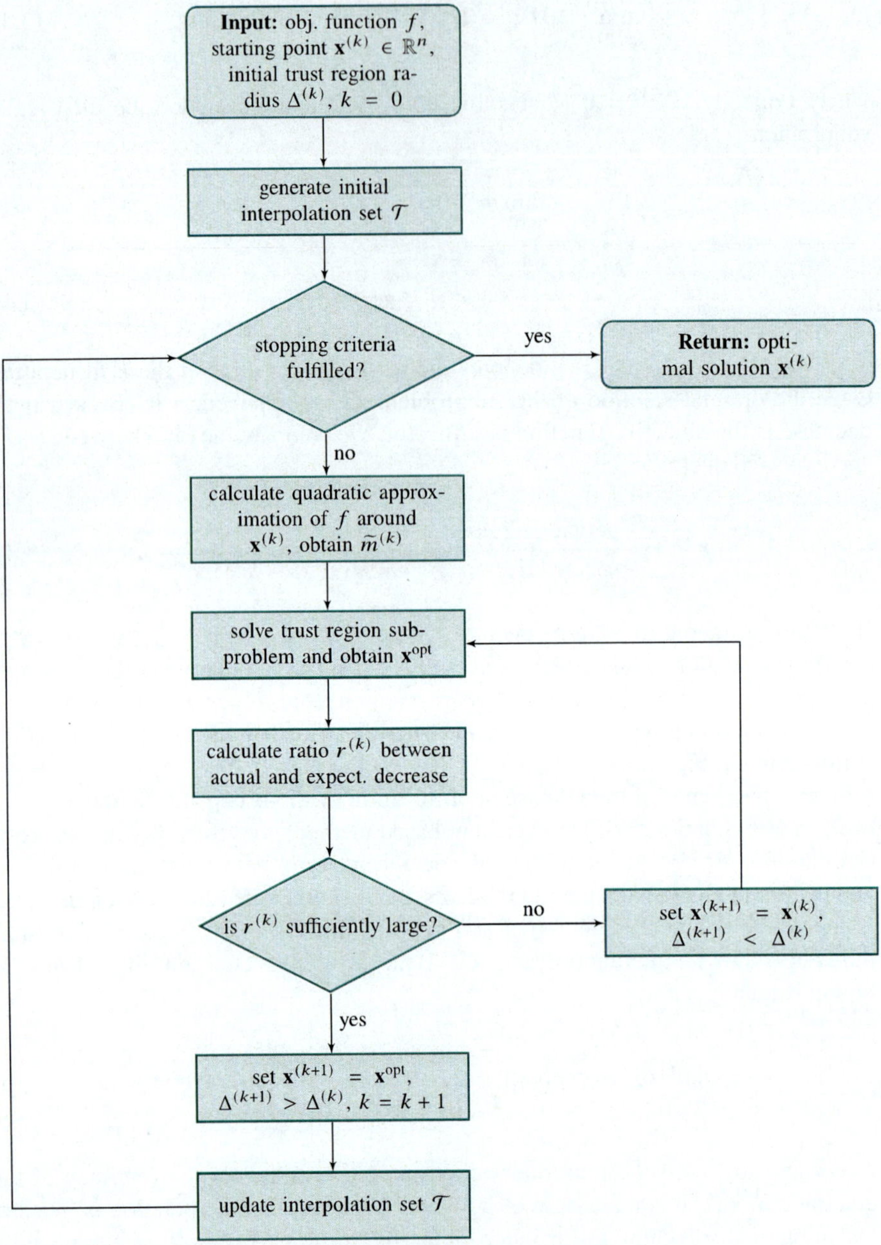

Fig. 3.3 Flowchart for the BOBYQA method, content based on [9]

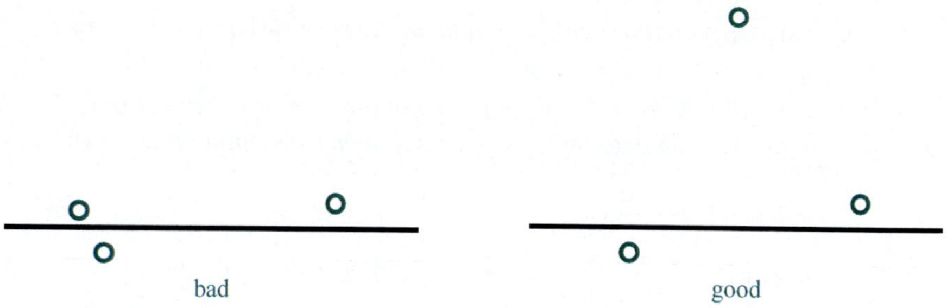

Fig. 3.4 Illustration of the poisedness of an interpolation set in a 2-dimensional space. Left a poorly poised set, right a well poised set

Definition 3.1 (*Poisedness*) Let an interpolation set $\mathcal{T}$ be defined as in (3.37), let $\Phi = \{\Phi_0(\mathbf{x}), \ldots, \Phi_p(\mathbf{x})\}$ be a polynomial basis of maximum degree d, let interpolation conditions be given as defined in (3.39) and let $\mathbf{M}$ denote the system matrix of the resulting interpolation problem. Then, the interpolation set $\mathcal{T}$ is poised for polynomial interpolation of degree d if and only if the matrix $\mathbf{M}$ is nonsingular.

This implies that an interpolation set is poised if and only if the solution of the interpolating polynomial exists and is unique [7, Lemma 3.2]. Please note that the definition of poisedness is independent of the chosen basis.

Definition 3.2 (Λ-*poisedness in the interpolation sense*) Let a constant $\Lambda > 0$, a set $\mathcal{B} \subset \mathbb{R}^n$, a polynomial basis $\Phi = \{\Phi_0(\mathbf{x}), \ldots, \Phi_p(\mathbf{x})\}$ of maximum degree d, $\boldsymbol{\Phi}(\mathbf{x}) = \left(\Phi_0(\mathbf{x}), \ldots, \Phi_p(\mathbf{x})\right)^\top$ and a poised interpolation set $\mathcal{T}$ as defined in (3.37). Then the training data set $\mathcal{T}$ is Λ-poised in $\mathcal{B}$ (in the interpolation sense) if and only if

$$\forall \mathbf{x} \in \mathcal{B}\ \exists \mathbf{l}(\mathbf{x}) \in \mathbb{R}^{p+1} \text{ s.t. } \sum_{i=0}^{p} l^i(\mathbf{x})\boldsymbol{\Phi}(\mathbf{y}^i) = \boldsymbol{\Phi}(\mathbf{x}) \quad \text{with} \quad ||\mathbf{l}(\mathbf{x})||_\infty \leq \Lambda. \tag{3.44}$$

Remark 3.1 When using the monomial basis, i.e., for $d = 2$ we use $\Phi = \left\{1, x_1, \ldots, x_n, \frac{1}{2}x_1^2, x_1x_2, x_1x_3, \ldots, x_{n-1}x_n, \frac{1}{2}x_n^2\right\}$, the equality in Definition 3.2 can be rewritten as $\mathbf{M}^\top\mathbf{l}(\mathbf{x}) = \boldsymbol{\Phi}(\mathbf{x})$. Then, the $l^i(\mathbf{x})$, $i = 0, \ldots, p$ are uniquely defined by the Lagrange polynomials and can be calculated by

$$\mathbf{M}\boldsymbol{\lambda}^i = \mathbf{e}^{i+1}, \tag{3.45}$$

where $\mathbf{e}^i \in \mathbb{R}^{p_1}$ denotes the i-th unit vector and the entries of $\boldsymbol{\lambda}_i$ are the coefficients of the polynomial l^i.

For the underdetermined interpolation case in BOBYQA, the following definition is needed.

Definition 3.3 (*Λ-poisedness in the minimum-norm sense*) Let a constant $\Lambda > 0$, a set $\mathcal{B} \subset \mathbb{R}^n$, the monomial basis $\Phi = \{\Phi_0(\mathbf{x}), \ldots, \Phi_q(\mathbf{x})\}$ of maximum degree d, $\mathbf{\Phi}(\mathbf{x}) = \left(\Phi_0(\mathbf{x}), \ldots, \Phi_q(\mathbf{x})\right)^\top$, a poised interpolation set $\mathcal{T}$ as defined in (3.37) and $p < q$. Then the training data set $\mathcal{T}$ is Λ-poised in $\mathcal{B}$ (in the minimum-norm sense) if and only if

$$\forall \mathbf{x} \in \mathcal{B}\ \exists \mathbf{l}(\mathbf{x}) \in \mathbb{R}^{p+1} \text{ s.t. } \sum_{i=0}^{p} l^i(\mathbf{x}) \mathbf{\Phi}(\mathbf{y}^i) \overset{\text{l.s.}}{=} \mathbf{\Phi}(\mathbf{x}) \text{ with } ||\mathbf{l}(\mathbf{x})||_\infty \leq \Lambda, \tag{3.46}$$

where l.s. indicates that the equations system is solved in the least squares sense.

Remark 3.2 The $l^i(\mathbf{x})$, $i = 0, \ldots, p$, correspond to the Lagrange polynomials (in the minimum norm sense), cf. [7, Definition 5.1]. Analogously to Remark 3.1 they can be calculated by the minimum norm solution of

$$\mathbf{M}\boldsymbol{\lambda}^i \overset{\text{m.n.}}{=} \mathbf{e}^{i+1}, \tag{3.47}$$

using the elements of $\boldsymbol{\lambda}^i$ as coefficients for the polynomial l^i.

In [7, Chap. 3] it is shown that the poisedness constant Λ, or rather $1/\Lambda$ can be understood as the distance to singularity of the system matrix of the interpolation problem. Definitions 3.2 and 3.3 show that the poisedness constant Λ corresponds to the quality of the interpolation set, and from (3.43) follows that in order to apply the convergence theory for gradient based methods to DFO methods, it is required that Λ remains uniformly bounded during the optimization procedure, i.e., for all interpolation sets used within the algorithm. For the maintenance of the interpolation set during the optimization follows that the poisedness constant Λ is a crucial factor. An accepted solution is added to the interpolation set, i.e., $\mathcal{T} = \mathcal{T} \cup \{(\mathbf{y}^{\text{add}}, f(\mathbf{y}^{\text{add}}))\}$ with $\mathbf{y}^{\text{add}} = \mathbf{x}^{\text{opt}}$. Since the size of the interpolation set is fixed, another interpolation point has to leave the set. Hereby, the aim is to reduce the poisedness constant Λ as much as possible. Let l^i, $i = 0, \ldots, p$, be the Lagrange polynomials obtained by evaluating (3.45) or (3.47) and

$$i^{\text{go}} = \arg\max_{i=0,\ldots,p} \left(|l^i(\mathbf{y}^{\text{add}})| \max\left\{1, \frac{||\mathbf{y}^i - \mathbf{y}^{\text{add}}||_2^4}{||\Delta^{(k)}||^4}\right\}\right). \tag{3.48}$$

Per construction, $\mathbf{y}^{i^{\text{go}}}$ is the point with the worst (largest) value of the corresponding Lagrange polynomial, evaluated at the new iterate solution. Hence it has the worst impact on the Λ-poisedness of the interpolation set and is replaced by the new solution, i.e., the updated interpolation set is $\mathcal{T} = \mathcal{T} \cup \{(\mathbf{y}^{\text{add}}, f(\mathbf{y}^{\text{add}}))\} \setminus \{(\mathbf{y}^{i^{\text{go}}}, f(\mathbf{y}^{i^{\text{go}}}))\}$. Sometimes in the algorithm, it is required to improve the interpolation set by exchanging points, regardless of the subproblem's optimal solution. Then, a new point is determined as follows. Let $\mathbf{y}^i$ be an interpolation point which shall leave the interpolation set based on (3.48) and l^i the corresponding Lagrange polynomial. Then, the new point is chosen by solving

$$\mathbf{y}^{\text{new}} = \max_{\mathbf{y}\in\mathcal{B}} |l^i(\mathbf{y})| \qquad (3.49)$$
$$\text{s.t. } \mathbf{x_{lb}} \leq \mathbf{y} \leq \mathbf{x_{ub}}$$
$$||\mathbf{y} - \mathbf{x}^{\text{opt}}|| \leq \Delta^{(k)}.$$

For more details we refer to the original work by Powell [9]. The author also proposed two other methods, LINCOA [13] (LINearly Constrained Optimization Algorithm), which allows linearly constrained optimization problems to be solved, and COBYLA [14] (Constrained Optimization BY Linear Approximations), which allows general constraints but approximates the objective function only linearly.

Scalarization Methods for Multi-objective Optimization

A common technique for solving MOO problems is to transform them into a SOO problem. This approach is called scalarization. In weighted sum scalarization [6, Chap. 3], the objective functions are combined into a weighted sum, which then yields the new objective function. We obtain

$$\min_{\mathbf{x}\in\mathbb{X}} \sum_{i=1}^{k} w_i\, f_i(\mathbf{x}), \qquad (3.50)$$

with weighting parameters $w_i > 0, i = 1, \ldots, k$. The ε-constraint method [6, Chap. 4] is another scalarization technique. Only one of the objective functions is considered as the objective function, while all the remaining ones are provided with an upper bound and included into the constraints. Without limitation of generality, we assume that f_1 is the objective function which will be kept in the ε-constraint formulation. This yields

$$\min_{\mathbf{x}\in\mathbb{X}} f_1(\mathbf{x})$$
$$\text{s.t.} f_i(\mathbf{x}) \leq \varepsilon_i, \quad i = 2, \ldots, k, \qquad (3.51)$$

with parameters $\varepsilon_i, i = 2, \ldots, k$. The advantage of transforming the MOO problem into a SOO problem is that it can be solved efficiently with standard SOO methods. However, the solution of (3.50) or (3.51) is only one pareto-optimal solution, depending on the starting point and the choice of the parameters w_i or ε_i, respectively. In order to obtain the pareto-front, the SOO problems have to be solved several times with different values for weights or bounds, respectively. Further, the feasible set needs to fulfill some convexity requirements to give the weighted sum method the chance to find each pareto-optimal solution. In Fig. 3.1 for example, the solutions $\mathbf{x}_1$ and $\mathbf{x}_3$ could be found, however the pareto-optimal solution $\mathbf{x}_2$ could not be found using the weighted sum approach, cf. [6, Chap. 3].

Genetic Algorithms

Genetic algorithms solve MOO problems directly and the solution is a set of solutions converging to the pareto-front [16, Chap. 4]. The term genetic algorithm is motivated by the natural evolution process of organisms and covers various algorithms, e.g. the Non-dominated Sorting Genetic Algorithms I-III (NSGA I-III) [17–19]. The main idea is briefly sketched in the following. In the beginning, an initial set of solutions is generated. Each solution is called an individual, the set is called population. All objective functions are evaluated on each individual. Based on this information, the so-called fitness value is determined, which is a metric for the quality of an individual. Based on the fitness value, in each iteration, the population is updated and hence, a new generation is created. This update includes three main elements: selection (survival of the fittest), crossover (reproduction process where genetic traits of the individuals are propagated) and mutation (variation of genetic traits). The procedure is continued until a stopping criterion is reached, e.g. the maximum number of generations is exceeded or the improvement in the solution or objective space stagnates. An illustration of the simplified genetic algorithm is provided in Fig. 3.5. Compared to solving a MOO once with a scalarization technique, the genetic algorithms are computationally more expensive since the objective functions are solved for each individual of each generation. Considering the aim of maximizing the yield, which involves the repeated solving of probably expensive quantities of interest (QoIs), we see that this may become computationally challenging. However, genetic algorithms approximate directly the complete pareto-front, not only one solution. Further, they do not require convexity or other previous knowledge about the problem. Also, the drawback of manually setting parameters like weights (weighted sum) or bounds (ε-constraint) is eliminated.

3.3 Uncertainty Quantification

In Sect. 2.3 we discussed high fidelity models, e.g. based on the FEM, and their approximation error. However, even if the computer model is highly accurate, the actually manufactured device may differ from it. This is not necessarily caused by erroneous computer models, but by manufacturing or environmental uncertainties, so-called aleatory uncertainties. These may lead to deviations in the original design parameters or other system parameters. To account for these uncertainties, we model the affected parameters as random variables. Let $\boldsymbol{\xi}$ denote the multivariate random variable representing the uncertain parameter and let us assume that $\boldsymbol{\xi}$ follows a given probability density function $\varphi(\boldsymbol{\xi})$. A realization of $\boldsymbol{\xi}$ is denoted by $\boldsymbol{\xi}^{(i)} \in \mathbb{R}^n$. Let us consider a QoI $Q : \mathbb{R}^n \to \mathbb{R}$, which maps the uncertain input $\boldsymbol{\xi}$ to an uncertain output $Q(\boldsymbol{\xi})$. The QoI Q can be assumed to be sufficiently smooth to be well approximated by GPs or polynomial interpolation. Estimating the impact of these deviations and considering them already in the design process based on computer models helps to increase the reliability of manufacturing processes. The field dealing with this is called UQ. In this section we recall several approximation methods frequently used in UQ.

Genetic algorithms

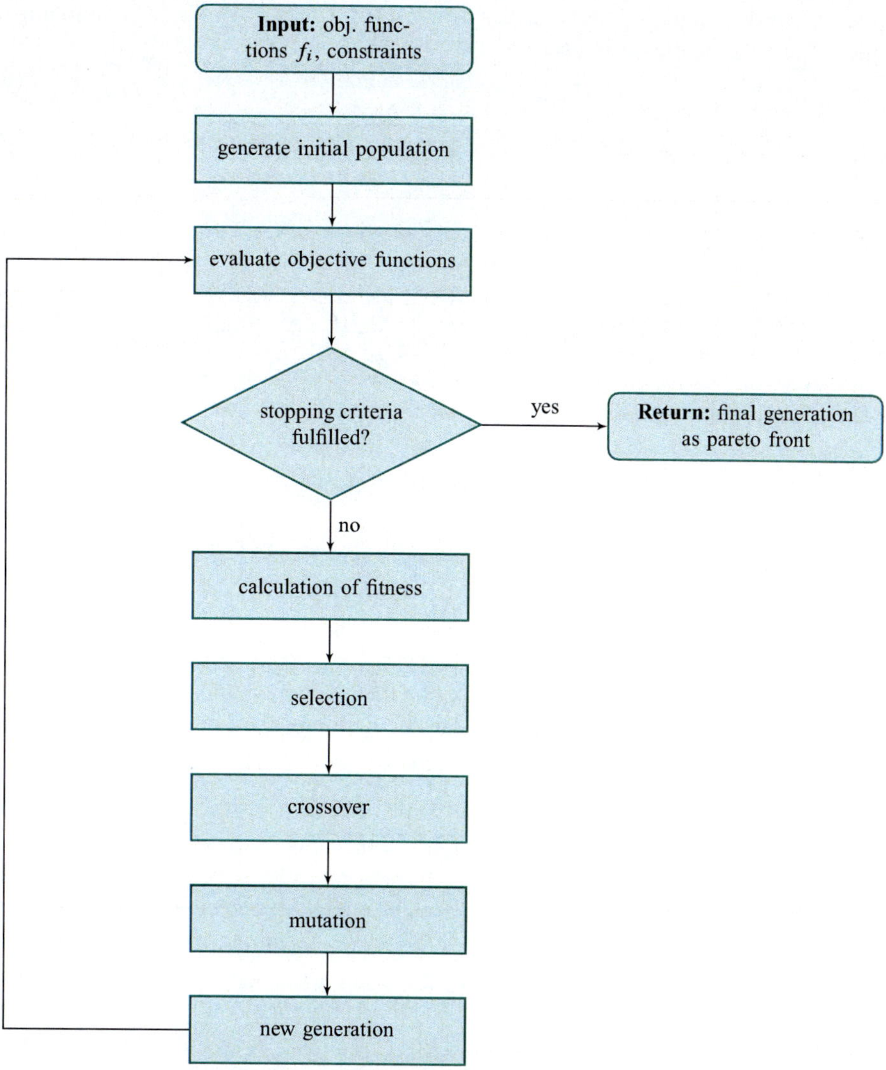

Fig. 3.5 Flowchart for the genetic MOO approach, based on [15, Fig. 3]

3.3.1 *Monte Carlo Analysis*

A widely used method in UQ is MC analysis [20, Chap. 5]. It is used to estimate statistical measures of the function Q, e.g., the mean value, the variance or failure probabilities as we will see in Sect. 4.2. Therefore, a set of MC sample points

$\{\boldsymbol{\xi}^{(i)}\}_{i=1,\ldots,N_{\mathrm{MC}}}$ of the uncertain parameter is generated according to the corresponding probability density function. According to [2, Chap. 3.4], the MC estimator of the mean value is then given by

$$\mathbb{E}[Q(\boldsymbol{\xi})] = \int_{\mathbb{R}^n} Q(\boldsymbol{\xi})\,\varphi(\boldsymbol{\xi})\,\mathrm{d}\boldsymbol{\xi} \approx \frac{1}{N_{\mathrm{MC}}}\sum_{i=1}^{N_{\mathrm{MC}}} Q(\boldsymbol{\xi}^{(i)}) = \widetilde{\mathbb{E}}_{\mathrm{MC}}[Q(\boldsymbol{\xi})] \tag{3.52}$$

and the variance by

$$\mathbb{V}[Q(\boldsymbol{\xi})] \approx \widetilde{\mathbb{V}}_{\mathrm{MC}}[Q(\boldsymbol{\xi})] = \frac{1}{N_{\mathrm{MC}}-1}\sum_{i=1}^{N_{\mathrm{MC}}}\left(Q(\boldsymbol{\xi}^{(i)}) - \widetilde{\mathbb{E}}_{\mathrm{MC}}[Q(\boldsymbol{\xi})]\right)^2 . \tag{3.53}$$

The MC estimator is unbiased, i.e., for all $N_{\mathrm{MC}} \geq 1$

$$\mathbb{E}[\widetilde{\mathbb{E}}_{\mathrm{MC}}[Q(\boldsymbol{\xi})]] = \mathbb{E}[Q(\boldsymbol{\xi})]. \tag{3.54}$$

For the variance of the MC estimator of the mean value follows

$$\mathbb{V}[\widetilde{\mathbb{E}}_{\mathrm{MC}}[Q(\boldsymbol{\xi})]] = \frac{\mathbb{V}[Q(\boldsymbol{\xi})]}{N_{\mathrm{MC}}}. \tag{3.55}$$

By taking the square root of (3.55) we obtain the MC error indicator $\varepsilon_{\mathrm{MC}}$ [21], which coincides with the RMSE and the standard deviation of the estimator

$$\varepsilon_{\mathrm{MC}} = \mathrm{Std}[\widetilde{\mathbb{E}}_{\mathrm{MC}}[Q(\boldsymbol{\xi})]] = \underbrace{\left(\mathbb{E}\left[\left|\mathbb{E}[Q(\boldsymbol{\xi})] - \widetilde{\mathbb{E}}_{\mathrm{MC}}[Q(\boldsymbol{\xi})]\right|^2\right]\right)^{\frac{1}{2}}}_{\mathrm{RMSE}} = \frac{\sqrt{\mathbb{V}[Q(\boldsymbol{\xi})]}}{\sqrt{N_{\mathrm{MC}}}} = \frac{\mathrm{Std}[Q(\boldsymbol{\xi})]}{\sqrt{N_{\mathrm{MC}}}}. \tag{3.56}$$

From (3.56) follows that the MC analysis provides high accuracy for large sample sets, but slow convergence with $\mathcal{O}\left(1/\sqrt{N_{\mathrm{MC}}}\right)$. Assuming that the evaluation of $Q(\boldsymbol{\xi}^{(i)})$, $i = 1, \ldots, N_{\mathrm{MC}}$ requires solving a complex model numerically, as introduced in Chap. 2, this approach quickly becomes computationally prohibitive if high accuracy is desired. On the other hand, Eq. (3.56) also shows that the accuracy of the MC estimator does not directly depend on the number of uncertain parameters n. Besides the simple implementation and applicability, this is an important advantage of the MC analysis. An alternative is a surrogate based MC analysis, i.e., a cheap to evaluate surrogate model $\widetilde{Q}$ of Q is built and then $\widetilde{Q}$ is used in (3.52) and (3.53). In the following, we will introduce two possible surrogate models. In contrast to a pure MC analysis, the discussed surrogate models require that the QoI depends smoothly on the input parameter. Provided that this is fulfilled, this is also an advantage of the surrogate methods: they exploit the information about the smoothness of the QoI, which MC does not.

3.3.2 Sparse Grid Stochastic Collocation via Interpolation

Stochastic collocation refers to methods with the goal of building an accurate approximation of the considered function using sampling [22]. However, when talking about SC in the following, we refer to SC via polynomial interpolation; more concretely, to sparse grid SC based on Leja nodes [23]. Following [22, 23], a brief introduction is given.

Again, let the QoI $Q : \mathbb{R}^n \to \mathbb{R}$ be the function to be approximated, let

$$\mathcal{T}_{\mathrm{SC}} = \left\{ \left(\boldsymbol{\xi}^{(1)}, Q(\boldsymbol{\xi}^{(1)}) \right), \ldots, \left(\boldsymbol{\xi}^{(p)}, Q(\boldsymbol{\xi}^{(p)}) \right) \right\} \tag{3.57}$$

be a given training data set with $|\mathcal{T}_{\mathrm{SC}}| = p$ and let $\Phi = \{\phi_1, \ldots, \phi_p\}$ be a multivariate polynomial basis, i.e., $\phi_j : \mathbb{R}^n \to \mathbb{R}$. Assuming that Q is well-defined and sufficiently smooth, the surrogate model $\widetilde{Q}(\boldsymbol{\xi})$ is defined by

$$\widetilde{Q}(\boldsymbol{\xi}) = \sum_{j=1}^{p} \alpha_j \phi_j(\boldsymbol{\xi}), \tag{3.58}$$

where the coefficients α_j are determined by solving the linear equations system which originates by enforcing the interpolation conditions

$$\widetilde{Q}(\boldsymbol{\xi}^{(j)}) = Q(\boldsymbol{\xi}^{(j)}), \qquad \forall j = 1, \ldots, p. \tag{3.59}$$

Additionally, all or some derivatives can be used to extend the conditions (3.59) with

$$\frac{\partial}{\partial \xi_i} \widetilde{Q}(\boldsymbol{\xi}^{(j)}) = \frac{\partial}{\partial \xi_i} Q(\boldsymbol{\xi}^{(j)}), \qquad j \in \{1, \ldots, p\}, i \in \{1, \ldots, n\}, . \tag{3.60}$$

This approach is called Hermite interpolation [24].

For the basis of (3.58) a Lagrange interpolation approach can be chosen [25, Chap. 2.2]. From univariate interpolation it is well-known that the structure of the training data points is crucial for the quality of the surrogate model. The fact that polynomial interpolation on unstructured or equidistant points becomes unstable is known as the Runge phenomena [24, Chap. 6.2]. For approximations with high accuracy and robustness, training data points should be chosen on a grid with higher density toward the boundaries of the interval. A common choice are Chebyshev nodes [26, Chap. 20]. Unfortunately, those nodes have a specific construction and are not nested, i.e., the surrogate model cannot simply be updated later with additional training data points. For an improvement of the model, a new structured training data set has to be generated and evaluated, which requires the high computational effort of building a complete surrogate model.

Transferring the findings from univariate interpolation to multivariate interpolation leads to a tensor grid approach. Let $\mathcal{T}_{\mathrm{SC},i}$ with $|\mathcal{T}_{\mathrm{SC},i}| = p_i, i = 1, \ldots, n$ denote

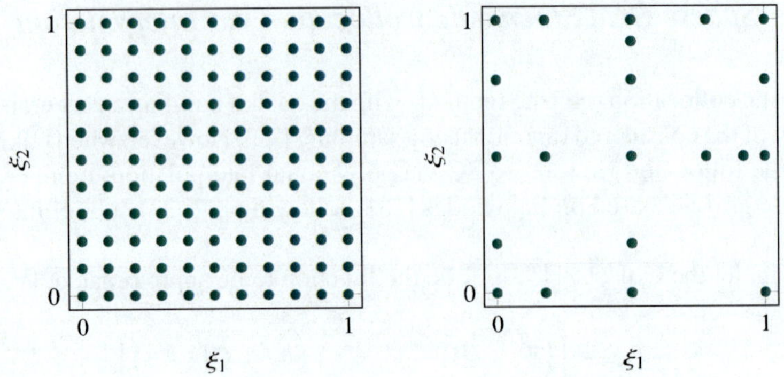

Fig. 3.6 Illustration of training data points for SC in two dimensions. Left tensor grid with equidistant nodes, right sparse grid with adaptive Leja nodes

the training data set for the univariate surrogate model of the i-th dimension and let $\widetilde{Q}_i$ denote this surrogate model. Then, the multivariate surrogate model is given by

$$\widetilde{Q}(\boldsymbol{\xi}) = \widetilde{Q}_1(\boldsymbol{\xi}) \otimes \cdots \otimes \widetilde{Q}_n(\boldsymbol{\xi}) \tag{3.61}$$

and the set of training data points by

$$\mathcal{T}_{\mathrm{SC}} = \mathcal{T}_{\mathrm{SC},1} \times \cdots \times \mathcal{T}_{\mathrm{SC},n}. \tag{3.62}$$

This approach results in a total number of training data points $p = p_1 \times \cdots \times p_n$. Employing all these points of the grid is often computationally intractable. For dimensions $n \gg 1$ the convergence rate deteriorates rapidly, which is often referred to as the *curse of dimensionality* [22].

An alternative is (Smolyak) sparse grids SC [27]. The idea is to neglect a subset of the training data set $\mathcal{T}_{\mathrm{SC}}$, which does not contribute significantly to the approximation accuracy. A visualization of training data sets in the tensor grid and the sparse grid approach is given in Fig. 3.6. There exist different approaches to choose the training data points which shall be considered, e.g., Clenshaw Curtis nodes, which are the extrema of Chebyshev polynomials [28], or weighted Leja nodes [23], which will be introduced in the following. For each dimension $i = 1, \ldots, n$, the weighted Leja nodes are defined recursively by an optimization problem. Let $p_{\mathrm{L},i}$ denote the number of Leja nodes in the i-th dimension and $\varphi(\xi_i)$ the probability density function of the corresponding input parameter ξ_i. For $j = 2, \ldots, p_{\mathrm{L},i}$ the j-th Leja node is calculated by

$$\xi_i^{(j)} = \underset{\xi_i \in \mathbb{R}}{\arg\max} \sqrt{\varphi(\xi_i)} \prod_{k=1}^{j-1} |\xi_i - \xi_i^{(k)}|. \tag{3.63}$$

The first node in each dimension can be chosen arbitrarily, and is typically set $\xi_i^{(1)} = 0$. The sparse grids can also be constructed adaptively, detecting highly sensitive parameters ξ_i, cf. [23, 29].

Adjoint Error Indicator for Stochastic Collocation

As the original model we consider the high fidelity solution of the QoI, i.e., $Q_h(\boldsymbol{\xi})$, obtained for example with FEM. We introduce an adjoint error indicator for SC [30, 31]. Therefore, we recall the primal (2.19) and the dual (2.23) problem from Sect. 2.3 with their corresponding systems of linear equations

$$\mathbf{K}\mathbf{u}(\boldsymbol{\xi}) = \mathbf{f}, \tag{3.64}$$

and

$$\mathbf{K}^{\dagger}\mathbf{z}(\boldsymbol{\xi}) = \mathbf{q}, \tag{3.65}$$

respectively. We are interested in the error between the original solution Q_h and the SC approximation $\widetilde{Q}_h$. Therefore, besides the SC approximation of the QoI $\widetilde{Q}_h$, we also build SC approximations of the solutions $\mathbf{u}$ and $\mathbf{z}$ of the original problem (3.64) and the dual problem (3.65), respectively. The same set of training data points is employed, cf. [32]. For $\mathbf{u}$ no additional computing effort is caused, for $\mathbf{z}$ problem (3.65) needs to be solved for each training data point. However, the additional costs for the adjoint problem can often be mitigated, e.g. by re-using the LU decomposition of the primal problem, see [33, Chap. 4.3]. The resulting approximations are denoted by $\widetilde{\mathbf{u}}$ and $\widetilde{\mathbf{z}}$. Following [32], the SC error is given by

$$\begin{aligned}
\varepsilon_{\mathrm{sc}}(\boldsymbol{\xi}) &= \left|Q_h(\boldsymbol{\xi}) - \widetilde{Q}_h(\boldsymbol{\xi})\right| \\
&= \left|\left(\mathbf{q}, \mathbf{u}(\boldsymbol{\xi})\right)_D - \left(\mathbf{q}, \widetilde{\mathbf{u}}(\boldsymbol{\xi})\right)_D\right| \\
&= \left|\left(\mathbf{q}, \mathbf{u}(\boldsymbol{\xi}) - \widetilde{\mathbf{u}}(\boldsymbol{\xi})\right)_D\right| \\
&\overset{(3.65)}{=} \left|\left(\mathbf{K}^{\dagger}\, \mathbf{z}(\boldsymbol{\xi}), \mathbf{u}(\boldsymbol{\xi}) - \widetilde{\mathbf{u}}(\boldsymbol{\xi})\right)_D\right| \\
&\overset{(2.17)}{=} \left|\left(\mathbf{z}(\boldsymbol{\xi}), \mathbf{K}\left(\mathbf{u}(\boldsymbol{\xi}) - \widetilde{\mathbf{u}}(\boldsymbol{\xi})\right)\right)_D\right| \\
&\overset{(3.64)}{=} \left|\left(\mathbf{z}(\boldsymbol{\xi}), \mathbf{f} - \mathbf{K}\,\widetilde{\mathbf{u}}(\boldsymbol{\xi})\right)_D\right|.
\end{aligned} \tag{3.66}$$

Using $\widetilde{\mathbf{z}}$ instead of $\mathbf{z}$ yields the SC error indicator

$$\widetilde{\varepsilon}_{\mathrm{sc}}(\boldsymbol{\xi}) := \left|\left(\widetilde{\mathbf{z}}(\boldsymbol{\xi}), \mathbf{f} - \mathbf{K}\,\widetilde{\mathbf{u}}(\boldsymbol{\xi})\right)_D\right|. \tag{3.67}$$

3.3.3 Gaussian Process Regression

Another surrogate-based method is GPR [4, Chap. 2.2]. In contrast to SC, the training data points can be chosen arbitrarily and their number does not automatically scale with the dimension. The structure of this section is based on our work in [34, Sect. 3.1]. Again, let $Q : \mathbb{R}^n \to \mathbb{R}$ be the QoI to be approximated. GPR can be divided into four mandatory and one optional step. These steps are described in the following.

1. Prior: Before we have seen any data, we start with some prior assumptions about the function we aim to approximate, for example, about the smoothness or the mean value. These assumptions are called the prior and are formulated as a GP $\{Q_{\boldsymbol{\xi}}\}_{\boldsymbol{\xi}\in\mathcal{X}}$, with $\mathcal{X} \subset \mathbb{R}^n$. The GP is completely defined by its mean function $m(\boldsymbol{\xi})$ and its kernel function $k(\boldsymbol{\xi}, \boldsymbol{\xi}')$, cf. Sect. 3.1.2. In other words, according to the assumptions, in this step a mean function and a kernel function are chosen. We write

$$\{Q_{\boldsymbol{\xi}}\}_{\boldsymbol{\xi}\in\mathcal{X}} \sim \mathcal{GP}\left(m(\boldsymbol{\xi}), k\left(\boldsymbol{\xi}, \boldsymbol{\xi}'\right)\right). \tag{3.68}$$

2. Training data: In this step, training data is collected by running simulations on the original model or by taking measurements. The training data set is defined by

$$\mathcal{T}_{\text{GPR}} = \left\{\left(\boldsymbol{\xi}^{(1)}, Q(\boldsymbol{\xi}^{(1)})\right), \ldots, \left(\boldsymbol{\xi}^{(p)}, Q(\boldsymbol{\xi}^{(p)})\right)\right\}, \tag{3.69}$$

where for $i = 1, \ldots, p$, $\boldsymbol{\xi}^{(i)} \in \mathbb{R}^n$ denote the input data and $Q(\boldsymbol{\xi}^{(i)}) \in \mathbb{R}$ the output data (observation), respectively. Further we denote the matrices of input and output data by

$$\mathcal{X} = \left(\boldsymbol{\xi}^{(1)}, \ldots, \boldsymbol{\xi}^{(p)}\right) \in \mathbb{R}^{n\times p} \quad \text{and} \quad \mathcal{Q} = \left(Q(\boldsymbol{\xi}^{(1)}), \ldots, Q(\boldsymbol{\xi}^{(p)})\right) \in \mathbb{R}^p. \tag{3.70}$$

3. Posterior: The information from the prior assumptions and the collected training data is combined in order to obtain a new GP, with updated mean and kernel function. Using

$$\mathcal{K} = \begin{pmatrix} k(\boldsymbol{\xi}^{(1)}, \boldsymbol{\xi}^{(1)}) & \ldots & k(\boldsymbol{\xi}^{(1)}, \boldsymbol{\xi}^{(p)}) \\ \vdots & & \vdots \\ k(\boldsymbol{\xi}^{(p)}, \boldsymbol{\xi}^{(1)}) & \ldots & k(\boldsymbol{\xi}^{(p)}, \boldsymbol{\xi}^{(p)}) \end{pmatrix} \quad \text{and} \quad \mathbf{m} = \begin{pmatrix} m(\boldsymbol{\xi}^{(1)}) \\ \vdots \\ m(\boldsymbol{\xi}^{(p)}) \end{pmatrix} \tag{3.71}$$

and applying Bayes' rule, cf. Sect. 3.1.1, the posterior distribution of the output $Q_{\boldsymbol{\xi}}$ depending on the training data set is given by

$$Q_{\boldsymbol{\xi}} | \mathcal{X}, \mathcal{Q} \sim \mathcal{N}(\mathbf{m}, \mathcal{K}). \tag{3.72}$$

In Fig. 3.7 example realizations for the prior and the posterior are illustrated.

4. Predictions: Depending on the prior assumptions and the training data, for an arbitrary test point $\boldsymbol{\xi}^\star$ the predicted distribution is

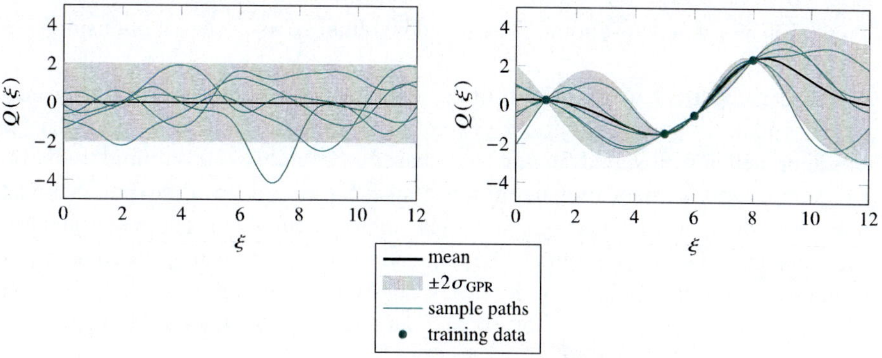

Fig. 3.7 Visualization of Gaussian process regression with mean value (i.e., prediction), $2\sigma_{\text{GPR}}$ interval, training data points and sample paths drawn from the prior (left) and the posterior (right)

$$Q_{\boldsymbol{\xi}^\star}|\boldsymbol{\xi}^\star, \mathcal{X}, \mathcal{Q} \sim \mathcal{N}(\underbrace{m(\boldsymbol{\xi}^\star) + \mathbf{k}(\boldsymbol{\xi}^\star, \mathcal{X})\mathcal{K}^{-1}(\mathcal{Q} - \mathbf{m})}_{\text{mean}}, \underbrace{k(\boldsymbol{\xi}^\star, \boldsymbol{\xi}^\star) - \mathbf{k}(\boldsymbol{\xi}^\star, \mathcal{X})\mathcal{K}^{-1}\mathbf{k}(\mathcal{X}, \boldsymbol{\xi}^\star)}_{\text{variance}}), \tag{3.73}$$

with

$$\mathbf{k}(\boldsymbol{\xi}^\star, \mathcal{X}) = \left(k(\boldsymbol{\xi}^\star, \boldsymbol{\xi}^{(1)}), \ldots, k(\boldsymbol{\xi}^\star, \boldsymbol{\xi}^{(p)})\right), \tag{3.74}$$

$$\mathbf{k}(\mathcal{X}, \boldsymbol{\xi}^\star) = \left(k(\boldsymbol{\xi}^{(1)}, \boldsymbol{\xi}^\star), \ldots, k(\boldsymbol{\xi}^{(p)}, \boldsymbol{\xi}^\star)\right)^\top. \tag{3.75}$$

In order to predict the function value of Q in $\boldsymbol{\xi}^\star$, the mean value of (3.73) in this point is taken, i.e., $\widetilde{Q}_{\text{GPR}}(\boldsymbol{\xi}^\star)$. In order to predict the error or uncertainty of the GPR model in this specific point, the standard deviation of (3.73) is considered, i.e., $\varepsilon_{\text{GPR}}(\boldsymbol{\xi}^\star) = \sigma_{\text{GPR}}(\boldsymbol{\xi}^\star)$.

5. Model update (optional): An existing GPR model can be updated with a new data point $(\boldsymbol{\xi}^{\text{add}}, Q(\boldsymbol{\xi}^{\text{add}}))$ by adding it to the training data set by

$$\mathcal{T}_{\text{GPR,new}} = \mathcal{T}_{\text{GPR}} \cup \left\{\left(\boldsymbol{\xi}^{\text{add}}, Q(\boldsymbol{\xi}^{\text{add}})\right)\right\}, \tag{3.76}$$

and extending the input and output matrices by

$$\mathcal{X}_{\text{new}} = \left(\mathcal{X}, \boldsymbol{\xi}^{\text{add}}\right) \in \mathbb{R}^{n \times p+1} \quad \text{and} \quad \mathcal{Q}_{\text{new}} = \left(\mathcal{Q}, Q(\boldsymbol{\xi}^{\text{add}})\right) \in \mathbb{R}^{p+1}. \tag{3.77}$$

We update (3.71) according to

$$\mathcal{K}_{\text{new}} = \begin{pmatrix} \mathcal{K} & \mathbf{k}(\mathcal{X}, \boldsymbol{\xi}^{\text{add}}) \\ \mathbf{k}(\boldsymbol{\xi}^{\text{add}}, \mathcal{X}) & k(\boldsymbol{\xi}^{\text{add}}, \boldsymbol{\xi}^{\text{add}}) \end{pmatrix} \quad \text{and} \quad \mathbf{m}_{\text{new}} = \begin{pmatrix} \mathbf{m} \\ m(\boldsymbol{\xi}^{\text{add}}) \end{pmatrix}. \tag{3.78}$$

Predictions for a new test point $\boldsymbol{\xi}^\star$ can then be obtained as in step 4, but using (3.77) and (3.78).

Since in contrast to SC, the training data points are chosen arbitrarily and a model update is possible without generating a completely new training data set, this approach is well suited for surrogate based MC analysis. If required, individual MC sample points can be used to improve the GPR model at any time. For the update, a linear system with matrix $\mathcal{K}_{\text{new}}$ has to be solved, which is of cubic complexity in the worst case, i.e., $\mathcal{O}(p^3)$, but can be reduced to $\mathcal{O}(r^2 p)$ by using low-rank approximations, where the rank of the low-rank approximation is denoted by r [4, Chap. 8]. Thus, the computational effort of such an update is assumed to be negligible in comparison to one evaluation of the original function Q.

3.4 Summary

In this chapter we discussed the mathematical foundations in the areas probability theory, optimization and uncertainty quantification. Manufacturing uncertainties may lead to deviations in the geometry or material parameters. To account for these deviations, we model uncertain parameters as random variables. Therefore, the basic concepts of probability theory are required. Further, we recalled the law of large numbers, which is the foundation for all sampling based methods in uncertainty quantification. We described the widely used MC analysis as a basic sampling method. In preparation for the next chapter we introduced two surrogate models, SC and GPR. These will be part of the hybrid yield estimation methods proposed in Sects. 4.3 and 4.4. In the field of optimization, we recalled basic definitions and briefly described common optimization approaches: The gradient based Newton method, which will be adapted to tackle computationally expensive yield optimization in Sect. 5.3, the derivative free BOBYQA method, which will be modified to handle mixed derivative information in Sect. 5.4, and MOO approaches, which will be applied for simultaneous reliability and performance optimization in Chap. 6.

References

1. Brémaud P (2020). Probability theory and stochastic processes. Springer. https://doi.org/10.1007/978-3-030-40183-2
2. Eckle-Kohler J, Kohler M (2017) Eine Einführung in die Statistik und ihre Anwendungen. Springer
3. Insua D, Ruggeri F, Wiper M (2012) Bayesian analysis of stochastic process models. Wiley
4. Rasmussen CE, Williams CK (2006) Gaussian processes for machine learning. The MIT Press
5. Ulbrich M, Ulbrich S (2012) Nichtlineare Optimierung. Birkhäuser
6. Ehrgott M (2005) Multicriteria optimization. Springer
7. Conn AR, Scheinberg K, Vicente LN (2009) Introduction to derivative-free optimization. SIAM
8. Marti RC, Pardalos PM, Resende MG (2018) Handbook of heuristics. Springer

9. Powell MJ (2009) The BOBYQA algorithm for bound constrained optimization without derivatives. Cambridge NA Report NA2009/06, University of Cambridge, Cambridge 26, 26–46
10. Cartis C, Fiala J, Marteau B, Roberts L (2019) Improving the flexibility and robustness of model-based derivative-free optimization solvers. ACM Trans Math Softw 45(3). https://doi.org/10.1145/3338517
11. Fuhrländer M, Schöps S (2023) Hermite least squares optimization: a modification of BOBYQA for optimization with limited derivative information. Optim Eng. https://doi.org/10.1007/s11081-023-09795-y
12. Conn AR, Scheinberg K, Vicente LN (2008) Geometry of interpolation sets in derivative free optimization. Math Program 111(1):141–172
13. Powell MJ (2015) On fast trust region methods for quadratic models with linear constraints. Math Program Comput 7(3):237–267
14. Powell MJ (1994) A direct search optimization method that models the objective and constraint functions by linear interpolation. Springer, pp 51–67
15. Fuhrländer M, Schöps S (2021) Yield optimization using hybrid Gaussian process regression and a genetic multi-objective approach. Adv Radio Sci 19(B):41–48
16. Audet C, Hare W (2017) Derivative-free and blackbox optimization. Springer. https://doi.org/10.1007/978-3-319-68913-5
17. Srinivas N, Deb K (1994) Multiobjective optimization using nondominated sorting in genetic algorithms. Evolut Comput 2(3):221–248. https://doi.org/10.1162/evco.1994.2.3.221
18. Deb K, Pratap A, Agarwal S, Meyarivan T (2002) A fast and elitist multiobjective genetic algorithm: NSGA-II. IEEE Trans Evolut Comput 6(2):182–197
19. Blank J, Deb K, Roy PC (2019) Investigating the normalization procedure of NSGA-III. In: International conference on evolutionary multi-criterion optimization. Springer, pp 229–240
20. Hammersley JM, Handscomb DC (1964) Monte Carlo methods. Methuen & Co Ltd
21. Giles MB (2015) Multilevel Monte Carlo methods. Acta Numerica 24. https://doi.org/10.1017/S09624929
22. Xiu D (2015) Stochastic collocation methods: a survey. Handbook of uncertainty quantification, pp 1–18
23. Narayan A, Jakeman JD (2014) Adaptive Leja sparse grid constructions for stochastic collocation and high-dimensional approximation. SIAM J Sci Comput 36(6)
24. Hermann M (2011) Numerische Mathematik. Oldenbourg Wissenschaftsverlag
25. Georg N, Loukrezis D, Römer U, Schöps S (2020) Enhanced adaptive surrogate models with applications in uncertainty quantification for nanoplasmonics. Int J Uncertain Quantif 10(2)
26. Stewart GW (1996) Afternotes on numerical analysis. SIAM
27. Bungartz HJ, Griebel M (2004) Sparse grids. Acta Numerica 13:147–269. https://doi.org/10.1017/S0962492904000182
28. Clenshaw CW, Curtis AR (1960) A method for numerical integration on an automatic computer. Numerische Mathematik 2(1):197–205
29. Loukrezis D (2019) Adaptive approximations for high-dimensional uncertainty quantification in stochastic parametric electromagnetic field simulations. PhD thesis, Technische Universität Darmstadt
30. Butler T, Constantine P, Wildey T (2012) A posteriori error analysis of parameterized linear systems using spectral methods. SIAM J Matrix Anal Appl 33(1):195–209
31. Butler T, Dawson C, Wildey T (2013) Propagation of uncertainties using improved surrogate models. SIAM/ASA J Uncertain Quantif 1(1):164–191
32. Jakeman JD, Wildey T (2015) Enhancing adaptive sparse grid approximations and improving refinement strategies using adjoint-based a posteriori error estimates. J Comput Phys 280:54–71
33. Georg N (2021) Surrogate modeling and uncertainty quantification for radio frequency and optical applications. PhD thesis, Technische Universität Darmstadt
34. Fuhrländer M, Schöps S (2020) A blackbox yield estimation workflow with Gaussian process regression applied to the design of electromagnetic devices. J Math Ind 10(1):1–17

Chapter 4
Yield Estimation

This chapter is dedicated to yield (or failure probability) estimation. First, the yield is introduced formally and an overview of existing approaches for yield estimation is provided. Then, we propose two new hybrid methods for efficient yield estimation. One of them is based on SC, the other one on GPR. The aim of these methods is to maintain the high estimation accuracy achieved with classic MC, while the computational effort is reduced by evaluating most of the sample points on surrogate models. The content and structure of this chapter follow our work in [1, 2].

4.1 Yield Formulation

In the design process a device or component is optimized over its design parameters, i.e. geometry and material parameters, such that some predefined performance requirements are fulfilled. However, manufacturing or environmental uncertainties may lead to deviations in the geometry and material parameters and this may lead to a violation of the requirements. The probability, that the requirements are fulfilled, under consideration of uncertainties, is called the *yield*. This implies the following relation between the yield (probability of success) and the so-called failure probability: yield = 1 − failure probability. In this section we will formally introduce the yield, following the definitions in [3]. Therefore, we require certain preparations.

We define three kinds of parameters: uncertain design parameters, deterministic design parameters and range parameters. Please note that we use the term design parameters for any kind of geometry or material parameters of the device. However, some of these parameters cannot be influenced in the design process. For example, the magnetization of a permanent magnet (PM) might be affected by uncertainties due to natural material deviations, but is not subject to optimization when designing the permanent magnet synchronous machine (PMSM). The uncertain parameters $\boldsymbol{\xi}$

M. Fuhrländer, *Design Methods for Reducing Failure Probabilities with Examples from Electrical Engineering*, Springer Theses,
https://doi.org/10.1007/978-3-031-37019-9_4

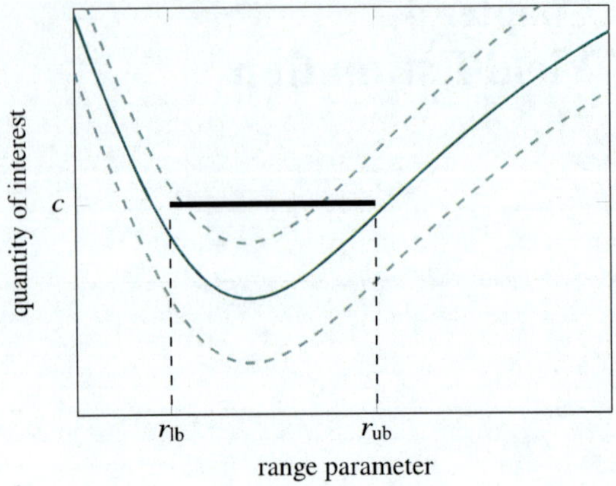

Fig. 4.1 Illustration of a QoI plotted over a range parameter. The black bar indicates the upper bound of the PFS. The solid curve corresponds to the QoI evaluated in $\overline{\boldsymbol{\xi}}$, the dashed curves illustrate deviations in the QoI due to design uncertainties

are modeled as random variables, where $\overline{\boldsymbol{\xi}} \in \mathbb{R}^{n_\xi}$ indicates the mean or expectation value and $\varphi(\boldsymbol{\xi})$ the corresponding probability density function. The deterministic parameters are given by $\mathbf{d} \in \mathbb{R}^{n_\mathrm{d}}$. The range parameter describes the environment in which the performance requirements have to be fulfilled and is denoted by $r \in T_\mathrm{r} \subset \mathbb{R}$. Typical range parameters are temperature or frequency. We introduce n_Q quantities of interest (QoIs), cf. (2.73), in form of a function $Q : \mathbb{R}^{n_\xi + n_\mathrm{d} + 1} \to \mathbb{R}^{n_\mathrm{Q}}$, and define its solution vector evaluated in $(\boldsymbol{\xi}, \mathbf{d}, r)$ by $\mathbf{Q}_r(\boldsymbol{\xi}, \mathbf{d}) \in \mathbb{R}^{n_\mathrm{Q}}$ and a bound $\mathbf{c} \in \mathbb{R}^{n_\mathrm{Q}}$. Then the performance feature specifications (PFS) are given by

$$\mathbf{Q}_r(\boldsymbol{\xi}, \mathbf{d}) \leq \mathbf{c} \quad \forall r \in T_\mathrm{r} = [r_\mathrm{lb}, r_\mathrm{ub}], \tag{4.1}$$

where the inequality is understood componentwise. In Fig. 4.1 we see a QoI plotted over a range parameter. The black bar indicates the upper bound of the PFS. While the solid curve corresponds to the QoI evaluated in $\overline{\boldsymbol{\xi}}$, the dashed curves represent possible deviations due to uncertainties.

We define the safe domain as the set of all parameter combinations of the uncertain parameter fulfilling the PFS. It depends on the current value of the deterministic parameter. We write

$$\Omega \equiv \Omega_\mathbf{d} = \{\boldsymbol{\xi} : \mathbf{Q}_r(\boldsymbol{\xi}, \mathbf{d}) \leq \mathbf{c} \ \ \forall r \in T_\mathrm{r}\}\,. \tag{4.2}$$

Let $\mathbf{1}_{\Omega_\mathrm{d}}(\boldsymbol{\xi})$ denote the indicator function, i.e.,

$$\mathbf{1}_{\Omega_\mathrm{d}}(\boldsymbol{\xi}) = \begin{cases} 1, & \boldsymbol{\xi} \in \Omega_\mathbf{d} \\ 0, & \boldsymbol{\xi} \notin \Omega_\mathbf{d}, \end{cases} \tag{4.3}$$

and let $\mathbb{P}$ denote the probability and $\mathbb{E}$ the expectation value. We define the yield as

$$Y_\varphi(\overline{\boldsymbol{\xi}}, \mathbf{d}) \equiv Y(\overline{\boldsymbol{\xi}}, \mathbf{d}) := \mathbb{P}\left[\boldsymbol{\xi} \in \Omega_\mathbf{d}\right] = \mathbb{E}[\mathbf{1}_{\Omega_\mathrm{d}}(\boldsymbol{\xi})] := \int_{\mathbb{R}^{n_\xi}} \mathbf{1}_{\Omega_\mathrm{d}}(\boldsymbol{\xi})\, \varphi(\boldsymbol{\xi})\, \mathrm{d}\boldsymbol{\xi}. \tag{4.4}$$

Note that the yield depends on the probability distribution of the uncertain parameters. However, for simplicity of notation we omit the subscript φ in the remainder of this work. Further, if there are no deterministic parameters to be considered, we write only $Y(\overline{\boldsymbol{\xi}})$. Since the yield is a probability, $0 \le Y(\overline{\boldsymbol{\xi}}, \mathbf{d}) \le 1$ holds.

4.2 Method Review

As a result of the definition of the yield via the expectation value of the indicator function, the MC analysis introduced in Sect. 3.3.1 is a straightforward approach for yield estimation. Using definition (4.4) and applying (3.52), we obtain the MC yield estimator

$$Y_{\mathrm{MC}}(\overline{\boldsymbol{\xi}}, \mathbf{d}) = \widetilde{\mathbb{E}}_{\mathrm{MC}}[\mathbf{1}_{\Omega_{\mathbf{d}}}(\boldsymbol{\xi})] = \frac{1}{N_{\mathrm{MC}}} \sum_{i=1}^{N_{\mathrm{MC}}} \mathbf{1}_{\Omega_{\mathbf{d}}}(\boldsymbol{\xi}^{(i)}) \approx \mathbb{E}[\mathbf{1}_{\Omega_{\mathbf{d}}}(\boldsymbol{\xi})] = Y(\overline{\boldsymbol{\xi}}, \mathbf{d}), \tag{4.5}$$

where $\{\boldsymbol{\xi}^{(i)}\}_{i=1,\ldots,N_{\mathrm{MC}}}$ are the MC sample points of the uncertain parameter, generated according to the corresponding probability density function. In Fig. 4.2 an example with two uncertain parameters is shown. Since the variance of the indicator function is given by

$$\mathbb{V}[\mathbf{1}_{\Omega_{\mathbf{d}}}(\boldsymbol{\xi})] = \mathbb{E}[\mathbf{1}_{\Omega_{\mathbf{d}}}(\boldsymbol{\xi})] \left(1 - \mathbb{E}[\mathbf{1}_{\Omega_{\mathbf{d}}}(\boldsymbol{\xi})]\right) = Y(\overline{\boldsymbol{\xi}}, \mathbf{d}) \left(1 - Y(\overline{\boldsymbol{\xi}}, \mathbf{d})\right), \tag{4.6}$$

an error indicator of the MC yield estimator is obtained by inserting (4.6) into (3.56)

$$\varepsilon_{\mathrm{MC},Y} = \sqrt{\frac{Y(\overline{\boldsymbol{\xi}}, \mathbf{d})(1 - Y(\overline{\boldsymbol{\xi}}, \mathbf{d}))}{N_{\mathrm{MC}}}} \le \frac{0.5}{\sqrt{N_{\mathrm{MC}}}}. \tag{4.7}$$

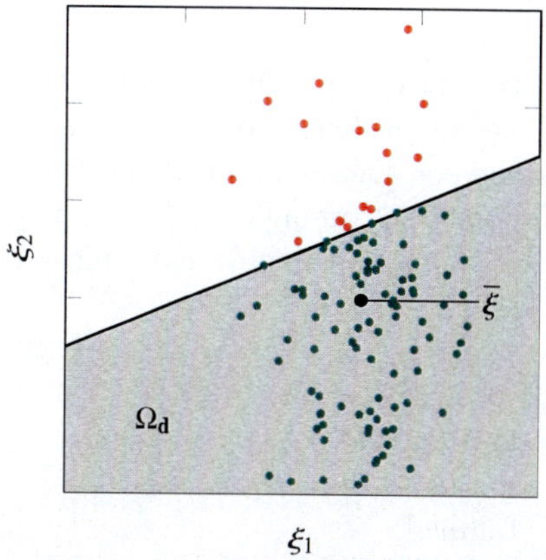

Fig. 4.2 Visualization of Monte Carlo analysis. Example with two uncertain parameters ξ_1 and ξ_2. Accepted sample points are given in green, rejected sample points in red color. Illustration based on [4, Fig. 4.2]

The inequality provides an upper bound for the error indicator which is met in case of $Y(\bar{\xi}, \mathbf{d}) = 0.5$.

In practice, calculating the QoI typically involves solving an expensive model, e.g. from finite element (FE) discretization, see Sect. 2.3. Since the QoI has to be evaluated for each sample point, and a large sample set is required for high accuracy, a classic MC analysis becomes computationally prohibitive. An alternative to MC are sampling-free methods such as the first order reliability method (FORM) and the second order reliability method (SORM) [5, 6]. The aim of FORM/SORM is a first or second order approximation of the limit state function, i.e. the surface between safe domain and failure domain. However, in the past years there has been much research on improving the efficiency of sample-based methods. Mainly, we can distinguish two ideas in order to reduce their computing effort. First, reducing the number of sample points, e.g., through importance sampling [7] or subset simulation [8–10]. Second, reducing the effort of evaluating one sample point, e.g. with model order reduction [11] or surrogate based approaches. In surrogate methods a small set of sample points, so-called training data points, is used to approximate the QoI. Then, a MC analysis is conducted using the approximation, i.e., the surrogate model, instead of the original model of the QoI, cf. [12]. Different methods for building the surrogate model have been employed, e.g., linear regression [13], SC [14, 15], GPR [16] and more recently neural networks [17]. The two ideas of reducing the computing effort can also be combined. In [18–20] importance sampling and GPR are employed. In industry, on the other hand, linearization approaches are frequently applied, assuming that the design parameter deviations are small enough [21, Online Help: Yield Analysis Overview]. Although this approach is very efficient, the linearity assumption is often not valid. In Sects. 3.3.2 and 3.3.3, SC and GPR have been introduced. In the following sections, two efficient and accurate yield estimation methods are proposed, using SC and GPR in a hybrid framework. Therefore, first we introduce and motivate the idea of a hybrid approach for yield estimation.

For the sake of simplicity in notation, in the following we assume $n_{\mathrm{Q}} = 1$, i.e., we consider one scalar-valued QoI $Q : \mathbb{R}^{n_\xi + n_\mathrm{d} + 1} \to \mathbb{R}$ with the PFS

$$Q_r(\boldsymbol{\xi}, \mathbf{d}) \leq c \ \ \forall r \in T_\mathrm{r}. \tag{4.8}$$

However, the following surrogate based approaches can be extended straightforwardly to the case of $n_{\mathrm{Q}} > 1$, e.g. by constructing one surrogate model for each QoI and considering all resulting PFS in $\Omega_\mathbf{d}$, and when determining the critical sample points. According to (4.8), the inequality needs to be fulfilled for all $r \in T_\mathrm{r}$. We discretize T_r and obtain the discretized range parameter set T_d. We reformulate the PFS as

$$Q_{r^{(j)}}(\boldsymbol{\xi}, \mathbf{d}) \leq c \ \ \forall j = 1, \ldots, |T_\mathrm{d}|. \tag{4.9}$$

If (4.9) holds, the PFS (4.8) is considered to be fulfilled. On the other hand, if the inequality in (4.9) is violated for one $r^{(j)} \in T_\mathrm{d}$, the PFS (4.8) is considered to be not fulfilled.

Hybrid Method

The reliability of a surrogate based yield estimator depends on the accuracy of the surrogate model, which depends, among other things, on the number of training data points. This leads to a trade-off between efficiency and accuracy. However, even if the surrogate model is highly accurate, measured by classical norms or pointwise, the yield estimation may still fail drastically. In [15] the following example is provided. Let a QoI be defined by the constant function $Q(\xi) = 0$ and the PFS by $Q(\xi) \geq 0$. Then the yield is $\mathbb{P}\left[Q(\xi) \geq 0\right] = 1$, since the requirement is fulfilled for each $\xi \in \mathbb{R}$. Let a surrogate model be given by

$$\widetilde{Q}_N(\xi) = -\frac{1}{N}, \quad N \geq 1, \tag{4.10}$$

which implies uniform convergence

$$\lim_{N\to\infty} \widetilde{Q}_N(\xi) = Q(\xi). \tag{4.11}$$

However, the yield with respect to the surrogate model is $\mathbb{P}[\widetilde{Q}_N(\xi) \geq 0] = 0$ for each finite N.

To overcome this problem, in [15] a hybrid approach is proposed, which follows the procedure illustrated in Fig. 4.3. The idea behind this procedure is to trust the surrogate model only if the predicted function value is far enough from the critical area, i.e., the limit between safe domain and failure domain. Otherwise, the original model is consulted. Note, that the definition of the term *far enough* is crucial, and typically depends on the prediction, an error indicator and a threshold γ. The MC sample points for which the original model is evaluated are referred to as *critical sample points*. Hybrid schemes have also been employed for example in [18, 22, 23]. Based on this key idea the following questions have to be considered when developing a hybrid approach:

1. Which kind of surrogate is used?
2. How is the term *far enough* defined?
3. (How) Is it possible to update the surrogate model during the estimation process?
4. Are there further steps in order to improve efficiency and/or accuracy of the estimation?

The first two questions are mandatory, the other two questions are optional features. In Sects. 4.3 and 4.4 we propose two hybrid approaches based on this idea, but differing in how they deal with each of these questions.

Excursion: Rare Event Simulation

A subfield of research on yield estimation deals with rare event simulation, i.e., very small probabilities of failure. In [7] it is shown that a minimum number of failure events must occur in order to provide accurate estimation results. More precisely, for failure probabilities of $\mathbb{P}_\mathrm{f} = 10^{-n}$ at least 10^{n+2} sample points are required. According to [15], a rule of thumb states that for a reliable estimation result, at least 10 failure

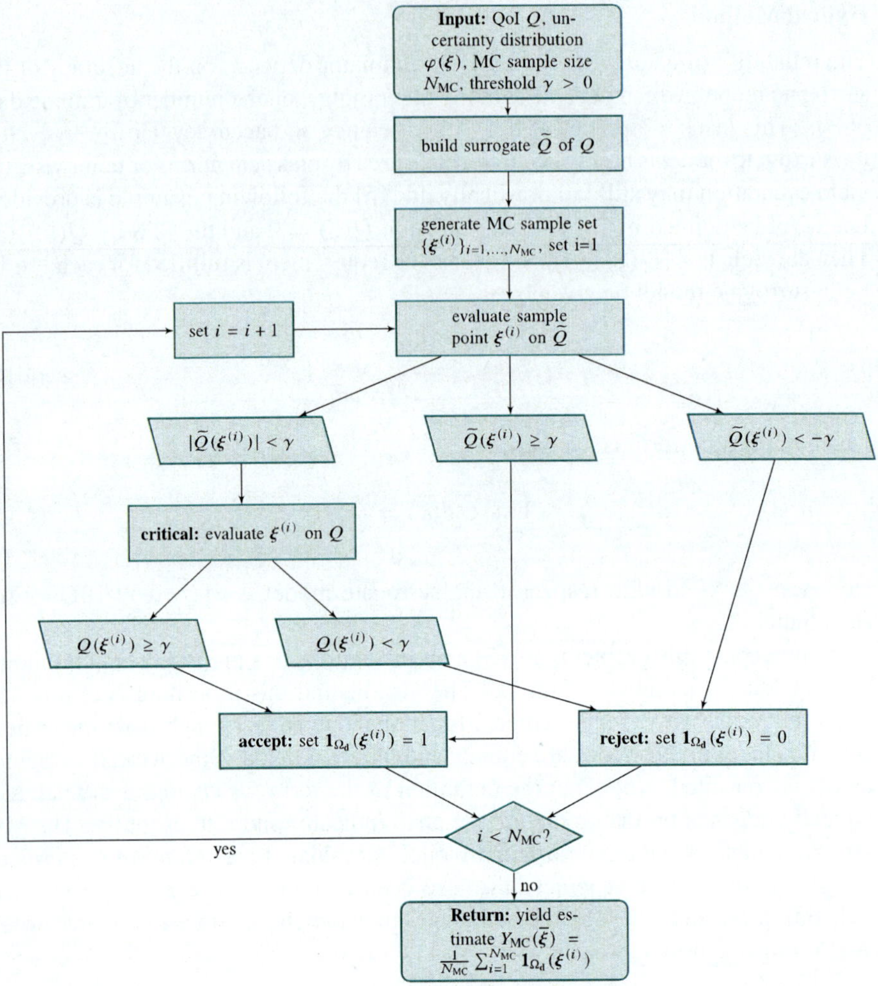

Fig. 4.3 Flowchart for the basic hybrid scheme proposed in [15] for $c = 0$

sample points have to be generated. Therefore, it is natural that methods suitable for the reduction of sample points are also utilized for rare event simulation. Importance sampling [7] reduces the number of sample points by concentrated sampling in the area of interest. This is achieved by sampling a modified probability density function and applying weight functions for each sample point when calculating the yield by (4.5). In subset simulation [8, 9, 24] a sequence of nested subsets is generated and sequential MC is conducted. In each step, conditional probabilities are calculated, each of them larger than the very low failure probability, and thus, less sample points are required. In the end, the failure probability (and thus the yield) can be calculated as a product of conditional probabilities. These are just two approaches of a very particular research field in the context of yield estimation. In the remainder of this work higher failure probabilities are assumed, i.e., $\mathbb{P}_\mathrm{f} \gtrsim 0.01$. This implies that we are not in the situation of rare events and choose the error bounds accordingly.

4.3 Stochastic Collocation Hybrid Approach

In this section we propose a hybrid approach based on the sparse grid SC introduced in Sect. 3.3.2. Before the method is described in detail, we summarize the main ideas. For the QoI a surrogate model is built using sparse grid SC based on adaptive, weighted Leja nodes, in the following denoted by SC. The univariate Leja nodes are generated according to (3.63) and the adaptive selection of the corresponding multivariate nodes is based on the adjoint error indicator from (3.67), cf. [25, 26]. A MC analysis is conducted on the surrogate model. For each MC sample point, the approximation error is estimated, also using an adjoint error indicator for the SC surrogate model and for the FE discretization, in the following referred to as SC error and FE error. Based on that, it is decided if the FE model is evaluated, and whether the FE discretization is refined. Using these evaluations, the MC yield estimator is calculated. The structure and content of this section follows our work in [1].

The proposed SC-Hybrid approach is based on the hybrid method in [15]. One main difference lies in the way the critical sample points are determined. The authors from [15] define a tube around the limit between safe domain and failure domain. Its size is either fixed in advance or determined by iteratively adding critical sample points until some error bound is reached. In our work, on the contrary, an adjoint error indicator is used in order to determine the critical sample points. An adjoint error indicator in combination with GPR has also been applied in [27], but contrary to the hybrid approach there, in our work the polynomial surrogate model is based on SC. Furthermore, we consider the model error, i.e., the FE discretization error, in addition to the SC error as hybrid distinction criterion and provide a strategy for FE model refinements, if required.

Error Indicator for SC and FEM

Let $\widetilde{Q}_h$ denote the SC approximation of the QoI, for which the training data points have been evaluated on a FE model with mesh size h, i.e., on Q_h. In order to determine the critical sample points, we aim to quantify the total error of the SC surrogate $\widetilde{Q}_h(\boldsymbol{\xi}, \mathbf{d}, r)$ compared to the *true* value $Q(\boldsymbol{\xi}, \mathbf{d}, r)$, which is the combination of the SC error estimation given in (3.67) and the FE error given in (2.25). However, for the computation of the FE error, it is required to solve problem (2.19) with FEM and the adjoint problem (2.23), this one even on a refined grid. If this would be done for each MC sample point, it would be computationally more expensive than a classic MC analysis on the original model. Therefore, a SC surrogate approximation of $\varepsilon^\mathrm{c}_{\mathrm{fe},h}(\boldsymbol{\xi}, \mathbf{d}, r)$ from (2.25) is used, denoted by $\widetilde{\varepsilon}_{\mathrm{fe},h}(\boldsymbol{\xi}, \mathbf{d}, r)$. Then, the total approximation error indicator can be defined by

$$\begin{aligned}\varepsilon_{\mathrm{tot},h}(\boldsymbol{\xi},\mathbf{d},r) &:= \left|Q(\boldsymbol{\xi},\mathbf{d},r)-\widetilde{Q}_h(\boldsymbol{\xi},\mathbf{d},r)\right| \\ &\leq \left|Q(\boldsymbol{\xi},\mathbf{d},r)-Q_h(\boldsymbol{\xi},\mathbf{d},r)\right| + \left|Q_h(\boldsymbol{\xi},\mathbf{d},r)-\widetilde{Q}_h(\boldsymbol{\xi},\mathbf{d},r)\right| \\ &\approx \widetilde{\varepsilon}_{\mathrm{fe},h}(\boldsymbol{\xi},\mathbf{d},r)+\widetilde{\varepsilon}_{\mathrm{sc}}(\boldsymbol{\xi},\mathbf{d},r) \\ &=: \widetilde{\varepsilon}_{\mathrm{tot},h}(\boldsymbol{\xi},\mathbf{d},r). \end{aligned} \tag{4.12}$$

Note that the separation of the FE error and the SC error by the triangle inequality is a rather conservative choice to define the total error.

The Algorithm

The first step of the SC-Hybrid approach is the construction of the SC surrogate model based on adaptive Leja nodes, as introduced in Sect. 3.3.2. For each range parameter point $r_j \in T_{\mathrm{d}}$ an own surrogate model is built. Then, a MC analysis is carried out based on the safe domain (4.2) using the SC surrogates. For each MC sample point $\boldsymbol{\xi}^{(i)}, i = 1, \ldots, N_{\mathrm{MC}}$, and each range parameter point $r^{(j)}, j = 1, \ldots, |T_{\mathrm{d}}|$, i.e., each surrogate model denoted by $\widetilde{Q}_h^{(j)}$, the total approximation error $\widetilde{\varepsilon}_{\mathrm{tot},h}(\boldsymbol{\xi}^{(i)}, \mathbf{d}, r^{(j)})$ is calculated according to (4.12). We define the interval

$$\mathcal{I}^{\varepsilon}_{\mathrm{tot}}(\boldsymbol{\xi}^{(i)},\mathbf{d},r^{(j)}) = \left[\widetilde{Q}_h^{(j)}(\boldsymbol{\xi}^{(i)},\mathbf{d}) - s\,\widetilde{\varepsilon}_{\mathrm{tot},h}(\boldsymbol{\xi}^{(i)},\mathbf{d},r^{(j)}),\ \widetilde{Q}_h^{(j)}(\boldsymbol{\xi}^{(i)},\mathbf{d}) + s\,\widetilde{\varepsilon}_{\mathrm{tot},h}(\boldsymbol{\xi}^{(i)},\mathbf{d},r^{(j)})\right], \tag{4.13}$$

where $s \geq 1$ indicates a safety factor. If for each $j = 1, \ldots, |T_{\mathrm{d}}|$ each element of the interval $\mathcal{I}^{\varepsilon}_{\mathrm{tot}}(\boldsymbol{\xi}^{(i)}, \mathbf{d}, r^{(j)})$ fulfills (or does not fulfill) the PFS (4.9), the i-th sample point is classified as *accepted* (or *not accepted*). If the PFS is only fulfilled for a subset of the interval $\mathcal{I}^{\varepsilon}_{\mathrm{tot}}(\boldsymbol{\xi}^{(i)}, \mathbf{d}, r^{(j)})$, the sample point is classified as *critical*. For all critical sample points, the original model, i.e., the FE model with mesh size h, will be evaluated. We obtain $Q_h(\boldsymbol{\xi}^{(i)}, \mathbf{d}, r^{(j)})$. For these sample points the SC error vanishes, while the FE error remains unchanged. We define the FE error interval by

$$\mathcal{I}^{\varepsilon}_{\mathrm{fe}}(\boldsymbol{\xi}^{(i)},\mathbf{d},r^{(j)}) = \left[Q_h(\boldsymbol{\xi}^{(i)},\mathbf{d},r^{(j)}) - s\,\widetilde{\varepsilon}_{\mathrm{fe},h}(\boldsymbol{\xi}^{(i)},\mathbf{d},r^{(j)}),\ Q_h(\boldsymbol{\xi}^{(i)},\mathbf{d},r^{(j)}) + s\,\widetilde{\varepsilon}_{\mathrm{fe},h}(\boldsymbol{\xi}^{(i)},\mathbf{d},r^{(j)})\right]. \tag{4.14}$$

Following the same decision rules as above, the sample point is again classified as accepted, not accepted, or critical. If the sample point is not critical, it is finally classified and we can continue with the next MC sample point. Otherwise we iteratively refine the FE model, e.g. by using a finer mesh $h = h/2$, and reevaluate the FE model and the corresponding FE error for this critical sample point until the sample point can be reliably classified or a maximum number of refinement steps is reached. Applying this procedure, we obtain an accuracy comparable to the classic MC analysis, using the finest refinement. A difference would only occur if the error for one sample point would be greatly underestimated, leading to a wrong classification of accepted or not accepted instead of critical. Since the adjoint error indicators are no strict upper bounds, the safety factor s is introduced in order to avoid wrong classification by error underestimation. The size of the safety factor is determined by evaluating a small random sample on the surrogate model $\widetilde{Q}_h$ and the original model Q_h with the finest mesh. Then we consider the maximum ratios between the actual error and the sum of the adjoint error indicators. Similarly to the application

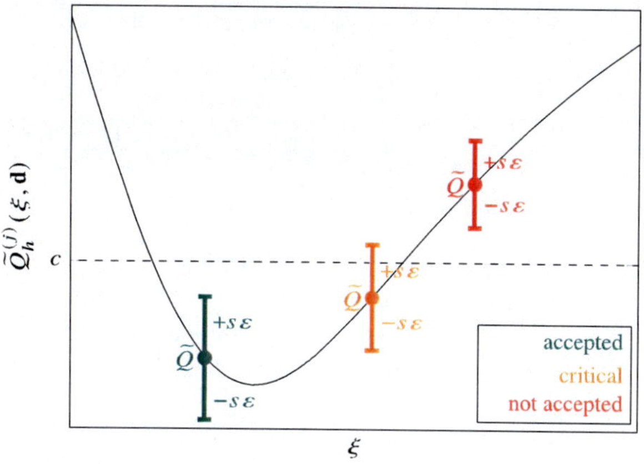

Fig. 4.4 Visualization for the classification of a sample point $\xi^{(i)}$ as accepted, not accepted or critical, based on model value $\widetilde{Q} \equiv \widetilde{Q}_h^{(j)}(\xi, \mathbf{d})$, adjoint error indicator $\varepsilon \equiv \widetilde{\varepsilon}_{\mathrm{tot},h}^{(j)}(\xi, \mathbf{d})$ and safety factor s

of the triangle inequality for the total approximation error (4.12), the safety factor is chosen rather conservatively in this work. A conservative choice may increase the computational effort due to too many sample points classified as critical, however it increases the accuracy by avoiding misclassification. A visualization of classifying a sample point as accepted, not accepted or critical is given in Fig. 4.4. The SC-Hybrid decision process for the i-th sample point is summarized in a flowchart in Fig. 4.5.

To further increase the efficiency of the yield estimation process, we propose a shortcut strategy for the consideration of the range parameter points $r^{(j)}$. Since the PFS must hold for each $r^{(j)}$, $j = 1, \ldots, |T_{\mathrm{d}}|$, each sample point $\boldsymbol{\xi}^{(i)}$, $i = 1, \ldots, N_{\mathrm{MC}}$ is evaluated for each range parameter point, i.e., on each surrogate model $\widetilde{Q}_h^{(j)}$. If the PFS are fulfilled for one range parameter point (on the corresponding surrogate model), the test is carried out for the next range parameter point. However, if the sample point fails to fulfill the PFS for a single arbitrary range parameter point, the sample point can be immediately classified as not accepted and the remaining range parameter points do not need to be considered. In the case that this sample point would be classified as critical for a later range parameter point, computational effort can be saved. Please note that this shortcut strategy is independent of the hybrid approach, i.e., it can also be applied in a classic MC analysis or a SC surrogate-based MC analysis without hybrid scheme. However, in the hybrid method we can further benefit from the fact that the QoI is evaluated in two stages. First on the surrogate model, which is computationally cheap, then on the FE models. Since the PFS are defined as upper bounds, we assume that range parameter points for which the surrogate model value $\widetilde{Q}_h^{(j)}(\boldsymbol{\xi}^{(i)}, \mathbf{d})$ is higher have a higher likelihood that the original model value $Q_h(\boldsymbol{\xi}^{(i)}, \mathbf{d}, r^{(j)})$ violates the requirements. Hence, we apply an ordering strategy, ordering the range parameter points for each sample point according to their surrogate model values and starting examination at the range parameter point $r^{(j_{\mathrm{start}})}$ where

$$j_{\mathrm{start}} = \underset{j=1,\ldots,|T_{\mathrm{d}}|}{\arg\max}\; \widetilde{Q}_h^{(j)}(\boldsymbol{\xi}^{(i)}, \mathbf{d}) \tag{4.15}$$

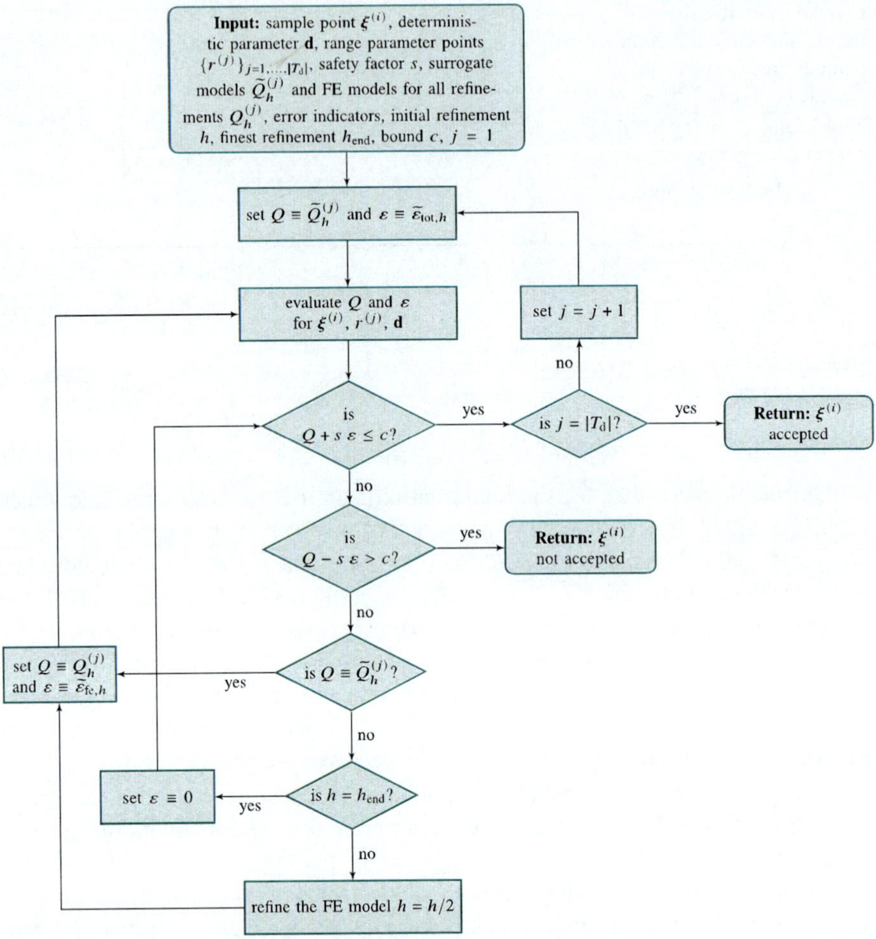

Fig. 4.5 Flowchart summarizing the hybrid decision process for one MC sample point in the SC-Hybrid approach. Content based on [1, Algorithm 1]

holds. Please note that for the sake of clarity this ordering strategy is not reflected in Fig. 4.5.

Comments and Summary

The proposed SC-Hybrid approach allows a reduction in the number of high fidelity function evaluations, while maintaining high accuracy comparable with classic MC. All relevant error sources are controlled within this algorithm: the MC error, the FE error and the surrogate error, which is the SC error in our case. The MC error is controlled by defining a target accuracy and determining the sample size according to (4.7). The FE error is controlled by calculating the adjoint FE error indicator and adjusting the mesh size, if required. The SC error is controlled by calculating the adjoint SC error indicator and evaluating the sample point on the FE model, if necessary. At this point it should be mentioned that the SC-Hybrid approach is an intrusive method. For the calculation of the adjoint error indicators FEM solutions and system matrices are required. Therefore, the method may not be applicable in combination with commercial software. Also, in its current form it is not possible to easily update the surrogate model. A variant of this approach could use regression instead of interpolation. Since the need for structured training data would be eliminated, SC model updates would be possible, similar to the updates described for the GPR-Hybrid approach in the next section. On the other hand, the absence of SC model updates allows for straightforward parallelization when evaluating critical sample points, just as in classic MC analysis. Finally, we summarize the SC-Hybrid method by answering the questions regarding hybrid approaches, posed on Sect. 4.2:

1. Surrogate: Sparse grid stochastic collocation based on adaptive, weighted Leja nodes.
2. Definition of *far enough*: Based on adjoint error indicators for FE error and SC error.
3. Updates: No. If an improvement of the surrogate is required, a completely new SC model with new training data points needs to be built.
4. Further features: Control of the FE error for possible model refinements. Straightforward parallelization.

4.4 Gaussian Process Regression Hybrid Approach

In this section we propose a hybrid approach based on GPR as introduced in Sect. 3.3.3. Similar to the SC-Hybrid approach, a surrogate model for the QoI is built, here the approximation is a GP. Then, a MC analysis is performed on the GPR surrogate. The mean value of the GP in a MC sample point is considered as a QoI value prediction, and its standard deviation is considered as an error indicator. Based on that, the MC sample point is classified. In contrast to the SC-Hybrid approach, for GPR the training data is not required to be structured. This advantage is used to update the GPR model during the estimation process with critical MC sample points, i.e., with sample points which have been evaluated on the original high fidelity model during the estimation process. Further, sorting and parallelization strategies for the MC analysis are investigated. GPR is a non-intrusive method, which allows the use of commercial or proprietary software as a blackbox FEM solver. Analogously to the SC-Hybrid approach, a FE model refinement strategy could be implemented. In order to preserve the blackbox character of the GPR-Hybrid method, it is recommended to use a non-intrusive FE error estimator or an error estimator included in the software. The structure and content of this section follows our work in [2].

Research on yield or failure probability estimation using surrogate models based on GPR has already been proposed in the past, e.g. [9, 19, 20, 23, 27]. In [27] a hybrid approach using GPR for the surrogate model combined with adjoint error indicators for determining critical sample points is proposed. In the GPR-Hybrid approach, we generalize and extend this method using GPR for both, building the surrogate model and obtaining an error indicator for determining the critical sample points. In [23], the authors discuss and compare various sorting strategies applied to the GPR training data points, with the aim of requiring less training data points, and thus less evaluations on the high fidelity model. In [9], small failure probabilities are calculated with a limited budget of function evaluations on the high fidelity model, i.e., a limited budget of training data points. An adaptive GPR surrogate model is used, but it is not combined with a hybrid approach, hence there are no critical sample points to improve the GPR surrogate model. Instead, the authors distinguish between sample points generated by Bayesian subset simulation (sequential MC) for the estimation of the failure probability and sample points adaptively generated as training data for the GPR model based on a stepwise uncertainty reduction technique. In [19, 20], combined surrogate model approaches based on GPR with importance sampling are proposed. The GPR model and the importance sampling density are improved adaptively by adding MC sample points from the last iteration to the training data set. These points are selected by a learning function and evaluated on the high fidelity model. A hybrid approach as proposed in [15] is not applied in [19, 20].

The Algorithm

A set of training data points for the initial GPR model is generated. Since no specific structure is required for the training data, it can be generated for example randomly according to the probability density function of the uncertain parameters. As in the SC-Hybrid approach, for each range parameter point $r^{(j)} \in T_\mathrm{d}$ an own surrogate model is built. The MC analysis is started. In contrast to the SC-Hybrid method, the MC analysis is not conducted completely, since we aim to improve the GPR model during the estimation process by updates with critical sample points. The MC sample is generated, and then, for each sample point $\boldsymbol{\xi}^{(i)}$, $i = 1, \ldots, N_\mathrm{MC}$, and each range parameter point $r^{(j)}$, $j = 1, \ldots, |T_\mathrm{d}|$, the predicted value of the QoI $\widetilde{Q}^{(j)}(\boldsymbol{\xi}^{(i)}, \mathbf{d})$ and the predicted standard deviation $\sigma_\mathrm{GPR}^{(j)}(\boldsymbol{\xi}^{(i)}, \mathbf{d}) \equiv \varepsilon_\mathrm{GPR}^{(j)}(\boldsymbol{\xi}^{(i)}, \mathbf{d})$ are obtained. Following the concept of sigma levels [28], again a safety factor $s \geq 1$ is introduced, which is multiplied with the standard deviation to define the GPR error interval

$$\mathcal{I}_\mathrm{GPR}^{\varepsilon}(\boldsymbol{\xi}^{(i)}, \mathbf{d}, r^{(j)}) = \left[\widetilde{Q}^{(j)}(\boldsymbol{\xi}^{(i)}, \mathbf{d}) - s\,\varepsilon_\mathrm{GPR}^{(j)}(\boldsymbol{\xi}^{(i)}, \mathbf{d}),\ \widetilde{Q}^{(j)}(\boldsymbol{\xi}^{(i)}, \mathbf{d}) + s\,\varepsilon_\mathrm{GPR}^{(j)}(\boldsymbol{\xi}^{(i)}, \mathbf{d})\right]. \tag{4.16}$$

As described in Sect. 4.3, the choice of s depends on the problem and can be determined by evaluating a small test set on the high fidelity and the GPR model and calculating the ratio of true and predicted error. If the PFS are fulfilled (or not fulfilled) for a MC sample point $\boldsymbol{\xi}^{(i)}$, i.e., for each $j = 1, \ldots, |T_\mathrm{d}|$, each element of the interval $\mathcal{I}_\mathrm{GPR}^{\varepsilon}(\boldsymbol{\xi}^{(i)}, \mathbf{d}, r^{(j)})$ fulfills (or does not fulfill) the PFS, the i-the sample point is classified as *accepted* (or *not accepted*). If the PFS are fulfilled for a subset of the interval $\mathcal{I}_\mathrm{GPR}^{\varepsilon}(\boldsymbol{\xi}^{(i)}, \mathbf{d}, r^{(j)})$ the i-th sample point is classified as *critical*. If the sample point is critical, the high fidelity model is evaluated for this point, then it is classified. Before we continue with the next MC sample point, the GPR model is updated, i.e., the evaluated critical sample point is added to the training data set and the GP is recalculated. As described in Sect. 3.3.3, the computational effort of updating the GPR model is negligible compared to one evaluation of the high fidelity model. And since the critical sample points are evaluated in the hybrid method anyway, this update requires almost no additional computational effort. In the following we distinguish between *offline* and *online* evaluations of the high fidelity model. Typically the computational effort of surrogate based approaches lies in the offline evaluations, i.e., the evaluation of training data before for example the MC analysis is started. The hybrid approach based on GPR allows us to start with a rather small initial training data set to keep the offline costs low. The resulting less accurate GPR model does not pose a problem for yield estimation accuracy, since by reevaluating critical sample points on the high fidelity model, the hybrid approach ensures the correct classification as accepted or not accepted. The only difference is that there may be more sample points classified as critical in the beginning, if the initial GPR model is built with a smaller training data set. The evaluations of critical sample points used for accurate classification and for model updates during the estimation process are referred to as online costs. Since the accuracy of the GPR model increases by updates during the yield estimation, the number of critical sample points decreases towards the end of the process.

The procedure for classifying one MC sample point within the GPR-Hybrid approach is provided in Algorithm 4.1. As in the SC-Hybrid approach, a shortcut strategy is applied, i.e., if a sample point fails to fulfill the PFS for one range parameter point, the remaining range parameter points do not need to be considered. Further, the range parameter points could be ordered according to (4.15). This ordering is not represented in Algorithm 4.1.

Parallelization

A classic MC analysis is well-suited for parallelization, since each sample point is evaluated independently. The same holds for a hybrid approach without model updates. However, if the GPR model is to be updated with each critical sample point, the evaluation of critical sample points and also the evaluation of all sample points on the GPR model is required to be performed sequentially. In order to enable parallel computing for hybrid approaches with model updates, we introduce so-called *batches*. Let N_{B} denote the size of a batch. Only after the evaluation of N_{B} critical sample points on the high fidelity model (possibly in parallel) is the GPR model updated. Setting $N_{\mathrm{B}} = 1$ indicates that no batches are used and the model update is conducted immediately, i.e., no parallelization is possible.

After evaluating a batch of critical sample points, all or a subset of the critical sample points can be used to update the GPR model. If all critical sample points are used for the update, all critical points are added to the training data set of the corresponding j-th surrogate model $\widetilde{Q}^{(j)}$, $j = 1, \dots, |T_{\mathrm{d}}|$, the Gaussian processes (GPs) are updated and we continue with the classification of the remaining MC sample points. If only a subset is used, they are chosen in a greedy way. The critical sample points of the batch corresponding to the j-th range parameter point are

Algorithm 4.1 Classification of i-th MC sample point in the GPR-Hybrid approach, based on [2, Algorithm 1]

Input: MC sample point $\boldsymbol{\xi}^{(i)}$, range parameter points $r^{(j)} \in T_\mathrm{d}$ and corresponding GPR models $\widetilde{Q}^{(j)}$, $j = 1, \dots, |T_\mathrm{d}|$, PFS threshold c, safety factor s, high fidelity model Q
for $j = 1, \dots, |T_\mathrm{d}|$ **do**
  Evaluate the GPR model and obtain $\widetilde{Q} \equiv \widetilde{Q}^{(j)}(\boldsymbol{\xi}^{(i)}, \mathbf{d})$ and $\varepsilon_\mathrm{GPR} \equiv \varepsilon^{(j)}_\mathrm{GPR}(\boldsymbol{\xi}^{(i)}, \mathbf{d})$
  if $\widetilde{Q} + s\,\varepsilon_\mathrm{GPR} \leq c$ **then**
    PFS fulfilled for $r^{(j)}$
  else if $\widetilde{Q} - s\,\varepsilon_\mathrm{GPR} > c$ **then**
    Classify $\boldsymbol{\xi}^{(i)} \notin \Omega$ (not accepted) and stop
  else
    Classify $\boldsymbol{\xi}^{(i)}$ as critical sample point
    Evaluate the high fidelity model and obtain $Q \equiv Q(\boldsymbol{\xi}^{(i)}, \mathbf{d}, r^{(j)})$
    Update the GPR model $\widetilde{Q}$
    if $Q \leq c$ **then**
      PFS fulfilled for $r^{(j)}$
    else
      Classify $\boldsymbol{\xi}^{(i)} \notin \Omega$ (not accepted) and stop
    end if
  end if
  if $j = |T_\mathrm{d}|$ **then**
    Classify $\boldsymbol{\xi}^{(i)} \in \Omega$ (accepted)
  else
    Continue with $j = j + 1$
  end if
end for

collected in the set $\mathcal{C}^{(j)}$. The sample point in $\mathcal{C}^{(j)}$ with the maximum difference between predicted value and real value of the QoI is added to the training set first, i.e.,

$$\boldsymbol{\xi}^{\mathrm{add}} = \underset{\boldsymbol{\xi}^{(i)} \in \mathcal{C}^{(j)}}{\arg\max} \left| \widetilde{Q}^{(j)}(\boldsymbol{\xi}^{(i)}, \mathbf{d}) - Q(\boldsymbol{\xi}^{(i)}, \mathbf{d}, r^{(j)}) \right| . \tag{4.17}$$

The GP is updated with this additional point and the remaining critical sample points in $\mathcal{C}^{(j)}$ are evaluated on the updated GPR model in order to obtain a new, better prediction and (4.17) is evaluated again. This procedure is repeated until the error is below a given tolerance ε_B. When solving in batches, it can be advisable to include only a subset of the critical sample points in order to avoid too many, closely neighboring training data points. In a sequential updating procedure without batches, this problem cannot occur, since a sample point is unlikely to be classified as critical, if a very closely neighboring point is already in the training data set. Setting $\varepsilon_\mathrm{B} = 0$ indicates using all critical sample points for model updates. After the updates, like before, the hybrid yield estimation proceeds with evaluating the remaining MC sample points on the updated GPR models until again N_B sample points have been identified as critical and evaluated on the high fidelity model (possibly in parallel). Note that without much extra cost, after each model update it is also possible to reevaluate all already considered non-critical MC sample points on the updated GPR model. The procedure for yield estimation with the GPR-Hybrid approach is represented in Algorithm 4.2. The combination of Algorithms 4.1 and 4.2 is illustrated in a flowchart in Fig. 4.6.

Sorting Strategies

The efficiency of the hybrid approach with model updates can be improved by considering the MC sample points in an optimized order. The sample points are sorted, such that we start the classification with the most promising sample points in the sense that they contribute most to the improvement of the GPR model. This shall lead to an earlier improvement of the GPR model and hence, to more sample points classified correctly as accepted or not accepted without being evaluated on the high fidelity model. The sorting of the MC sample points requires information about the model value and the prediction error. Therefore, first, the whole MC sample set is evaluated on the initial GPR surrogate models to obtain $\widetilde{Q}^{(j)}(\boldsymbol{\xi}^{(i)}, \mathbf{d})$ and $\varepsilon_{\mathrm{GPR}}^{(j)}(\boldsymbol{\xi}^{(i)}, \mathbf{d})$, for $i = 1, \ldots, N_{\mathrm{MC}}$ and $j = 1, \ldots, |T_{\mathrm{d}}|$. Then, a sorting criterion is applied to order the MC sample points. In [23] a collection and comparison of various sorting criteria can be found. We focus on two criteria. The first one is introduced by Echard, Gayton and Lemaire in [29], it is referred to as the EGL criterion in the following. For the i-th sample point, the EGL criterion is calculated by

---

Algorithm 4.2 Yield estimation with the GPR-Hybrid approach, based on [2, Algorithm 2]

Input: MC sample set $\left\{\boldsymbol{\xi}^{(i)}\right\}_{i=1,\ldots,N_{\mathrm{MC}}}$, range parameter points $r^{(j)} \in T_{\mathrm{d}}$ and corresponding GPR models $\widetilde{Q}^{(j)}$, $j = 1, \ldots, |T_{\mathrm{d}}|$, PFS threshold c, safety factor s, high fidelity model Q, error tolerance $\varepsilon_{\mathrm{B}} \geq 0$, batch size $N_{\mathrm{B}} \in \mathbb{N}$

for $i = 1, \ldots, N_{\mathrm{MC}}$ **do**
  Classify $\boldsymbol{\xi}^{(i)}$ according to Algorithm 4.1
  Count number of online high fidelity evaluations $\#\mathrm{HF}_{\mathrm{GPR\text{-}H}}^{\mathrm{online}}$
  for $j = 1, \ldots, |T_{\mathrm{d}}|$ **do**
    Define $\mathcal{C}^{(j)} = \left\{\boldsymbol{\xi}^{(i)} : \boldsymbol{\xi}^{(i)} \text{ classified as critical for } r^{(j)} \text{ in last } N_{\mathrm{B}} \text{ MC evaluations}\right\}$
  end for
  if $\#\mathrm{HF}_{\mathrm{GPR\text{-}H}}^{\mathrm{online}}$ is an integer multiplier of N_{B} **then**
    for $j = 1, \ldots, |T_{\mathrm{d}}|$ **do**
      Initialize $\varepsilon = \infty$
      while $\varepsilon > \varepsilon_{\mathrm{B}}$ **do**
        Set $\boldsymbol{\xi}^{\mathrm{add}} = \underset{\boldsymbol{\xi}^{(i)} \in \mathcal{C}^{(j)}}{\arg\max} \left|\widetilde{Q}^{(j)}(\boldsymbol{\xi}^{(i)}, \mathbf{d}) - Q(\boldsymbol{\xi}^{(i)}, \mathbf{d}, r^{(j)})\right|$
        Update $\widetilde{Q}^{(j)}$ with sample point $\boldsymbol{\xi}^{\mathrm{add}}$
        Evaluate updated GPR model and obtain updated $\widetilde{Q}^{(j)}(\boldsymbol{\xi}^{(i)}, \mathbf{d})$ for all $\boldsymbol{\xi}^{(i)} \in \mathcal{C}^{(j)}$
        Calculate $\varepsilon = \max_{\boldsymbol{\xi}^{(i)} \in \mathcal{C}^{(j)}} \left|\widetilde{Q}^{(j)}(\boldsymbol{\xi}^{(i)}, \mathbf{d}) - Q(\boldsymbol{\xi}^{(i)}, \mathbf{d}, r^{(j)})\right|$
      end while
    end for
  end if
end for
Estimate the yield with $Y(\overline{\boldsymbol{\xi}}) = \frac{|\Omega|}{N_{\mathrm{MC}}}$

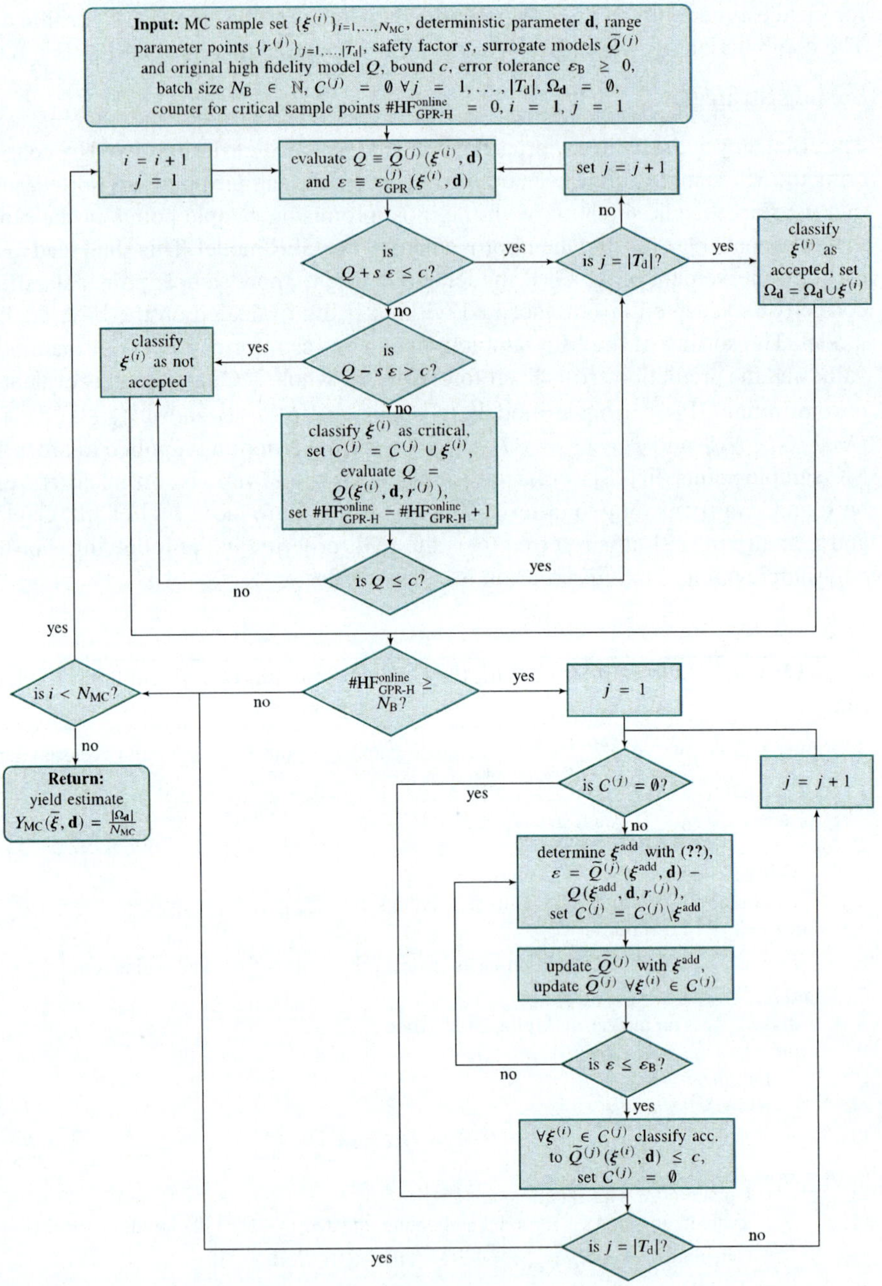

Fig. 4.6 Flowchart summarizing the yield estimation process in the GPR-Hybrid approach. Content based on [2, Algorithms 1, 2]

$$C_{\mathrm{EGL}}(\boldsymbol{\xi}^{(i)}) := \min_{j=1,\ldots,|T_{\mathrm{d}}|} \frac{\left|\widetilde{Q}^{(j)}(\boldsymbol{\xi}^{(i)},\mathbf{d}) - c\right|}{\left|\varepsilon_{\mathrm{GPR}}^{(j)}(\boldsymbol{\xi}^{(i)},\mathbf{d})\right|}, \tag{4.18}$$

where c denotes the upper bound from the PFS (4.8). Then, all MC sample points are sorted beginning with the smallest value of the EGL criterion, i.e.,

$$i^{\mathrm{start}} = \operatorname*{arg\,min}_{i=1,\ldots,N_{\mathrm{MC}}} C_{\mathrm{EGL}}(\boldsymbol{\xi}^{(i)}). \tag{4.19}$$

We propose a second criterion, based on the hybrid decision criterion (4.16), it is referred to as the hybrid criterion and defined by

$$\begin{aligned} C_{\mathrm{H}}(\boldsymbol{\xi}^{(i)}) := \max_{j=1,\ldots,|T_{\mathrm{d}}|} &\left(c - \left(\widetilde{Q}^{(j)}(\boldsymbol{\xi}^{(i)},\mathbf{d}) - s\,\varepsilon_{\mathrm{GPR}}^{(j)}(\boldsymbol{\xi}^{(i)},\mathbf{d})\right)\right) \\ &\times \left(\left(\widetilde{Q}^{(j)}(\boldsymbol{\xi}^{(i)},\mathbf{d}) + s\,\varepsilon_{\mathrm{GPR}}^{(j)}(\boldsymbol{\xi}^{(i)},\mathbf{d})\right) - c\right). \end{aligned} \tag{4.20}$$

Per definition it holds that

$$C_{\mathrm{H}}(\boldsymbol{\xi}^{(i)}) \begin{cases} > 0, & \text{if } \boldsymbol{\xi}^{(i)} \text{ is critical} \\ \leq 0, & \text{else.} \end{cases} \tag{4.21}$$

For the hybrid criterion, the MC sample points are sorted beginning with the largest value of the hybrid criterion, i.e.,

$$i^{\mathrm{start}} = \operatorname*{arg\,max}_{i=1,\ldots,N_{\mathrm{MC}}} C_{\mathrm{H}}(\boldsymbol{\xi}^{(i)}). \tag{4.22}$$

Algorithm 4.3 is a modification of Algorithm 4.2, applying the sorting strategy. Although, in this work we focus on sampling strategies based on the EGL and the hybrid criterion, any other criterion could be used instead, for example one from [23]. Once a GPR model is updated for one batch of critical sample points, the remaining MC sample points are reevaluated on the updated GPR models and are sorted again, according to the chosen criterion. This procedure is repeated until each sample point is classified as accepted or not accepted and the MC yield estimator can be calculated.

Comments and Summary

Using the GPR-Hybrid approach for yield estimation enables the reduction of high fidelity evaluations by maintaining high accuracy comparable with classic MC. The GPR-Hybrid approach has three important advantages over the SC-Hybrid approach. First, the training data set is not required to be structured and updating the GPR model with a new training data point is computationally negligible. This allows initialization with a rather small initial training data set and updates of the GPR model during the MC estimation procedure. Second, GPR provides an error indicator in form of the standard deviation of the GP. Hence, no additional computational effort nor infor-

Algorithm 4.3 Sorting strategy for yield estimation in the GPR-Hybrid approach, based on [2, Algorithm 3]

Input: input from Algorithm 4.2, sorting criterion
Evaluate all sample points $\xi^{(i)}, i = 1, \ldots, N_{\mathrm{MC}}$ on the GPR models
Sort all sample points according to the chosen sorting criterion
for $i = 1, \ldots, N_{\mathrm{MC}}$ **do**
Run lines 3–7 from Algorithm 4.2 (i.e., classify $\xi^{(i)}$ and define $\mathcal{C}^{(j)}$)
if $\#\mathrm{HF}_{\text{GPR-H}}^{\text{online}}$ is an integer multiplier of N_{B} **then**
Run lines 9–17 from Algorithm 4.2 (i.e., update GPR models)
Evaluate remaining sample points $\xi^{(k)}, k = i + 1, \ldots, N_{\mathrm{MC}}$
Sort the remaining sample points according to the chosen sorting criterion
end if
end for
Estimate the yield with $Y(\overline{\xi}) = \frac{|\Omega|}{N_{\mathrm{MC}}}$

mation is required to calculate an error indicator for the hybrid distinction criterion. This leads to the third point, neither for the generation of the surrogate model nor for the calculation of the error indicator are any FEM solutions or system matrices required, only the value of the QoI. This makes the GPR-Hybrid approach a black-box method easily usable with proprietary software. Additionally, the application of parallelization and sorting strategies is possible. Finally, we summarize the GPR-Hybrid method by answering the questions regarding hybrid approaches posed on Sect. 4.2:

1. Surrogate: GPR.
2. Definition of *far enough*: Based on the standard deviation of the GPR model.
3. Updates: Yes. Since there is no need of structured training data, arbitrary points, e.g. critical MC sample points, can be added to the training data set and the GPR model can be updated with negligible computational costs.
4. Further features: The usage of batches provides a parallelization strategy. The option of sorting MC sample points may further increase efficiency.

4.5 Summary

In the beginning of this chapter we formally introduced the yield as the fraction of realizations in a manufacturing process with uncertainties fulfilling all PFS. We provided a review of existing yield estimation methods, before we proposed two new yield estimation approaches. Both are hybrid methods, which means that most of the MC sample points are evaluated on a surrogate model while only a small subset of critical sample points is evaluated on the high fidelity model. This saves computing time while high accuracy standards from classic MC are maintained. The SC-Hybrid approach relies on SC as surrogate model. For determining the critical sample points, an adjoint SC error indicator is employed. This requires knowledge of the FEM matrices and solutions, which limits the use of proprietary software. The GPR-Hybrid approach relies on GPR as the surrogate model and utilizes the standard deviation of the GP as error indicator for determining critical sample points. The blackbox character enables the straightforward application of proprietary software for solving the QoI (here FEM software) and the unstructured training data points allow the use of critical sample points for GP model updates during the estimation process. Parallelization strategies were also discussed. In both methods a FE model refinement strategy can be applied. For the SC-Hybrid approach this has been implemented based on an adjoint FE error indicator. The next chapter uses those algorithms within the maximization of the yield.

References

1. Fuhrländer M, Georg N, Römer U, Schöps S (2020) Yield optimization based on adaptive Newton-Monte Carlo and polynomial surrogates. Int J Uncertain Quantif 10(4):351–373. https://doi.org/10.1615/Int.J.UncertaintyQuantification.2020033344
2. Fuhrländer M, Schöps S (2020) A blackbox yield estimation workflow with Gaussian process regression applied to the design of electromagnetic devices. J Math Ind 10(1):1–17
3. Gräb HE (2007) Analog design centering and sizing. Springer (2007)
4. Huber M, Fuhrländer M, Schöps S (2023) Multi-objective yield optimization for electrical machines using Gaussian processes to learn faulty designs. IEEE Trans Ind Appl 59(2):1340–1350
5. Choi CK, Yoo HH (2012) Uncertainty analysis of nonlinear systems employing the first-order reliability method. J Mech Sci Technol 26(1):39–44. https://doi.org/10.1007/s12206-011-1011-x
6. Breitung K (1984) Asymptotic approximations for multinormal integrals. J Eng Mech 110(3):357–366
7. Gallimard L (2019) Adaptive reduced basis strategy for rare-event simulations. Int J Numer Meth Eng 1:1–20. https://doi.org/10.1002/nme.6135
8. Au SK, Beck JL (2001) Estimation of small failure probabilities in high dimensions by subset simulation. Probab Eng Mech 16:263–277. https://doi.org/10.1016/S0266-8920(01)00019-4
9. Bect J, Li L, Vazquez E (2017) Bayesian subset simulation. SIAM/ASA J Uncert Quantif 5(1):762–786. https://doi.org/10.1137/16m1078276
10. Kouassi A, Bourinet J, Lalléchère S, Bonnet P, Fogli M (2016) Reliability and sensitivity analysis of transmission lines in a probabilistic EMC context. IEEE Trans Electromagn Compat 58(2):561–572. https://doi.org/10.1109/TEMC.2016.2520205
11. Hess MW, Benner P (2013) Fast evaluation of time-harmonic Maxwell's equations using the reduced basis method. IEEE Trans Microw Theory Tech 61(6):2265–2274. https://doi.org/10.1109/TMTT.2013.2258167
12. Bogoclu C, Roos D (2016). A benchmark of contemporary metamodeling algorithms. https://doi.org/10.7712/100016.2039.7645
13. Rao CR, Toutenburg H (1999) Linear models: least squares and alternatives, 2 edn. Springer (1999)
14. Babuška I, Nobile F, Tempone R (2007) A stochastic collocation method for elliptic partial differential equations with random input data. SIAM J Numer Anal 45(3):1005–1034. https://doi.org/10.1137/100786356
15. Li J, Xiu D (2010) Evaluation of failure probability via surrogate models. J Comput Phys 229(23):8966–8980. https://doi.org/10.1016/j.jcp.2010.08.022
16. Rasmussen CE, Williams CK (2006) Gaussian processes for machine learning. The MIT Press (2006)

17. Goodfellow I, Bengio Y, Courville A (2016) Deep learning. The MIT Press
18. Tyagi A, Jonsson X, Beelen T, Schilders W (2018) Hybrid importance sampling Monte Carlo approach for yield estimation in circuit design. J Math Ind 8. https://doi.org/10.1186/s13362-018-0053-4
19. Xiao S, Oladyshkin S, Nowak W (2020) Reliability analysis with stratified importance sampling based on adaptive kriging. Reliab Eng Syst Safety 197 (2020)
20. Zhang J, Taflanidis AA (2019) Accelerating MCMC via kriging-based adaptive independent proposals and delayed rejection. Comput Methods Appl Mech Eng 355:1124–1147
21. Dassault Systèmes Deutschland GmbH: CST Studio Suite ® (2018). www.3ds.com
22. Li J, Li J, Xiu D (2011) An efficient surrogate-based method for computing rare failure probability. J Comput Phys 230(24):8683–8697
23. Bect J, Ginsbourger D, Li L, Picheny V, Vazquez E (2012) Sequential design of computer experiments for the estimation of a probability of failure. Stat Comput 22(3):773–793
24. Dhulipala SL, Shields MD, Spencer BW, Bolisetti C, Slaughter AE, Laboure VM, Chakroborty P (2022) Active learning with multifidelity modeling for efficient rare event simulation. J Comput Phys
25. Butler T, Constantine P, Wildey T (2012) A posteriori error analysis of parameterized linear systems using spectral methods. SIAM J Matrix Anal Appl 33(1):195–209
26. Butler T, Dawson C, Wildey T (2013) Propagation of uncertainties using improved surrogate models. SIAM/ASA J Uncert Quantif 1(1):164–191
27. Butler T, Wildey T (2018) Utilizing adjoint-based error estimates for surrogate models to accurately predict probabilities of events. Int J Uncertain Quantif 8(2):143–159. https://doi.org/10.1615/Int.J.UncertaintyQuantification.2018020911
28. Kumar U, Crocker J, Chitra T, Saranga H (2006) Reliability and six sigma. EngineeringPro collection. Springer (2006). https://books.google.td/books?id=5_amcGFkhEIC
29. Echard B, Gayton N, Lemaire M (2011) AK-MCS: an active learning reliability method combining kriging and Monte Carlo simulation. Struct Saf 33(2):145–154

Chapter 5
Yield Optimization

We introduced the yield as the fraction of realizations in a manufacturing process fulfilling the performance feature specifications (PFS), see (4.4). Besides the estimation of the yield, e.g. with one of the methods presented in Chap. 4, the maximization of the yield is the next natural task. This chapter deals with different methods for yield optimization, depending on the specific problem and the information available.

The adaptive Newton-MC method is a modification of the globalized Newton method specialized on keeping the computational effort of yield optimization low. The Hermite least squares and Hermite BOBYQA methods are modifications of Powell's BOBYQA method, for the case that some partial derivatives are available and are to be used. This is a general optimization framework and not limited to yield optimization. However, in Sect. 5.1 we will see that the situation of mixed gradient information is particularly relevant if a yield is optimized over uncertain and deterministic parameters at the same time.

For the yield defined in (4.4) the general formulation of the optimization problem is given by

$$\max_{\overline{\boldsymbol{\xi}},\mathbf{d}} Y(\overline{\boldsymbol{\xi}}, \mathbf{d}), \tag{5.1}$$

where $\overline{\boldsymbol{\xi}} \in \mathbb{R}^{n_\xi}$ and $\mathbf{d} \in \mathbb{R}^{n_\mathbf{d}}$. In this formulation, the yield is maximized over both types of design parameters—deterministic and uncertain parameters—and the problem is unconstrained. In Sect. 5.2 we will introduce different kinds of constraints.

5.1 Derivative Calculation

In Sect. 3.2.2 we summarized common optimization methods dividing them into gradient based and derivative free optimization methods. In order to apply and modify gradient based optimization techniques in the following sections, we derive the first and second order derivatives of the yield based on [1, Sect. 7.1].

M. Fuhrländer, *Design Methods for Reducing Failure Probabilities with Examples from Electrical Engineering*, Springer Theses,
https://doi.org/10.1007/978-3-031-37019-9_5

5.1.1 Derivatives with Respect to Mean of Uncertain Parameters

We recall the definition of the yield from (4.4) and derive first the derivative with respect to the mean value of the uncertain parameter $\overline{\boldsymbol{\xi}}$, i.e.,

$$\nabla_{\overline{\boldsymbol{\xi}}} Y(\overline{\boldsymbol{\xi}}, \mathbf{d}) \overset{(4.4)}{=} \nabla_{\overline{\boldsymbol{\xi}}} \int_{\mathbb{R}^{n_\xi}} \mathbf{1}_{\Omega_\mathrm{d}}(\boldsymbol{\xi})\, \varphi(\boldsymbol{\xi})\, \mathrm{d}\boldsymbol{\xi} \tag{5.2}$$

$$\overset{(\star)}{=} \int_{\mathbb{R}^{n_\xi}} \nabla_{\overline{\boldsymbol{\xi}}} \left(\mathbf{1}_{\Omega_\mathrm{d}}(\boldsymbol{\xi})\, \varphi(\boldsymbol{\xi})\right) \mathrm{d}\boldsymbol{\xi} \tag{5.3}$$

$$\overset{(\star\star)}{=} \int_{\mathbb{R}^{n_\xi}} \mathbf{1}_{\Omega_\mathrm{d}}(\boldsymbol{\xi})\, \nabla_{\overline{\boldsymbol{\xi}}}\varphi(\boldsymbol{\xi})\, \mathrm{d}\boldsymbol{\xi}. \tag{5.4}$$

The equality $(\star)$ holds since the integral is built over $\boldsymbol{\xi}$ and not over $\overline{\boldsymbol{\xi}}$, the equality $(\star\star)$ holds since $\overline{\boldsymbol{\xi}}$ does not appear in the indicator function. Hence, only the gradient of the probability density function needs to be calculated. Let us assume, that all uncertain parameters are Gaussian distributed, cf. (3.5), then, $\varphi(\boldsymbol{\xi})$ is the probability density function of a multivariate Gaussian distribution, and its derivative is given by

$$\nabla_{\overline{\boldsymbol{\xi}}}\varphi(\boldsymbol{\xi}) \equiv \nabla_{\overline{\boldsymbol{\xi}}}\varphi_{\mathcal{N}(\overline{\boldsymbol{\xi}}, \boldsymbol{\Sigma}_\xi)}(\boldsymbol{\xi}) \tag{5.5}$$

$$= \nabla_{\overline{\boldsymbol{\xi}}} \det(2\pi\,\boldsymbol{\Sigma}_\xi)^{-\frac{1}{2}} \exp\left(-\frac{1}{2}(\boldsymbol{\xi}-\overline{\boldsymbol{\xi}})^\top \boldsymbol{\Sigma}_\xi^{-1}(\boldsymbol{\xi}-\overline{\boldsymbol{\xi}})\right) \tag{5.6}$$

$$= \boldsymbol{\Sigma}_\xi^{-1}(\boldsymbol{\xi}-\overline{\boldsymbol{\xi}})\varphi_{\mathcal{N}(\overline{\boldsymbol{\xi}}, \boldsymbol{\Sigma}_\xi)}(\boldsymbol{\xi}), \tag{5.7}$$

where $\boldsymbol{\Sigma}_\xi$ is the covariance matrix of the uncertain parameter. Inserting (5.7) into (5.4) yields

$$\nabla_{\overline{\boldsymbol{\xi}}} Y(\overline{\boldsymbol{\xi}}, \mathbf{d}) = \int_{\mathbb{R}^{n_\xi}} \mathbf{1}_{\Omega_\mathrm{d}}(\boldsymbol{\xi})\, \boldsymbol{\Sigma}_\xi^{-1}(\boldsymbol{\xi}-\overline{\boldsymbol{\xi}})\varphi_{\mathcal{N}(\overline{\boldsymbol{\xi}}, \boldsymbol{\Sigma}_\xi)}(\boldsymbol{\xi})\, \mathrm{d}\boldsymbol{\xi}. \tag{5.8}$$

In the same way, the Hessian with respect to $\overline{\boldsymbol{\xi}}$ can be obtained

$$\nabla^2_{\overline{\boldsymbol{\xi}}} Y(\overline{\boldsymbol{\xi}}, \mathbf{d}) = \int_{\mathbb{R}^{n_\xi}} \mathbf{1}_{\Omega_\mathrm{d}}(\boldsymbol{\xi})\, \boldsymbol{\Sigma}_\xi^{-1}\left((\boldsymbol{\xi}-\overline{\boldsymbol{\xi}})(\boldsymbol{\xi}-\overline{\boldsymbol{\xi}})^\top - \boldsymbol{\Sigma}_\xi\right)\boldsymbol{\Sigma}_\xi^{-1}\varphi_{\mathcal{N}(\overline{\boldsymbol{\xi}}, \boldsymbol{\Sigma}_\xi)}(\boldsymbol{\xi})\, \mathrm{d}\boldsymbol{\xi}. \tag{5.9}$$

Hence, in case of Gaussian distributed uncertain parameters, closed form formulations of the gradient and the Hessian with respect to $\overline{\boldsymbol{\xi}}$ exist. In general, i.e., for arbitrary distributions, this is not the case, as we will see in the end of this section.

Simplification and Estimation of Derivatives

We aim to simplify the formulations (5.8) and (5.9) in order to show that the gradient and the Hessian with respect to the mean value of Gaussian distributed uncertain parameters can be estimated without additional computing effort when the yield itself has been estimated with a MC analysis. The required simplification is stated in Theorem 5.1, but first we need some preparation. We follow [1, Sect. 7.1].

Definition 5.1 (*Acceptance-truncated distribution (ATD)*) Let $Y(\overline{\boldsymbol{\xi}}, \mathbf{d})$ be the yield as defined in (4.4) with deterministic parameter $\mathbf{d}$ and uncertain parameter $\boldsymbol{\xi} \sim \mathcal{N}(\overline{\boldsymbol{\xi}}, \boldsymbol{\Sigma}_{\xi})$. Further let $\varphi_{\mathcal{N}(\overline{\xi}, \boldsymbol{\Sigma}_{\xi})}(\boldsymbol{\xi})$ denote the corresponding probability density function. Then,

$$\varphi_{\mathrm{ATD}}(\boldsymbol{\xi}) = \frac{1}{Y(\overline{\boldsymbol{\xi}}, \mathbf{d})} \mathbf{1}_{\Omega_{\mathbf{d}}}(\boldsymbol{\xi}) \varphi_{\mathcal{N}(\overline{\xi}, \boldsymbol{\Sigma}_{\xi})}(\boldsymbol{\xi}) \tag{5.10}$$

defines the acceptance-truncated distribution with respect to $\varphi_{\mathcal{N}(\overline{\xi}, \boldsymbol{\Sigma}_{\xi})}(\boldsymbol{\xi})$.

The distribution (5.10) represents the uncertain parameter combinations fulfilling the PFS, i.e., lying inside the safe domain $\Omega_{\mathbf{d}}$. Please note, that in general, this is not a Gaussian distribution. The mean and the covariance of $\varphi_{\mathrm{ATD}}(\boldsymbol{\xi})$ are given by

$$\overline{\boldsymbol{\xi}}_{\mathrm{ATD}} = \mathbb{E}_{\varphi_{\mathrm{ATD}}}[\boldsymbol{\xi}] = \frac{1}{Y(\overline{\boldsymbol{\xi}}, \mathbf{d})} \int_{\mathbb{R}^{n_\xi}} \boldsymbol{\xi}\, \mathbf{1}_{\Omega_{\mathbf{d}}}(\boldsymbol{\xi}) \varphi_{\mathcal{N}(\overline{\xi}, \boldsymbol{\Sigma}_{\xi})}(\boldsymbol{\xi})\, \mathrm{d}\boldsymbol{\xi}, \tag{5.11}$$

$$\begin{aligned} \boldsymbol{\Sigma}_{\mathrm{ATD}} &= \mathbb{E}_{\varphi_{\mathrm{ATD}}}\left[(\boldsymbol{\xi} - \overline{\boldsymbol{\xi}}_{\mathrm{ATD}})(\boldsymbol{\xi} - \overline{\boldsymbol{\xi}}_{\mathrm{ATD}})^{\top}\right] \\ &= \frac{1}{Y(\overline{\boldsymbol{\xi}}, \mathbf{d})} \int_{\mathbb{R}^{n_\xi}} (\boldsymbol{\xi} - \overline{\boldsymbol{\xi}}_{\mathrm{ATD}})(\boldsymbol{\xi} - \overline{\boldsymbol{\xi}}_{\mathrm{ATD}})^{\top} \mathbf{1}_{\Omega_{\mathbf{d}}}(\boldsymbol{\xi}) \varphi_{\mathcal{N}(\overline{\xi}, \boldsymbol{\Sigma}_{\xi})}(\boldsymbol{\xi})\, \mathrm{d}\boldsymbol{\xi}. \end{aligned} \tag{5.12}$$

We obtain a simplified and calculable formulation of the gradient and the Hessian of the yield.

Theorem 5.1 *Let $Y(\overline{\boldsymbol{\xi}}, \mathbf{d})$ be the yield as defined in (4.4) with deterministic parameter $\mathbf{d}$ and uncertain parameter $\boldsymbol{\xi} \sim \mathcal{N}(\overline{\boldsymbol{\xi}}, \boldsymbol{\Sigma}_{\xi})$. Further let $\varphi_{\mathcal{N}(\overline{\xi}, \boldsymbol{\Sigma}_{\xi})}(\boldsymbol{\xi})$ denote the probability density function and $\varphi_{\mathrm{ATD}}(\boldsymbol{\xi})$ the corresponding acceptance-truncated distribution with mean $\overline{\boldsymbol{\xi}}_{\mathrm{ATD}}$ and covariance matrix $\boldsymbol{\Sigma}_{\mathrm{ATD}}$. Then, the gradient and the Hessian of the yield with respect to $\overline{\boldsymbol{\xi}}$ are given by*

$$\nabla_{\overline{\xi}} Y(\overline{\boldsymbol{\xi}}, \mathbf{d}) = Y(\overline{\boldsymbol{\xi}}, \mathbf{d}) \boldsymbol{\Sigma}_{\xi}^{-1} (\overline{\boldsymbol{\xi}}_{\mathrm{ATD}} - \overline{\boldsymbol{\xi}}), \tag{5.13}$$

$$\nabla^2_{\overline{\xi}} Y(\overline{\boldsymbol{\xi}}, \mathbf{d}) = Y(\overline{\boldsymbol{\xi}}, \mathbf{d}) \boldsymbol{\Sigma}_{\xi}^{-1} (\boldsymbol{\Sigma}_{\mathrm{ATD}} + (\overline{\boldsymbol{\xi}}_{\mathrm{ATD}} - \overline{\boldsymbol{\xi}})(\overline{\boldsymbol{\xi}}_{\mathrm{ATD}} - \overline{\boldsymbol{\xi}})^{\top} - \boldsymbol{\Sigma}_{\xi}) \boldsymbol{\Sigma}_{\xi}^{-1}. \tag{5.14}$$

Proof We start with the proof of (5.13). We derive

$$\nabla_{\overline{\xi}} Y(\overline{\boldsymbol{\xi}}, \mathbf{d}) \overset{(5.8)}{=} \int_{\mathbb{R}^{n_\xi}} \mathbf{1}_{\Omega_{\mathbf{d}}}(\boldsymbol{\xi})\, \boldsymbol{\Sigma}_{\xi}^{-1} (\boldsymbol{\xi} - \overline{\boldsymbol{\xi}})\, \varphi_{\mathcal{N}(\overline{\xi}, \boldsymbol{\Sigma}_{\xi})}(\boldsymbol{\xi})\, \mathrm{d}\boldsymbol{\xi} \tag{5.15}$$

$$= \boldsymbol{\Sigma}_{\xi}^{-1} \left(\int_{\mathbb{R}^{n_\xi}} \mathbf{1}_{\Omega_{\mathbf{d}}}(\boldsymbol{\xi})\, \boldsymbol{\xi}\, \varphi_{\mathcal{N}(\overline{\xi}, \boldsymbol{\Sigma}_{\xi})}(\boldsymbol{\xi})\, \mathrm{d}\boldsymbol{\xi} - \overline{\boldsymbol{\xi}} \int_{\mathbb{R}^{n_\xi}} \mathbf{1}_{\Omega_{\mathbf{d}}}(\boldsymbol{\xi})\, \varphi_{\mathcal{N}(\overline{\xi}, \boldsymbol{\Sigma}_{\xi})}(\boldsymbol{\xi})\, \mathrm{d}\boldsymbol{\xi} \right) \tag{5.16}$$

$$\overset{(5.11),(4.4)}{=} \boldsymbol{\Sigma}_{\xi}^{-1} \left(Y(\overline{\boldsymbol{\xi}}, \mathbf{d}) \overline{\boldsymbol{\xi}}_{\mathrm{ATD}} - \overline{\boldsymbol{\xi}} Y(\overline{\boldsymbol{\xi}}, \mathbf{d}) \right) \tag{5.17}$$

$$= Y(\overline{\boldsymbol{\xi}}, \mathbf{d}) \boldsymbol{\Sigma}_{\xi}^{-1} \left(\overline{\boldsymbol{\xi}}_{\mathrm{ATD}} - \overline{\boldsymbol{\xi}} \right). \tag{5.18}$$

For the Hessian we show (5.14)

$$\nabla^2_{\xi} Y(\bar{\xi}, \mathbf{d}) \overset{(5.9)}{=} \int_{\mathbb{R}^{n_\xi}} \mathbf{1}_{\Omega_\mathbf{d}}(\xi)\, \boldsymbol{\Sigma}_\xi^{-1} \left((\xi - \bar{\xi})(\xi - \bar{\xi})^\top - \boldsymbol{\Sigma}_\xi \right) \boldsymbol{\Sigma}_\xi^{-1} \varphi_{\mathcal{N}(\bar{\xi}, \boldsymbol{\Sigma}_\xi)}(\xi)\, \mathrm{d}\xi \tag{5.19}$$

$$= \boldsymbol{\Sigma}_\xi^{-1} \Big(\underbrace{\int_{\mathbb{R}^{n_\xi}} \mathbf{1}_{\Omega_\mathbf{d}}(\xi)\, (\xi - \bar{\xi})(\xi - \bar{\xi})^\top \varphi_{\mathcal{N}(\bar{\xi}, \boldsymbol{\Sigma}_\xi)}(\xi)\, \mathrm{d}\xi}_{=:(\star)} \tag{5.20}$$

$$- \underbrace{\int_{\mathbb{R}^{n_\xi}} \mathbf{1}_{\Omega_\mathbf{d}}(\xi)\, \boldsymbol{\Sigma}_\xi\, \varphi_{\mathcal{N}(\bar{\xi}, \boldsymbol{\Sigma}_\xi)}(\xi)\, \mathrm{d}\xi}_{=\boldsymbol{\Sigma}_\xi Y(\bar{\xi}, \mathbf{d})} \Big) \boldsymbol{\Sigma}_\xi^{-1}. \tag{5.21}$$

We transform $(\star)$ by replacing $(\xi - \bar{\xi})$ by $(\xi - \bar{\xi}_{\mathrm{ATD}}) + (\bar{\xi}_{\mathrm{ATD}} - \bar{\xi})$ and applying (4.4), (5.11) and (5.12)

$$(\star) = \int_{\mathbb{R}^{n_\xi}} \mathbf{1}_{\Omega_\mathbf{d}}(\xi) \left((\xi - \bar{\xi}_{\mathrm{ATD}}) + (\bar{\xi}_{\mathrm{ATD}} - \bar{\xi}) \right) \left((\xi - \bar{\xi}_{\mathrm{ATD}}) + (\bar{\xi}_{\mathrm{ATD}} - \bar{\xi}) \right)^\top \varphi_{\mathcal{N}(\bar{\xi}, \boldsymbol{\Sigma}_\xi)}(\xi)\, \mathrm{d}\xi \tag{5.22}$$

$$\overset{\text{binom.}}{=} \underbrace{\int_{\mathbb{R}^{n_\xi}} \mathbf{1}_{\Omega_\mathbf{d}}(\xi)\, (\xi - \bar{\xi}_{\mathrm{ATD}})(\xi - \bar{\xi}_{\mathrm{ATD}})^\top \varphi_{\mathcal{N}(\bar{\xi}, \boldsymbol{\Sigma}_\xi)}(\xi)\, \mathrm{d}\xi}_{\overset{(5.12)}{=} \boldsymbol{\Sigma}_{\mathrm{ATD}} Y(\bar{\xi}, \mathbf{d})} \tag{5.23}$$

$$+2 \int_{\mathbb{R}^{n_\xi}} \mathbf{1}_{\Omega_\mathbf{d}}(\xi)\, (\xi - \bar{\xi}_{\mathrm{ATD}})(\bar{\xi}_{\mathrm{ATD}} - \bar{\xi})^\top \varphi_{\mathcal{N}(\bar{\xi}, \boldsymbol{\Sigma}_\xi)}(\xi)\, \mathrm{d}\xi \tag{5.24}$$

$$+ \underbrace{\int_{\mathbb{R}^{n_\xi}} \mathbf{1}_{\Omega_\mathbf{d}}(\xi)\, (\bar{\xi}_{\mathrm{ATD}} - \bar{\xi})(\bar{\xi}_{\mathrm{ATD}} - \bar{\xi})^\top \varphi_{\mathcal{N}(\bar{\xi}, \boldsymbol{\Sigma}_\xi)}(\xi)\, \mathrm{d}\xi}_{\overset{(4.4)}{=} Y(\bar{\xi}, \mathbf{d})(\bar{\xi}_{\mathrm{ATD}} - \bar{\xi})(\bar{\xi}_{\mathrm{ATD}} - \bar{\xi})^\top} \tag{5.25}$$

$$= \boldsymbol{\Sigma}_{\mathrm{ATD}} Y(\bar{\xi}, \mathbf{d}) \tag{5.26}$$

$$+2 \Big(\underbrace{\int_{\mathbb{R}^{n_\xi}} \mathbf{1}_{\Omega_\mathbf{d}}(\xi)\, \xi\, \varphi_{\mathcal{N}(\bar{\xi}, \boldsymbol{\Sigma}_\xi)}(\xi)\, \mathrm{d}\xi}_{\overset{(5.11)}{=} Y(\bar{\xi}, \mathbf{d}) \bar{\xi}_{\mathrm{ATD}}} - \underbrace{\bar{\xi}_{\mathrm{ATD}} \int_{\mathbb{R}^{n_\xi}} \mathbf{1}_{\Omega_\mathbf{d}}(\xi)\, \varphi_{\mathcal{N}(\bar{\xi}, \boldsymbol{\Sigma}_\xi)}(\xi)\, \mathrm{d}\xi}_{\overset{(4.4)}{=} \bar{\xi}_{\mathrm{ATD}} Y(\bar{\xi}, \mathbf{d})} \Big) (\bar{\xi}_{\mathrm{ATD}} - \bar{\xi})^\top \tag{5.27}$$

$$+ Y(\bar{\xi}, \mathbf{d})(\bar{\xi}_{\mathrm{ATD}} - \bar{\xi})(\bar{\xi}_{\mathrm{ATD}} - \bar{\xi})^\top \tag{5.28}$$

and insert $(\star)$ back into (5.21)

$$\nabla^2_{\xi} Y(\bar{\xi}, \mathbf{d}) = \boldsymbol{\Sigma}_\xi^{-1} \left(\boldsymbol{\Sigma}_{\mathrm{ATD}} Y(\bar{\xi}, \mathbf{d}) + Y(\bar{\xi}, \mathbf{d})(\bar{\xi}_{\mathrm{ATD}} - \bar{\xi})(\bar{\xi}_{\mathrm{ATD}} - \bar{\xi})^\top - \boldsymbol{\Sigma}_\xi Y(\bar{\xi}, \mathbf{d}) \right) \boldsymbol{\Sigma}_\xi^{-1} \tag{5.29}$$

$$= Y(\bar{\xi}, \mathbf{d}) \boldsymbol{\Sigma}_\xi^{-1} (\boldsymbol{\Sigma}_{\mathrm{ATD}} + (\bar{\xi}_{\mathrm{ATD}} - \bar{\xi})(\bar{\xi}_{\mathrm{ATD}} - \bar{\xi})^\top - \boldsymbol{\Sigma}_\xi) \boldsymbol{\Sigma}_\xi^{-1}. \tag{5.30}$$

When the yield is estimated with a sample based approach, e.g. classic MC analysis (see Sect. 3.3.1), SC-Hybrid approach (see Sect. 4.3) or GPR-Hybrid approach (see

Sect. 4.4), then the mean and covariance of the acceptance-truncated distribution (5.11)–(5.12) can be estimated by [1]

$$\overline{\boldsymbol{\xi}}_{\text{ATD}} \approx \widetilde{\overline{\boldsymbol{\xi}}}_{\text{ATD}} = \frac{1}{N_{\Omega_{\mathbf{d}}}} \sum_{i=1}^{N_{\text{MC}}} \mathbf{1}_{\Omega_{\mathbf{d}}}(\boldsymbol{\xi}^{(i)})\boldsymbol{\xi}^{(i)}, \tag{5.31}$$

$$\boldsymbol{\Sigma}_{\text{ATD}} \approx \widetilde{\boldsymbol{\Sigma}}_{\text{ATD}} = \frac{1}{N_{\Omega_{\mathbf{d}}} - 1} \sum_{i=1}^{N_{\text{MC}}} \mathbf{1}_{\Omega_{\mathbf{d}}}(\boldsymbol{\xi}^{(i)})(\boldsymbol{\xi}^{(i)} - \widetilde{\overline{\boldsymbol{\xi}}}_{\text{ATD}})(\boldsymbol{\xi}^{(i)} - \widetilde{\overline{\boldsymbol{\xi}}}_{\text{ATD}})^{\top}, \tag{5.32}$$

where $N_{\Omega_{\mathbf{d}}} \equiv |\Omega_{\mathbf{d}}| \leq N_{\text{MC}}$ denotes the number of sample points inside the safe domain $\Omega_{\mathbf{d}}$ and $\boldsymbol{\xi}^{(i)}$, $i = 1, \ldots, N_{\text{MC}}$ the i-th MC sample point. Since $\mathbf{1}_{\Omega_{\mathbf{d}}}(\boldsymbol{\xi}^{(i)})$ has already been evaluated for all $i = 1, \ldots, N_{\text{MC}}$ within the yield estimation and also $\Omega_{\mathbf{d}}$ and thus $N_{\Omega_{\mathbf{d}}}$ are known, (5.31) and (5.32) are obtained without any additional computational effort. It follows, that once the yield is estimated, its gradient and Hessian can also be calculated without additional computational effort.

Corollary 5.1 *Assume we have the same setting as in Theorem 5.1 and note that mean value $\overline{\boldsymbol{\xi}}$ and covariance matrix $\boldsymbol{\Sigma}_{\boldsymbol{\xi}}$ of the uncertain parameter $\boldsymbol{\xi}$ are given. Let $\widetilde{Y}(\overline{\boldsymbol{\xi}}, \mathbf{d})$ denote a MC based estimation of the yield and $\widetilde{\overline{\boldsymbol{\xi}}}_{\text{ATD}}$ and $\widetilde{\boldsymbol{\Sigma}}_{\text{ATD}}$ the approximations from (5.31) and (5.32). Then, the gradient and the Hessian of the yield with respect to $\overline{\boldsymbol{\xi}}$ can be estimated by*

$$\nabla_{\overline{\boldsymbol{\xi}}} Y(\overline{\boldsymbol{\xi}}, \mathbf{d}) \approx \nabla_{\overline{\boldsymbol{\xi}}} \widetilde{Y}(\overline{\boldsymbol{\xi}}, \mathbf{d}) = \widetilde{Y}(\overline{\boldsymbol{\xi}}, \mathbf{d}) \boldsymbol{\Sigma}_{\boldsymbol{\xi}}^{-1} (\widetilde{\overline{\boldsymbol{\xi}}}_{\text{ATD}} - \overline{\boldsymbol{\xi}}), \tag{5.33}$$

$$\nabla^2_{\overline{\boldsymbol{\xi}}} Y(\overline{\boldsymbol{\xi}}, \mathbf{d}) \approx \nabla^2_{\overline{\boldsymbol{\xi}}} \widetilde{Y}(\overline{\boldsymbol{\xi}}, \mathbf{d}) = \widetilde{Y}(\overline{\boldsymbol{\xi}}, \mathbf{d}) \boldsymbol{\Sigma}_{\boldsymbol{\xi}}^{-1} (\widetilde{\boldsymbol{\Sigma}}_{\text{ATD}} + (\widetilde{\overline{\boldsymbol{\xi}}}_{\text{ATD}} - \overline{\boldsymbol{\xi}})(\widetilde{\overline{\boldsymbol{\xi}}}_{\text{ATD}} - \overline{\boldsymbol{\xi}})^{\top} - \boldsymbol{\Sigma}_{\boldsymbol{\xi}}) \boldsymbol{\Sigma}_{\boldsymbol{\xi}}^{-1}. \tag{5.34}$$

Non-Gaussian Uncertain Parameters

Let us now assume, that the uncertain parameters are not all Gaussian distributed. Then, in general, derivatives of their probability density functions are not available in closed form. This implies that the simplified formulation of the yield derivatives and the computability without additional effort cannot be ensured. For illustration, let us consider a uniformly distributed one-dimensional uncertain parameter. In (5.4) the gradient of the probability density function is required. Let $\overline{\xi}$ be the mean value of the uniform distribution and $\Delta_{\xi} > 0$. Then, the probability density function is given by

$$\varphi_{\mathcal{U}(\overline{\xi}-\Delta_{\xi}, \overline{\xi}+\Delta_{\xi})}(\xi) = \begin{cases} \frac{1}{2\Delta_{\xi}}, & \text{if } \overline{\xi} - \Delta_{\xi} \leq \xi \leq \overline{\xi} + \Delta_{\xi} \\ 0, & \text{else.} \end{cases} \tag{5.35}$$

This can also be written using an indicator function

$$\varphi_{\mathcal{U}(\overline{\xi}-\Delta_{\xi}, \overline{\xi}+\Delta_{\xi})}(\xi) = \frac{1}{2\Delta_{\xi}} \mathbf{1}_{[\overline{\xi}-\Delta_{\xi}, \overline{\xi}+\Delta_{\xi}]}(\xi) + \underbrace{\mathbf{1}_{<\overline{\xi}-\Delta_{\xi}}(\xi) \cdot 0 + \mathbf{1}_{>\overline{\xi}+\Delta_{\xi}} \cdot 0}_{=0}. \tag{5.36}$$

Calculating the gradient of (5.36) involves calculating the derivative of the indicator function. For this, we refer to the following section.

In Sect. 3.1 we introduced the truncated Gaussian distribution. This distribution is *almost* Gaussian, but ensures physical behavior of the parameters. Although the corresponding gradient and Hessian are not exactly given by (5.13) and (5.14), these are good approximations. In Sect. 6.3.1 we show that the resulting errors are negligible compared to the MC error. Hence, in the numerical tests in Chap. 6 we consider the derivatives as available, when the parameter is truncated Gaussian distributed and use (5.13) and (5.14), respectively.

5.1.2 *Derivatives with Respect to Deterministic Parameters*

In Problem (5.1) the yield is optimized over the mean of the uncertain parameter and over the deterministic parameter. In the last section we discussed the derivation of the gradient with respect to the former, this section deals with the derivation of the gradient with respect to the latter. Analogously to (5.2), we obtain

$$\nabla_{\mathbf{d}} Y(\overline{\boldsymbol{\xi}}, \mathbf{d}) \overset{(4.4)}{=} \nabla_{\mathbf{d}} \int_{\mathbb{R}^{n_\xi}} \mathbf{1}_{\Omega_{\mathbf{d}}}(\boldsymbol{\xi})\, \varphi(\boldsymbol{\xi})\, \mathrm{d}\boldsymbol{\xi} \tag{5.37}$$

$$\overset{(\star)}{=} \int_{\mathbb{R}^{n_\xi}} \nabla_{\mathbf{d}} \left(\mathbf{1}_{\Omega_{\mathbf{d}}}(\boldsymbol{\xi})\, \varphi(\boldsymbol{\xi})\right) \mathrm{d}\boldsymbol{\xi} \tag{5.38}$$

$$\overset{(\star\star)}{=} \int_{\mathbb{R}^{n_\xi}} \nabla_{\mathbf{d}} \mathbf{1}_{\Omega_{\mathbf{d}}}(\boldsymbol{\xi})\, \varphi(\boldsymbol{\xi})\, \mathrm{d}\boldsymbol{\xi}. \tag{5.39}$$

Again, the gradient can be drawn into the integral, cf. ($\star$). This time, the probability density function does not depend on the optimization variable for which we calculate the gradient, but the safe domain $\Omega_{\mathbf{d}}$ does, cf. ($\star\star$). Hence, the derivative of an indicator function is required. Since the gradient of the indicator function is related to the Dirac distribution [2], this derivative exists only in the distributional sense. Therefore, a numerical computation of (5.39) is also not straightforward. Hence, in the remainder of this work, the gradients including derivatives of the indicator function are considered as unavailable.

5.2 Optimization Problem Formulations

We summarize that once the yield is estimated, the first and second order derivatives of the yield with respect to the mean value of (almost) Gaussian distributed uncertain parameters are available without any noteworthy additional computing effort. This statement does not hold for derivatives with respect to deterministic or non-Gaussian distributed uncertain parameters. Although we discussed in Sect. 5.1.2 that approximations of these gradients are also possible, there are several impediments: especially the gradient of the QoI must be available and problem dependent approximations of the Dirac distribution must be found. To our best knowledge, for finding a well suited approximation of the Dirac distribution there are research achievements and rules, but no tuning algorithms or established criteria. For that reason, we decided not to approximate the derivatives with respect to deterministic or non-Gaussian distributed uncertain parameters. Instead, we assume them to be unavailable. This motivates the formulation of different optimization problems, depending on the type of optimization variables, i.e., depending on the derivative information available.

5.2.1 Single Objective Optimization Problem for Gradient Based Solvers

We formulate an unconstrained SOO problem which allows the usage of gradient based optimization algorithms without the necessity of approximating derivatives. The problem reads

$$\max_{\mathbf{x}\in\mathbb{R}^n} f(\mathbf{x}), \tag{5.40}$$

where $f : \mathbb{R}^n \to \mathbb{R}$ is twice continuously differentiable and $\nabla_{\mathbf{x}} f(\mathbf{x})$ and $\nabla^2_{\mathbf{x}} f(\mathbf{x})$ are available. In the context of yield optimization the problem reads

$$\max_{\overline{\xi}\in\mathbb{R}^{\xi}} Y(\overline{\boldsymbol{\xi}}, \mathbf{d}), \tag{5.41}$$

where the yield is defined as in (4.4) and $\boldsymbol{\xi} \sim \mathcal{N}(\overline{\boldsymbol{\xi}}, \boldsymbol{\Sigma}_{\xi})$ is a Gaussian distributed uncertain parameter. From Sect. 5.1.1 we know that the gradient and the Hessian of the yield are then available. In Sect. 5.3 we propose the adaptive Newton-MC algorithm which is an efficient modification of the Newton method to solve (5.41).

5.2.2 General Single Objective Optimization Problem

We generalize (5.40). We still assume that the objective function is twice continuously differentiable, but now we permit that some or all partial derivatives are unavailable, e.g. because they are too expensive to compute. In addition, we allow bound constraints. The optimization problem reads

$$\begin{aligned} &\max_{\mathbf{x}\in\mathbb{R}^n} f(\mathbf{x}) \\ &\text{s.t. } \mathbf{x}_{\mathbf{lb}} \leq \mathbf{x} \leq \mathbf{x}_{\mathbf{ub}} \end{aligned} \tag{5.42}$$

with $f : \mathbb{R}^n \to \mathbb{R}$ twice continuously differentiable, but now $\frac{\partial}{\partial x_i} f(\mathbf{x})$ is only available for a subset of $\{1, \ldots, n\}$.

Let the yield be defined as in (4.4). Let $\mathbf{d} \in \mathbb{R}^{\mathbf{d}}$ denote a deterministic design parameter, $\boldsymbol{\xi}_{\mathrm{G}} \sim \mathcal{N}(\overline{\boldsymbol{\xi}}_{\mathrm{G}}, \boldsymbol{\Sigma}_{\boldsymbol{\xi}_{\mathrm{G}}})$ a Gaussian distributed uncertain parameter and $\boldsymbol{\xi}_{\mathrm{nG}}$ a non-Gaussian distributed uncertain parameter with $\overline{\boldsymbol{\xi}}_{\mathrm{nG}} = \mathbb{E}[\boldsymbol{\xi}_{\mathrm{nG}}]$. Then, the general, bound constrained single-objective yield optimization problem reads

$$
\begin{aligned}
\max_{\overline{\boldsymbol{\xi}}_{\mathrm{G}}, \overline{\boldsymbol{\xi}}_{\mathrm{nG}}, \mathbf{d}} \quad & Y(\overline{\boldsymbol{\xi}}_{\mathrm{G}}, \overline{\boldsymbol{\xi}}_{\mathrm{nG}}, \mathbf{d}) \\
\text{s.t.} \quad & \overline{\boldsymbol{\xi}}_{\mathbf{G,lb}} \leq \overline{\boldsymbol{\xi}}_{\mathrm{G}} \leq \overline{\boldsymbol{\xi}}_{\mathbf{G,ub}} \\
& \overline{\boldsymbol{\xi}}_{\mathbf{nG,lb}} \leq \overline{\boldsymbol{\xi}}_{\mathrm{nG}} \leq \overline{\boldsymbol{\xi}}_{\mathbf{nG,ub}} \\
& \mathbf{d}_{\mathbf{lb}} \leq \mathbf{d} \leq \mathbf{d}_{\mathbf{ub}}.
\end{aligned}
\tag{5.43}
$$

Following Sects. 5.1.1 and 5.1.2, the derivatives with respect to $\overline{\boldsymbol{\xi}}_{\mathrm{G}}$ are available, the others are not. In (5.43), the yield depends on all parameter types $\overline{\boldsymbol{\xi}}_{\mathrm{G}}, \overline{\boldsymbol{\xi}}_{\mathrm{nG}}$ and $\mathbf{d}$ and all of them are optimization variables. But this is not necessary. The dependence and/or the optimization over types of parameters is allowed to be omitted. Problem (5.41) for example is an unconstrained special case of (5.43), where the yield depends only on $\overline{\boldsymbol{\xi}}_{\mathrm{G}} \equiv \overline{\boldsymbol{\xi}}$ and $\mathbf{d}$, and the only optimization variable is $\overline{\boldsymbol{\xi}}_{\mathrm{G}} \equiv \overline{\boldsymbol{\xi}}$. We formulate two further special cases, which are interesting for the remainder of this work. First, we consider the problem where the yield depends on deterministic and arbitrarily distributed uncertain parameters, but we only optimize over the deterministic parameters. Thus, only the derivatives with respect to $\mathbf{d}$ are of interest. Following Sect. 5.1.2 they are considered as unknown. The problem reads

$$
\begin{aligned}
\max_{\mathbf{d}} \; & Y(\overline{\boldsymbol{\xi}}_{\mathrm{G}}, \overline{\boldsymbol{\xi}}_{\mathrm{nG}}, \mathbf{d}) \\
\text{s.t.} \; & \mathbf{d}_{\mathbf{lb}} \leq \mathbf{d} \leq \mathbf{d}_{\mathbf{ub}}.
\end{aligned}
\tag{5.44}
$$

Second, we formulate a simplified version of (5.43) by leaving out the non-Gaussian distributed uncertain parameter. We obtain

$$
\begin{aligned}
\max_{\overline{\boldsymbol{\xi}}_{\mathrm{G}}, \mathbf{d}} \; & Y(\overline{\boldsymbol{\xi}}_{\mathrm{G}}, \mathbf{d}) \\
\text{s.t.} \; & \overline{\boldsymbol{\xi}}_{\mathbf{G,lb}} \leq \overline{\boldsymbol{\xi}}_{\mathrm{G}} \leq \overline{\boldsymbol{\xi}}_{\mathbf{G,ub}} \\
& \mathbf{d}_{\mathbf{lb}} \leq \mathbf{d} \leq \mathbf{d}_{\mathbf{ub}}.
\end{aligned}
\tag{5.45}
$$

The motivation for this formulation is to distinguish only two types of parameters. One type, for which the derivative information is available ($\boldsymbol{\xi}_{\mathrm{G}}$), and one type for which the derivative information is not available ($\mathbf{d}$). In Sect. 5.4 we propose an optimization strategy which is developed to efficiently handle objective functions for which some partial derivatives are known and others are unknown. These Hermite-type algorithms are general optimization approaches well suited for yield optimization as (5.45), but not limited to this application. In the Hermite-type method, non-Gaussian distributed uncertain parameters would be treated as deterministic parameters in the optimization procedure (certainly not in the estimation process).

5.2.3 *Multi Objective Optimization Problem*

Maximizing the yield means increasing the reliability of a design in a manufacturing process. In practice, however, there are often other performance features that also should be optimized, e.g. power, costs or size. In (3.32) we formulated a general MOO problem. Let f_1 be the yield. We obtain the general multi-objective yield optimization problem formulation

$$\begin{aligned} \max_{\overline{\boldsymbol{\xi}}_{\mathrm{G}},\overline{\boldsymbol{\xi}}_{\mathrm{nG}},\mathbf{d}} \quad & Y(\overline{\boldsymbol{\xi}}_{\mathrm{G}},\overline{\boldsymbol{\xi}}_{\mathrm{nG}},\mathbf{d}) \\ \max_{\overline{\boldsymbol{\xi}}_{\mathrm{G}},\overline{\boldsymbol{\xi}}_{\mathrm{nG}},\mathbf{d}} \quad & f_i(\overline{\boldsymbol{\xi}}_{\mathrm{G}},\overline{\boldsymbol{\xi}}_{\mathrm{nG}},\mathbf{d}) \quad i=2,\dots,k \\ \text{s.t.}\ & \overline{\boldsymbol{\xi}}_{\mathrm{G}} \in \mathbb{X}_{\mathrm{G}} \\ & \overline{\boldsymbol{\xi}}_{\mathrm{nG}} \in \mathbb{X}_{\mathrm{nG}} \\ & \mathbf{d} \in \mathbb{X}_{\mathrm{d}}. \end{aligned} \tag{5.46}$$

As in the previous sections, $\boldsymbol{\xi}_{\mathrm{G}}$ is a Gaussian distributed uncertain parameter for which the derivatives are assumed to be available, $\boldsymbol{\xi}_{\mathrm{nG}}$ and $\mathbf{d}$ are non-Gaussian distributed uncertain parameters and deterministic parameters, respectively, for which the derivatives are considered as unavailable. In contrast to the sections above, we formulate the constraints more generally, i.e., they are not limited to bound constraints. As in Sect. 5.2.2, different types of parameters could be left out which generates special cases analogously to (5.41), (5.44) and (5.45). We refrain from formulating these special cases. When solving (5.46) with a scalarization technique as described in Sect. 3.2.2, the problem is transformed into a SOO problem. If additionally, problem (5.46) is only bound constrained, the problem reduces to (5.43) (or one of the formulated special cases) and can be solved with the same techniques mentioned above. Otherwise, if scalarization methods are applied, but (5.46) is not only bound constrained, the methods LINCOA (for linear constraints) or COBYLA (for general constraints) can be applied, see Sect. 3.2.2. For solving (5.46) directly as MOO problem, genetic algorithms can be applied, see also Sect. 3.2.2.

5.3 Adaptive Newton-MC

In this section we will propose an algorithm that is called adaptive Newton-MC optimization. It is a modification of the Newton algorithm introduced in Sect. 3.2.2 and an efficient method to solve the yield optimization problem (5.41), i.e., yield optimization under the assumption that all first and second order derivatives are available. This section follows our work in [3].

The globalized Newton method illustrated in Fig. 3.2 is formulated for a minimization problem. Since the yield, i.e., the objective function, shall be maximized, we rewrite (5.41) as

$$\min_{\overline{\xi}\in\mathbb{R}^{\xi}} -Y(\overline{\xi}, \mathbf{d}). \tag{5.47}$$

Let us assume that the yield is estimated by a classic MC analysis, cf. Sects. 3.3.1 and 4.2. We recall the error estimator of the MC yield estimator from (4.7)

$$\varepsilon_{\mathrm{MC},Y} = \sqrt{\frac{Y(\overline{\xi}, \mathbf{d})(1 - Y(\overline{\xi}, \mathbf{d}))}{N_{\mathrm{MC}}}} \leq \frac{0.5}{\sqrt{N_{\mathrm{MC}}}}. \tag{5.48}$$

The accuracy of the MC yield estimator depends on the size N_{MC} of the MC sample set. On the other hand, so does the computational effort of the yield estimation. In each iteration of the Newton method, the yield has to be estimated—possibly several times. Thus, an accurate yield estimation in each step of the optimization is computationally challenging. However, the accuracy of the yield estimator at intermediate steps of the optimization routine is not essential to find an optimal solution in the end. It is sufficient that the gradient obtained in each step indicates the right descent direction. Only towards the termination of the algorithm, high accuracy of the yield and derivative estimation may be decisive to accurately determine the optimal solution.

The idea is to reduce accuracy and computational effort in the first Newton iterations by starting with a small size of the MC sample set. During the optimization procedure, the sample size is adaptively increased. A similar idea is also used in the stochastic gradient approach, see [4] for example. They also use approximated or inexact gradients. In contrast to this, our approach uses more sample points than typically used in the stochastic gradient approach. And since the yield gradient results directly from the calculation of the yield estimate, our approach uses the reduced sample size also for the estimation of the objective function, not only for the derivatives.

The proposed adaptive Newton-MC approach ensures that high, predefined, accuracy requirements are fulfilled at the final stages of the algorithm while computational effort can be reduced significantly. A pseudo code is provided in Algorithm 5.1. We begin with a relatively small sample size and run a few *fast* initial Newton iterations. If the yield improvement stagnates, the globalized Newton method illustrated in Fig. 3.2 would stop. Here, instead, we increase the sample size until an improvement of the yield is observed or a target accuracy is reached. Then, a new Newton iteration is started. Only if the target accuracy has been reached and the yield no longer improves, the algorithm terminates. Hence, in addition to the other parameters in the globalized Newton method, we require a definition of a target accuracy, an initial sample size and a rule for how and when to update the sample size.

Algorithm 5.1 Adaptive Newton-MC, based on [3, Algorithm 3]

1: **Input:** Starting point $\overline{\xi}^{(0)} \in \mathbb{R}^{n_\xi}$, target accuracy $\hat{\varepsilon}_{\mathrm{MC},Y}$, initial sample size $N_{\mathrm{MC}}^{\mathrm{start}}$, Newton method parameters $\beta \in (0,1), \gamma \in (0,1), \alpha_1, \alpha_2 > 0, q > 0$

2: **Output:** Optimal solution $\overline{\xi}^{\mathrm{opt}}$

3: **while** $\nabla Y_{\mathrm{MC}}(\overline{\xi}^{(k)}, \mathbf{d}) \neq 0$ **and** $\left\| \overline{\xi}^{(k)} - \overline{\xi}^{(k-1)} \right\| > 0$ **do**

4: Calculate $\mathbf{s}^{(k)}$ by solving Newton's equation $\nabla^2 Y_{\mathrm{MC}}(\overline{\xi}^{(k)}, \mathbf{d})\mathbf{s}^{(k)} = -\nabla Y_{\mathrm{MC}}(\overline{\xi}^{(k)}, \mathbf{d})$.

5: **if** "Calculation of $\mathbf{s}^{(k)}$ possible" **and** $-\nabla Y_{\mathrm{MC}}(\overline{\xi}^{(k)}, \mathbf{d})\mathbf{s}^{(k)} \geq \min\left(\alpha_1, \alpha_2 \|\mathbf{s}^{(k)}\|^q\right) \|\mathbf{s}^{(k)}\|^2$ **then**

6: Accept search direction $\mathbf{s}^{(k)}$.

7: **else**

8: Set search direction $\mathbf{s}^{(k)} = -\nabla Y_{\mathrm{MC}}(\overline{\xi}^{(k)}, \mathbf{d})$.

9: **end if**

10: Determine step size with Armijo rule, i.e., search for largest $\sigma^{(k)} \in \{\beta^0, \beta^1, \beta^2, \beta^3, \dots\}$, such that
$Y_{\mathrm{MC}}(\overline{\xi}^{(k)} + \sigma^{(k)} s^{(k)}, \mathbf{d}) - Y_{\mathrm{MC}}(\overline{\xi}^{(k)}, \mathbf{d}) \leq \sigma^{(k)} \gamma \nabla Y_{\mathrm{MC}}(\overline{\xi}^{(k)}, \mathbf{d})^\top \mathbf{s}^{(k)}$.

11: Set $\overline{\xi}^{(k+1)} = \overline{\xi}^{(k)} + \sigma^{(k)} \mathbf{s}^{(k)}$ and $k = k + 1$.

12: **end while**

13: Calculate MC error estimator $\varepsilon_{\mathrm{MC},Y} = \sqrt{\frac{Y_{\mathrm{MC}}(\overline{\xi}^{(k)}, \mathbf{d})(1 - Y_{\mathrm{MC}}(\overline{\xi}^{(k)}, \mathbf{d}))}{N_{\mathrm{MC}}}}$

14: **if** $\varepsilon_{\mathrm{MC},Y} > \hat{\varepsilon}_{\mathrm{MC},Y}$ **then**

15: **while** $\varepsilon_{\mathrm{MC},Y} > \hat{\varepsilon}_{\mathrm{MC},Y}$ **and** $\left| Y_{\mathrm{MC}}(\overline{\xi}^{(k-1)}, \mathbf{d}) - Y_{\mathrm{MC}}(\overline{\xi}^{(k)}, \mathbf{d}) \right| < \hat{\varepsilon}_{\mathrm{MC},Y}$ **do**

16: Increase sample size $N_{\mathrm{MC}}^{\mathrm{new}} = N_{\mathrm{MC}} + \mathrm{inc}\, N_{\mathrm{MC}}^{\mathrm{start}}$.

17: Calculate $Y_{\mathrm{MC}}(\overline{\xi}^{(k)}, \mathbf{d})$ and $\varepsilon_{\mathrm{MC},Y}$ with $N_{\mathrm{MC}}^{\mathrm{new}}$.

18: Set $N_{\mathrm{MC}} = N_{\mathrm{MC}}^{\mathrm{new}}$.

19: **end while**

20: Go back to line 3.

21: **else**

22: Return $\overline{\xi}^{\mathrm{opt}} = \overline{\xi}^{(k)}$.

23: **end if**

We define the target accuracy as an upper bound for the MC error estimator. According to (5.48) the error estimator $\varepsilon_{\mathrm{MC},Y}$ also depends on the size of the yield. In a non-adaptive strategy, a sample size would be chosen for the complete optimization procedure. As in yield estimation a target accuracy $\hat{\varepsilon}_{\mathrm{MC},Y}$ would be defined and the sample size would be calculated based on the worst case holding for $Y(\overline{\xi}, \mathbf{d}) = 0.5$ (see (4.7)), i.e.,

$$N_{\mathrm{MC}} = \frac{Y(\overline{\xi}, \mathbf{d})(1 - Y(\overline{\xi}, \mathbf{d}))}{\hat{\varepsilon}_{\mathrm{MC},Y}^2} \overset{Y(\overline{\xi}, \mathbf{d}) = 0.5}{\Longrightarrow} N_{\mathrm{MC}} \geq \frac{0.25}{\hat{\varepsilon}_{\mathrm{MC},Y}^2}. \tag{5.49}$$

In the adaptive Newton-MC method, the same target accuracy could be chosen, but it is only required to be fulfilled in the final stages. In the final stages, the yield estimate is typically larger than 0.5 and thus a smaller sample size is sufficient in order to reach the target accuracy. For example, a yield of size $Y(\overline{\xi}, \mathbf{d}) = 0.9$ requires only

$$N_{\mathrm{MC}} \geq \frac{0.09}{\hat{\varepsilon}^2_{\mathrm{MC},Y}} \tag{5.50}$$

sample points to fulfill the accuracy $\hat{\varepsilon}_{\mathrm{MC},Y}$. Let $\overline{\boldsymbol{\xi}}^{(k)}$ denote the solution of the k-th iteration. The sample size is updated, if the yield shows no improvement compared to the last iteration, i.e.,

$$\left| Y_{\mathrm{MC}}\left(\overline{\boldsymbol{\xi}}^{(k-1)}, \mathbf{d}\right) - Y_{\mathrm{MC}}\left(\overline{\boldsymbol{\xi}}^{(k)}, \mathbf{d}\right) \right| < \hat{\varepsilon}_{\mathrm{MC},Y} \tag{5.51}$$

and the target accuracy is not reached yet, i.e.,

$$\hat{\varepsilon}_{\mathrm{MC},Y} < \varepsilon_{\mathrm{MC},Y}\left(Y_{\mathrm{MC}}\left(\overline{\boldsymbol{\xi}}^{(k)}, \mathbf{d}\right)\right) = \sqrt{\frac{Y_{\mathrm{MC}}\left(\overline{\boldsymbol{\xi}}^{(k)}, \mathbf{d}\right)\left(1 - Y_{\mathrm{MC}}\left(\overline{\boldsymbol{\xi}}^{(k)}, \mathbf{d}\right)\right)}{N_{\mathrm{MC}}}}. \tag{5.52}$$

Let $N_{\mathrm{MC}}^{\mathrm{start}}$ denote the initial MC sample size. We define an incremental factor $\mathrm{inc} > 0$. Each time a sample size increase is required, the new sample size is calculated by

$$N_{\mathrm{MC}}^{\mathrm{new}} = N_{\mathrm{MC}}^{\mathrm{old}} + \mathrm{inc}\, N_{\mathrm{MC}}^{\mathrm{start}}. \tag{5.53}$$

Note that it is not necessary to generate and evaluate a completely new MC sample set of size $N_{\mathrm{MC}}^{\mathrm{new}}$ each time the sample size is increased. Only the additional $\mathrm{inc}\, N_{\mathrm{MC}}^{\mathrm{start}}$ sample points have to be evaluated and then they are fused with the previous sample set of size $N_{\mathrm{MC}}^{\mathrm{old}}$. The Newton-MC approach ensures that high, pre-defined, accuracy requirements are fulfilled at the final stages of the algorithm while computational effort can be reduced significantly.

Combination with Hybrid Estimation Approaches

In the beginning of this section we assumed that the yield is estimated with a classic MC analysis on the high fidelity model. However, this is not necessary. The optimization procedure remains unchanged for more involved MC based estimation methods like the SC-Hybrid approach (see Sect. 4.3) or the GPR-Hybrid approach (see Sect. 4.4). The surrogate models in the GPR-Hybrid approach can even be updated on the fly. In Sect. 4.4 we discussed updates after each evaluation of a critical sample point or—to enable parallelization—after a batch of critical sample points. In the adaptive Newton-MC framework another option would be to update after each Newton iteration. This would also allow parallelization, but not necessarily with optimal utilization of all cores. Please note that this optimization framework is different from the Bayesian optimization, which is often discussed in literature when GPR (or kriging) surrogate modeling is combined with optimization, see e.g. [5, 6]. There, typically the function which is approximated by the GP is optimized by choosing new training data points based on an acquisition function. This acquisition function aims to find a trade-off between exploration (improving the quality of the surrogate model) and exploitation (improving the objective function value of the approximated function). A well established acquisition function is for example the expected improvement (EI) [7]. Here instead, we maximize the yield. However, we do not learn the yield function itself, but the underlying QoI. This QoI defines the limit state function dividing all possible designs into accepted and not accepted designs and is assumed to be sufficiently smooth to be well approximated by a GP. By applying the GPR-Hybrid approach in the optimization, the accuracy of the surrogate model and the objective function value (here yield, not QoI) are also improved simultaneously.

The adaptive Newton-MC optimization can also be combined with the SC-Hybrid approach for yield estimation. In the SC-Hybrid approach, the surrogate model cannot be easily updated, a complete recalculation would be required. Hence, the collocation points for the SC surrogate model should be chosen such that a larger area is well approximated, since during the optimization the algorithm will move away from the initial solution. Alternatively, one could employ the regression variant mentioned in Sect. 4.3.

We like to emphasize that, in combination with the hybrid approaches, the adaptive Newton-MC method achieves an a-priori defined accuracy of the result. All error sources are controlled. The MC error is controlled by the adaptive adjustment of the MC sample set size, the surrogate model error (here SC or GPR error) is controlled by the hybrid scheme reevaluating critical sample points on the finite element (FE) model, and the FE error is controlled by the FE model refinement strategy (here only implemented for the SC-Hybrid approach).

5.4 Hermite-Type Modifications of BOBYQA

In this section we discuss optimization in the case that some partial derivatives are known, but others are not. Here, the objective function is assumed to be smooth, i.e., although some partial derivatives are unknown, they are assumed to exist. This is relevant for example for yield optimization with Gaussian distributed uncertain design parameters and deterministic design parameters, see problem (5.45). A common approach is the usage of gradient based solvers with finite differences approximations for the gradients and Broyden-Fletcher-Goldfarb-Shanno (BFGS) updates for the Hessians [8, Sect. 13.2]. In [9], we modified the adaptive Newton-MC method for yield optimization in this way. However, for a large number of unknown derivatives and computationally expensive problems, the additional objective function evaluations for the finite differences quickly become unattractive. In this section we propose two modifications of the originally derivative free BOBYQA method introduced in Sect. 3.2.2. These methods are well suited for mixed gradient information, and, moreover, they are not limited to yield optimization. The first approach will be referred to as Hermite least squares, the second as Hermite BOBYQA. This section follows our work in [10].

While BOBYQA was proposed by Powell [11] and originally implemented in Fortran, today there are several modifications and implementations. In this work we will mainly refer to two more recent versions. First, the Python implementation by Cartis et al., called PyBOBYQA [12]. And second, the algorithm proposed by Conn et al. [13, Algorithm 11.2], which is an (unconstrained) modification of BOBYQA such that global convergence to a stationary point can be proven. One key aspect for this is the fact that the interpolation set is always ensured to be Λ-poised, see [13, Sect. 3.3]. If it is not, the so-called model improvement algorithm [13, Algorithm 6.3] is applied which guarantees transformation of the set into a Λ-poised set within a finite number of steps. The existence of this model improvement algorithm is crucial for the proof of convergence. The implementations by Powell [11] and Cartis et al. [12] do not guarantee convergence. Although both methods update the interpolation set such that the poisedness constant Λ is *hopefully* reduced, they do not provide any bounds [14]. The model improvement algorithm would require checks of the Λ-poisedness more often and a new generation of the complete interpolation data set in situations of poorly balanced data points. However, to enable bound constrained optimization and in the sake of computational efficiency and thus, practical applicability, Powell and Cartis et al. refrain from provable convergence and rely on heuristics for when and how to control and improve the Λ-poisedness of the interpolation set.

Notation

Let $\mathbf{x} \in \mathbb{R}^n$ denote the optimization variable, where x_i, $i = 1, \ldots, n$ can be either deterministic or stochastic. Further, let $f : \mathbb{R}^n \to \mathbb{R}$ denote the objective function. As in BOBYQA we consider a bound constrained optimization problem, we recall from (3.27)

$$\begin{aligned} &\min_{\mathbf{x}\in\mathbb{R}^n} f(\mathbf{x}) \\ &\text{s.t. } \mathbf{x}_{\text{lb}} \le \mathbf{x} \le \mathbf{x}_{\text{ub}}. \end{aligned} \tag{5.54}$$

In BOBYQA the objective function is approximated via quadratic interpolation, in Hermite least squares and Hermite BOBYQA it is approximated via quadratic least squares regression. We formulate the approximation problem generally as an interpolation or least squares regression problem. Therefore, we recall the notation used in BOBYQA from Sect. 3.2.2. Let $\widetilde{m}(\mathbf{x})$ be a polynomial of degree d, $\Phi = \{\phi_0(\mathbf{x}), \ldots, \phi_q(\mathbf{x})\}$ a polynomial basis in $\mathcal{P}_n^d$ and $q_1 = q + 1$ the number of basis polynomials. For the vector of basis polynomials evaluated in $\mathbf{x}$ we write $\mathbf{\Phi}(\mathbf{x}) = (\phi_0(\mathbf{x}), \ldots, \phi_q(\mathbf{x}))^\top$. Although most of the results hold for any choice of a basis, in the following we consider the monomial basis of degree $d = 2$, i.e., if not specifically noted otherwise, for the remainder of this section, Φ is defined by the $(n+1)(n+2)/2$-dimensional basis

$$\Phi = \left\{1, x_1, \ldots, x_n, \frac{1}{2}x_1^2, x_1x_2, x_1x_3, \ldots, x_{n-1}x_n, \frac{1}{2}x_n^2\right\}. \tag{5.55}$$

The interpolation set is given by

$$\mathcal{T} = \{(\mathbf{y}^0, f(\mathbf{y}^0)), \dots, (\mathbf{y}^p, f(\mathbf{y}^p))\} \tag{5.56}$$

and the number of interpolation points is $p_1 = p + 1$. Then, the system matrix and the right hand side of the approximation problem are given by

$$\mathbf{M} \equiv \mathbf{M}(\Phi, \mathcal{T}) = \begin{pmatrix} \phi_0(\mathbf{y}^0) & \dots & \phi_q(\mathbf{y}^0) \\ \vdots & & \vdots \\ \phi_0(\mathbf{y}^p) & \dots & \phi_q(\mathbf{y}^p) \end{pmatrix} \text{ and } \mathbf{b} \equiv \mathbf{b}(\mathcal{T}) = \begin{pmatrix} f(\mathbf{y}^0) \\ \vdots \\ f(\mathbf{y}^p) \end{pmatrix}. \tag{5.57}$$

For $p_1 = q_1$ the system matrix $\mathbf{M}$ is quadratic and

$$\mathbf{M}\mathbf{v} = \mathbf{b} \tag{5.58}$$

yields an interpolation problem which has a unique solution $\mathbf{v} \in \mathbb{R}^{q_1}$ (note here $q_1 = p_1$) if and only if $\mathbf{M}$ is non-singular. According to Definition 3.1, the corresponding interpolation set is then said to be poised. For $p_1 > q_1$ the system matrix $\mathbf{M}$ lies in $\mathbb{R}^{p_1 \times q_1}$ and $\mathbf{v} \in \mathbb{R}^{q_1}$, which yields an overdetermined interpolation problem. It can be solved with least squares regression. We write

$$\mathbf{M}\mathbf{v} \overset{\text{l.s}}{=} \mathbf{b} \quad \Leftrightarrow \quad \min_{\mathbf{v} \in \mathbb{R}^{q_1}} ||\mathbf{M}\mathbf{v} - \mathbf{b}||^2 \quad \Leftrightarrow \quad \mathbf{M}^\top \mathbf{M}\mathbf{v} = \mathbf{M}^\top \mathbf{b}. \tag{5.59}$$

If $\mathbf{M}$ has full column rank, the linear system (5.59) can be uniquely solved. Analogously to the interpolation case and following [13, Chap. 4], the corresponding interpolation set is said to be poised for polynomial least squares regression. When speaking about interpolation sets in the following, we assume them to be poised, if not specifically noted otherwise.

The definitions of Λ-poisedness (Definitions 3.2, 3.3) can be extended to the regression case [13, Definition 4.7].

Definition 5.2 (*Λ-poisedness in the regression sense*) Let a constant $\Lambda > 0$, a set $\mathcal{B} \subset \mathbb{R}^n$, a polynomial basis $\Phi = \{\Phi_0(\mathbf{x}), \dots, \Phi_q(\mathbf{x})\}$ of maximum degree d, a poised set $\mathcal{T}$ as defined in (5.56) and let $p > q$. Then the training data set $\mathcal{T}$ is Λ-poised in $\mathcal{B}$ (in the regression sense) if and only if

$$\forall \mathbf{x} \in \mathcal{B}\ \exists \mathbf{l}(\mathbf{x}) \in \mathbb{R}^{p_1} \text{ s.t. } \sum_{i=0}^{p} l^i(\mathbf{x}) \boldsymbol{\Phi}(\mathbf{y}^i) = \boldsymbol{\Phi}(\mathbf{x}) \text{ with } ||\mathbf{l}(\mathbf{x})||_\infty \le \Lambda. \tag{5.60}$$

Remark 5.1 Since the system (5.60) is underdetermined, the coefficients $l^i(\mathbf{x})$, $i = 0, \dots, p$, are not uniquely defined. However, the minimum norm solution corresponds to the Lagrange polynomials (in the regression sense), cf. [13, Definition 4.4]. Analogously to Remarks 3.1 and 3.2 they can be calculated by solving

$$\mathbf{M}\boldsymbol{\lambda}^i \overset{\text{l.s.}}{=} \mathbf{e}^{i+1}, \tag{5.61}$$

using the elements of the solution vector $\boldsymbol{\lambda}^i$ as coefficients for the polynomial l^i.

Assumption for Hermite-Type Approaches

In contrast to the classic BOBYQA setting, now we assume that the partial derivatives with respect to some directions are known, and unknown for the others. With $\mathbf{x} \in \mathbb{R}^n$, let $\mathcal{I}_\mathcal{D} \subseteq \{1, \ldots, n\}$ denote the index set of known first order derivative directions. Then, the set of available first order derivatives is given by

$$\mathcal{D} := \left\{ \frac{\partial}{\partial x_i} f \right\}_{i \in \mathcal{I}_\mathcal{D}}. \tag{5.62}$$

Further, we define the set of available second order derivatives by

$$\mathcal{D}_2 := \left\{ \frac{\partial^2}{\partial x_i \partial x_j} f \right\}_{(i,j) \in \mathcal{I}_{\mathcal{D}_2}}, \tag{5.63}$$

where the tuple set $\mathcal{I}_{\mathcal{D}_2} \subseteq \{1, \ldots, n\} \times \{1, \ldots, n\}$ is the index set of known second order derivative directions. Higher order derivatives are not of concern, since we build quadratic approximations. For the sake of simplicity, in the following we focus on the practically more relevant case of given first order derivatives. However, in the end of the section we will show that the proposed methods can be straightforwardly adjusted for using second order derivatives. For better readability and without limitation of generality, we assume that the x_i are ordered, such that we can write

$$\mathcal{I}_\mathcal{D} = \{1, \ldots, n_{\text{kd}}\}, \; n_{\text{kd}} \leq n \tag{5.64}$$

for the index set of known first order derivative directions. The resulting interpolation set used for the Hermite-type modifications of BOBYQA is denoted by $\mathcal{T}_\text{H}$ and given by

$$\mathcal{T}_\text{H} = \left\{ \left(\mathbf{y}^0, f(\mathbf{y}^0), \frac{\partial}{\partial y_1} f(\mathbf{y}^0), \ldots, \frac{\partial}{\partial y_{n_\text{kd}}} f(\mathbf{y}^0) \right), \ldots, \left(\mathbf{y}^p, f(\mathbf{y}^p), \frac{\partial}{\partial y_1} f(\mathbf{y}^p), \ldots, \frac{\partial}{\partial y_{n_\text{kd}}} f(\mathbf{y}^p) \right) \right\}. \tag{5.65}$$

The Linear System of BOBYQA

The Hermite-type approaches are modifications of the original BOBYQA method. The aim is to use derivative information which we assume to be available in order to obtain better predictions of the quadratic model $\widetilde{m}(\mathbf{x})$. Before these modifications are discussed in detail, we recall the construction of the linear systems solved in BOBYQA to generate the approximation, cf. [11, 12]. Let $\mathbf{x}^\text{opt} \in \mathcal{T}$ denote the optimal solution of the current quadratic subproblem. Since this point is included into the interpolation set, cf. Sect. 3.2.2, without limitation of generality we assume $\mathbf{x}^\text{opt} = \mathbf{y}^p$.

In case of $p_1 = q_1 = (n+1)(n+2)/2$, the interpolation problem is uniquely solvable and reads

$$\mathbf{M}_{\mathrm{I}}\mathbf{v}_{\mathrm{I}}^{(k)} = \mathbf{b}_{\mathrm{I}} \tag{5.66}$$

with

$$\mathbf{M}_{\mathrm{I}} = \begin{pmatrix} \phi_1(\mathbf{y}^0 - \mathbf{x}^{\mathrm{opt}}) & \dots & \phi_q(\mathbf{y}^0 - \mathbf{x}^{\mathrm{opt}}) \\ \vdots & & \vdots \\ \phi_1(\mathbf{y}^{p-1} - \mathbf{x}^{\mathrm{opt}}) & \dots & \phi_q(\mathbf{y}^{p-1} - \mathbf{x}^{\mathrm{opt}}) \end{pmatrix} \in \mathbb{R}^{p\times q}, \tag{5.67}$$

$$\mathbf{v}_{\mathrm{I}}^{(k)} = \begin{pmatrix} \mathbf{g}^{(k)} \\ \mathbf{H}^{(k)\star} \end{pmatrix} \in \mathbb{R}^q \text{ and } \mathbf{b}_{\mathrm{full}} = \begin{pmatrix} f(\mathbf{y}^0) - f(\mathbf{x}^{\mathrm{opt}}) \\ \vdots \\ f(\mathbf{y}^{p-1}) - f(\mathbf{x}^{\mathrm{opt}}) \end{pmatrix} \in \mathbb{R}^p, \tag{5.68}$$

where $\mathbf{H}^{(k)\star}$ is a vector in $\mathbb{R}^{(n^2+n)/2}$ containing the lower triangular and the diagonal entries of the symmetric matrix $\mathbf{H}^{(k)}$. Further we set $c^{(k)} = f(\mathbf{x}^{\mathrm{opt}})$. Then, the quadratic model (see (3.38)) is formulated by

$$\widetilde{m}^{(k)}(\mathbf{x}) = c^{(k)} + \mathbf{g}^{(k)\top}(\mathbf{x} - \mathbf{x}^{(k)}) + \frac{1}{2}(\mathbf{x} - \mathbf{x}^{(k)})^\top \mathbf{H}^{(k)}(\mathbf{x} - \mathbf{x}^{(k)}). \tag{5.69}$$

Otherwise, if $n+2 \le p_1 < q_1 = (n+1)(n+2)/2$, further information is required, i.e., the system defined in (3.40) is solved, minimizing the change between $\mathbf{H}^{(k-1)}$ and $\mathbf{H}^{(k)}$. The resulting linear system reads

$$\mathbf{M}_{\mathrm{B}}\mathbf{v}_{\mathrm{B}}^{(k)} = \mathbf{b}_{\mathrm{B}}, \tag{5.70}$$

where

$$\mathbf{M}_{\mathrm{B}} = \begin{pmatrix} \mathbf{A} & \mathbf{B}^\top \\ \mathbf{B} & \mathbf{0} \end{pmatrix} \in \mathbb{R}^{p+n\times p+n}, \tag{5.71}$$

with $\mathbf{A} \in \mathbb{R}^{p\times p}$ given by

$$\mathbf{A} = \frac{1}{2}\begin{pmatrix} \left((\mathbf{y}^0 - \mathbf{x}^{\mathrm{opt}})^\top(\mathbf{y}^0 - \mathbf{x}^{\mathrm{opt}})\right)^2 & \dots & \left((\mathbf{y}^0 - \mathbf{x}^{\mathrm{opt}})^\top(\mathbf{y}^{p-1} - \mathbf{x}^{\mathrm{opt}})\right)^2 \\ \vdots & & \vdots \\ \left((\mathbf{y}^{p-1} - \mathbf{x}^{\mathrm{opt}})^\top(\mathbf{y}^0 - \mathbf{x}^{\mathrm{opt}})\right)^2 & \dots & \left((\mathbf{y}^{p-1} - \mathbf{x}^{\mathrm{opt}})^\top(\mathbf{y}^{p-1} - \mathbf{x}^{\mathrm{opt}})\right)^2 \end{pmatrix} \tag{5.72}$$

and

$$\mathbf{B} = \begin{pmatrix} | & & | \\ \mathbf{y}^0 - \mathbf{x}^{\mathrm{opt}} & \dots & \mathbf{y}^{p-1} - \mathbf{x}^{\mathrm{opt}} \\ | & & | \end{pmatrix} \in \mathbb{R}^{n\times p}. \tag{5.73}$$

Further,

$$\mathbf{v}_{\mathrm{B}}^{(k)} = \begin{pmatrix} v_1^{(k)} \\ \vdots \\ v_{p+n}^{(k)} \end{pmatrix} \in \mathbb{R}^{p+n}, \tag{5.74}$$

and

$$\mathbf{b}_{\mathrm{B}} = \begin{pmatrix} f(\mathbf{y}^0) - f(\mathbf{x}^{\mathrm{opt}}) \\ \vdots \\ f(\mathbf{y}^{p-1}) - f(\mathbf{x}^{\mathrm{opt}}) \\ \mathbf{0} \end{pmatrix} - \frac{1}{2} \begin{pmatrix} (\mathbf{y}^0 - \mathbf{x}^{\mathrm{opt}})^\top \mathbf{H}^{(k-1)} (\mathbf{y}^0 - \mathbf{x}^{\mathrm{opt}}) \\ \vdots \\ (\mathbf{y}^{p-1} - \mathbf{x}^{\mathrm{opt}})^\top \mathbf{H}^{(k-1)} (\mathbf{y}^{p-1} - \mathbf{x}^{\mathrm{opt}}) \\ \mathbf{0} \end{pmatrix} \in \mathbb{R}^{p+n}. \tag{5.75}$$

After solving the interpolation problem (5.70), the coefficients of the quadratic model (5.69) are determined by

$$c^{(k)} = f(\mathbf{x}^{\mathrm{opt}}), \quad \mathbf{g}^{(k)^\top} = (v_{p+1}^{(k)}, \ldots, v_{p+n}^{(k)}) \tag{5.76}$$

$$\mathbf{H}^{(k)} = \mathbf{H}^{(k-1)} + \sum_{i=0}^{p-1} v_{i+1}^{(k)} \left((\mathbf{y}^i - \mathbf{x}^{\mathrm{opt}})(\mathbf{y}^i - \mathbf{x}^{\mathrm{opt}})^\top \right). \tag{5.77}$$

5.4.1 Hermite Least Squares Method

The aim of Hermite interpolation is to find a polynomial approximation of a function by solving a linear system containing information on function values and partial derivative values in given training data points, cf. [15, Sect. 6.6] or [16]. We propose an approximation method using such a linear system, but with more information than necessary for a uniquely solvable interpolation problem and solve it with least squares regression. Hence, we call the optimization method based on this technique Hermite least squares.

In the Hermite least squares approach we extend the simple interpolation system (5.66) from BOBYQA with derivative information yielding an overdetermined system which will be solved by least squares regression. We introduce this method in two steps. First, we assume that the system (5.66) is uniquely solvable, i.e., the number of interpolation points p_1 coincides with the number of basis polynomials q_1. Second, we allow that the system (5.66) is underdetermined, i.e., $p_1 < q_1$. In the original BOBYQA this would not be solvable and would transform into (5.70), but by adding derivative information, still an overdetermined system is obtained. The number of rows of this new system is given by $||\mathcal{T}_{\mathrm{H}} = p_1(1 + n_{\mathrm{kd}})$, where p_1 is the number of interpolation points and $||\mathcal{T}_{\mathrm{H}}$ denotes the *number of information*, i.e., the number of objective function values and derivative values. For $p_1 = q_1$ global convergence to a stationary point can be proven under the same assumptions (e.g.

smoothness of the objective function, no bound constraints) as in Conn's version [13, Algorithm 11.2]. However, for $p_1 < q_1$ the method yields superior performance in practice, see Sect. 6.5.

Regression Based on Interpolation ($p_1 = q_1$)

Let us consider a training data set $\mathcal{T}_{\mathrm{I}}$ with $|\mathcal{T}_{\mathrm{I}}| = p_1 \equiv q_1$. Instead of determining the unique solution of (5.66), additionally we provide derivative information for the first n_{kd} partial derivatives of each training data point. We obtain the interpolation set $\mathcal{T}_{\mathrm{H}}$ from (5.65) with $||\mathcal{T}_{\mathrm{H}} = p_1(1 + n_{\mathrm{kd}})$. The gradient information is included in the linear system (5.66) by adding lines to the system matrix and the right hand side. We obtain

$$\mathbf{M}_{\mathrm{H}} = \begin{pmatrix} \phi_1(\mathbf{y}^0 - \mathbf{x}^{\mathrm{opt}}) & \dots & \phi_q(\mathbf{y}^0 - \mathbf{x}^{\mathrm{opt}}) \\ \vdots & & \vdots \\ \phi_1(\mathbf{y}^{p-1} - \mathbf{x}^{\mathrm{opt}}) & \dots & \phi_q(\mathbf{y}^{p-1} - \mathbf{x}^{\mathrm{opt}}) \\ \frac{\partial}{\partial y_1}\phi_1(\mathbf{y}^0 - \mathbf{x}^{\mathrm{opt}}) & \dots & \frac{\partial}{\partial y_1}\phi_q(\mathbf{y}^0 - \mathbf{x}^{\mathrm{opt}}) \\ \vdots & & \vdots \\ \frac{\partial}{\partial y_{n_{\mathrm{kd}}}}\phi_1(\mathbf{y}^0 - \mathbf{x}^{\mathrm{opt}}) & \dots & \frac{\partial}{\partial y_{n_{\mathrm{kd}}}}\phi_q(\mathbf{y}^0 - \mathbf{x}^{\mathrm{opt}}) \\ \frac{\partial}{\partial y_1}\phi_1(\mathbf{y}^1 - \mathbf{x}^{\mathrm{opt}}) & \dots & \frac{\partial}{\partial y_1}\phi_q(\mathbf{y}^1 - \mathbf{x}^{\mathrm{opt}}) \\ \vdots & & \vdots \\ \frac{\partial}{\partial y_{n_{\mathrm{kd}}}}\phi_1(\mathbf{y}^p - \mathbf{x}^{\mathrm{opt}}) & \dots & \frac{\partial}{\partial y_{n_{\mathrm{kd}}}}\phi_q(\mathbf{y}^p - \mathbf{x}^{\mathrm{opt}}) \end{pmatrix} \in \mathbb{R}^{p_1(1+n_{\mathrm{kd}})-1\times q} \tag{5.78}$$

and

$$\mathbf{b}_{\mathrm{H}} = \begin{pmatrix} f(\mathbf{y}^0) - f(\mathbf{x}^{\mathrm{opt}}) \\ \vdots \\ f(\mathbf{y}^{p-1}) - f(\mathbf{x}^{\mathrm{opt}}) \\ \frac{\partial}{\partial y_1} f(\mathbf{y}^0) \\ \vdots \\ \frac{\partial}{\partial y_{n_{\mathrm{kd}}}} f(\mathbf{y}^0) \\ \frac{\partial}{\partial y_1} f(\mathbf{y}^1) \\ \vdots \\ \frac{\partial}{\partial y_{n_{\mathrm{kd}}}} f(\mathbf{y}^p) \end{pmatrix} \in \mathbb{R}^{p_1(1+n_{\mathrm{kd}})-1}. \tag{5.79}$$

The resulting system

$$\mathbf{M}_{\mathrm{H}}\mathbf{v}_{\mathrm{H}}^{(k)} \overset{\text{l.s.}}{=} \mathbf{b}_{\mathrm{H}} \tag{5.80}$$

is solved with least squares regression in order to build the quadratic model for the trust region subproblem ($\mathbf{v}_{\mathrm{H}}^{(k)}$ is defined as $\mathbf{v}_{\mathrm{I}}^{(k)}$ in (5.4)). The formulation of the system using second order derivatives is provided at the end of this section. We will discuss how well this model approximates the original function. Therefore, we state the following theorem, which is a generalization of Theorem 4.1 in [17].

Theorem 5.2 *Let $\mathcal{T}_{\mathrm{I}}$ be a poised interpolation set as defined in (5.56), Φ the monomial basis with $|\mathcal{T}_{\mathrm{I}} = |\Phi$ and $\mathcal{B} \subset \mathbb{R}^n$. Further, let $\mathbf{M}_{\mathrm{I}}$ be the corresponding system matrix of the interpolation problem and $\mathbf{b}_{\mathrm{I}}$ the right hand side, respectively. Let $\mathcal{T}_{\mathrm{R}} \supset \mathcal{T}_{\mathrm{I}}$ be a training set containing further information, such that the corresponding system matrix $\mathbf{M}_{\mathrm{R}}$ has still full column rank. If $\mathcal{T}_{\mathrm{I}}$ is Λ-poised in $\mathcal{B}$ in the interpolation sense, then $\mathcal{T}_{\mathrm{R}}$ is at least Λ-poised in $\mathcal{B}$ in the regression sense.*

Proof The additional information can be added to the system matrix $\mathbf{M}_{\mathrm{I}}$ and the right hand side $\mathbf{b}_{\mathrm{I}}$ in form of additional rows. For the regression problem corresponding to $\mathcal{T}_{\mathrm{R}}$ we obtain

$$\mathbf{M}_{\mathrm{R}} = \begin{pmatrix} \mathbf{M}_{\mathrm{I}} \\ \mathbf{M}_{\mathrm{add}} \end{pmatrix} \text{and} \mathbf{b}_{\mathrm{R}} = \begin{pmatrix} \mathbf{b}_{\mathrm{I}} \\ \mathbf{b}_{\mathrm{add}} \end{pmatrix}. \tag{5.81}$$

The interpolation set $\mathcal{T}_{\mathrm{I}}$ is Λ-poised in the interpolation sense. Thus, by Definition 3.2 holds

$$\forall \mathbf{x} \in \mathcal{B}\ \exists \mathbf{l}_{\mathrm{I}}(\mathbf{x}) \in \mathbb{R}^{|\mathcal{T}_{\mathrm{I}}} \text{ s.t. } \sum_{\mathbf{y}^i \in \mathcal{T}_{\mathrm{I}}} l_{\mathrm{I}}^i(\mathbf{x}) \mathbf{m}_{\mathrm{I}}^i = \mathbf{\Phi}(\mathbf{x}) \quad \text{with} \quad \|\mathbf{l}_{\mathrm{I}}(\mathbf{x})\|_\infty \leq \Lambda, \tag{5.82}$$

where $\mathbf{m}_{\mathrm{I}}^i$ denotes the i-th column of the matrix $\mathbf{M}_{\mathrm{I}}^\top$. Let $\mathbf{l}_{\mathrm{R}}(\mathbf{x}) = (\mathbf{l}_{\mathrm{I}}(\mathbf{x}), \mathbf{0})^\top \in \mathbb{R}^{|\mathcal{T}_{\mathrm{R}}}$, then

$$\sum_{\mathbf{y}^i \in \mathcal{T}_{\mathrm{R}}} l_{\mathrm{R}}^i(\mathbf{x}) \mathbf{m}_{\mathrm{R}}^i = \mathbf{\Phi}(\mathbf{x}) \tag{5.83}$$

holds and $\|\mathbf{l}_{\mathrm{R}}(\mathbf{x})\|_\infty$ is bounded by Λ, since

$$\|\mathbf{l}_{\mathrm{R}}(\mathbf{x})\|_\infty = \max_{i=0,\ldots,|\mathcal{T}_{\mathrm{R}}} |l_{\mathrm{R}}^i(\mathbf{x}) \overset{\text{Definition } \mathbf{l}_{\mathrm{R}}}{=} \max_{i=0,\ldots,|\mathcal{T}_{\mathrm{I}}} |l_{\mathrm{I}}^i(\mathbf{x}) = \|\mathbf{l}_{\mathrm{I}}(\mathbf{x})\|_\infty \overset{(3.44)}{\leq} \Lambda. \tag{5.84}$$

In the beginning of Sect. 5.4 we stated that the model improvement algorithm is a necessary requirement for ensuring global convergence of Conn's version of the BOBYQA method. This algorithm is always able to generate a Λ-poised interpolation set in a finite number of steps. For regression it is not clear if such an algorithm exists [13, Chap. 6]. However, Theorem 5.2 shows that it is enough to ensure that a subset of $\mathcal{T}_{\mathrm{R}}$ is Λ-poised in the interpolation sense, and we can deduce that $\mathcal{T}_{\mathrm{R}}$ is Λ-poised in the regression sense. Hence, although there is no model improvement algorithm for regression, the model improvement algorithm for interpolation [13, Algorithm 6.3] can be applied to a subset of $\mathcal{T}_{\mathrm{R}}$ with $|\Phi = (n+1)(n+2)/2$ points. Then, the optimization procedure [13, Algorithm 11.2] can be applied, but instead of using interpolation, the quadratic model is built with $\mathcal{T}_{\mathrm{R}}$ and least squares regression. Further, we note that in the proof of Theorem 5.2 we set $l_{\mathrm{R}}^i(\mathbf{x}) = 0$ for $i > |\mathcal{T}_{\mathrm{I}}$. This implies that as long as the matrix $\mathbf{M}_{\mathrm{R}}$ has full column rank, the type of additional information in $\mathcal{T}_{\mathrm{R}}$ has no impact on the proof. Hence, instead of additional points and their function values, it is also possible to add derivative information for existing points and the set $\mathcal{T}_{\mathrm{R}}$ remains at least Λ-poised. Also the proof of convergence for the complete optimization procedure from [13] remains unaffected. However, in practice we expect that our modification converges faster due to better local quadratic approximations.

Regression Based on Underdetermined Interpolation ($p_1 < q_1$)

The number of required data points for building a uniquely solvable quadratic interpolation model is $p_1 = q_1 = (n+1)(n+2)/2$. Extending the system with additional gradient information allows the reduction of the number of data points and while still leading to a determined or overdetermined interpolation problem. The number of rows of the system matrix (5.78) in the Hermite least squares approach is $(1 + n_{\text{kd}})p_1 - 1$ and has to be greater than or equal to the number of columns of the system matrix, which is $q = |\Phi| - 1 = (n+1)(n+2)/2 - 1$. Thus, for the Hermite least squares approach only

$$p_1 \geq \left\lceil \frac{(n+1)(n+2)}{2(1+n_{\text{kd}})} \right\rceil. \tag{5.85}$$

interpolation points are required. The Hermite least squares system (5.78)–(5.80) can be built as before, the only difference is that $p_1 < q_1$. This implies that the data set $\mathcal{T}_{\text{H}}$ no longer contains a subset of Λ-poised (in the interpolation sense) data points. Thus, the model improvement algorithm cannot be applied to a subset of $\mathcal{T}_{\text{H}}$ and convergence cannot be proven.

Λ-Poisedness for Hermite Least Squares

We discuss how the derivative information can be considered for building and maintaining the interpolation set. We begin with the Hermite interpolation setting, i.e., let

$$p_1 = \frac{(n+1)(n+2)}{2(1+n_{\text{kd}})}, \tag{5.86}$$

s.t. $||\mathcal{T}_{\text{H}} = q_1$ and the Hermite interpolation problem (5.80) is uniquely solvable. We extend the definitions of Λ-poisedness (Definitions 3.2, 3.3, 5.2) to the Hermite interpolation case. Note that the following definition does not ensure that the error bounds required for provable convergence are provided (cf. [13, Sect. 6.1]). This leads to an algorithm without formal convergence proof, analogously to the original BOBYQA implementations [11, 12].

Definition 5.3 (*Λ-poisedness in the Hermite interpolation sense*) Let a constant $\Lambda > 0$, a set $\mathcal{B} \subset \mathbb{R}^n$, the monomial basis $\Phi = \{\Phi_0(\mathbf{x}), \ldots, \Phi_q(\mathbf{x})\}$ of maximum degree d, $\mathbf{\Phi}(\mathbf{x}) = \left(\Phi_0(\mathbf{x}), \ldots, \Phi_q(\mathbf{x})\right)^{\top}$ and a poised Hermite interpolation set $\mathcal{T}_{\text{H}}$ as defined in (5.65). Let $||\mathcal{T}_{\text{H}} = p_1(1+n_{\text{kd}}) = q_1 = ||\Phi$. Then the training data set $\mathcal{T}_{\text{H}}$ is Λ-poised in $\mathcal{B}$ (in the Hermite interpolation sense) if and only if

$$\forall \mathbf{x} \in \mathcal{B}\ \exists \mathbf{l}(\mathbf{x}) \in \mathbb{R}^{q_1} \text{ s.t. } \mathbf{M}_{\text{H}}^{\top}\mathbf{l}(\mathbf{x}) = \mathbf{\Phi}(\mathbf{x}) \quad \text{with} \quad ||\mathbf{l}(\mathbf{x})||_\infty \leq \Lambda. \tag{5.87}$$

Analogously to the simple interpolation and regression cases we define Lagrange-type polynomials for Hermite interpolation and show that they solve (5.87).

Definition 5.4 (*Lagrange-type polynomials for Hermite interpolation*) Let $\mathbf{M}_{\mathrm{H}} \in \mathbb{R}^{q_1 \times q_1}$ be the system matrix of a Hermite interpolation problem with respect to the basis Φ as defined in (5.78) and $\mathbf{e}^i \in \mathbb{R}^{q_1}$ the i-th unit vector. Let $\boldsymbol{\lambda}^i \in \mathbb{R}^{q_1}$ solve

$$\mathbf{M}_{\mathrm{H}} \boldsymbol{\lambda}^i = \mathbf{e}^{i+1}. \tag{5.88}$$

For $i = 0, \ldots, q$, let

$$t^i(\mathbf{x}) = \lambda_0^i \phi_0(\mathbf{x}) + \cdots + \lambda_q^i \phi_q(\mathbf{x}) \tag{5.89}$$

be the polynomial built with the basis Φ and the entries of $\boldsymbol{\lambda}^i$ as coefficients. Then, $t^i(\mathbf{x})$ is the i-th Lagrange-type polynomial for Hermite interpolation.

Lemma 5.1 *Let* $\mathbf{M}_{\mathrm{H}} \in \mathbb{R}^{q_1 \times q_1}$ *be the system matrix of a Hermite interpolation problem with respect to the basis* Φ *as defined in* (5.78) *and* $\mathbf{t}(\mathbf{x}) = \left(t^0(\mathbf{x}), \ldots, t^q(\mathbf{x})\right)^\top$ *be defined as in* (5.89)*. Then,* $\mathbf{t}(\mathbf{x})$ *solves* $\mathbf{M}_{\mathrm{H}}^\top \mathbf{l}(\mathbf{x}) = \boldsymbol{\Phi}(\mathbf{x})$ *(cf.* (5.87)*), i.e.,* $\mathbf{t}(\mathbf{x}) \equiv \mathbf{l}(\mathbf{x})$.

Proof Let $\mathbf{I}$ denote the $q_1 \times q_1$ identity matrix and define $\mathbf{T}$ by the solution vectors of (5.88), i.e.,

$$\mathbf{T} = \begin{pmatrix} | & & | \\ \boldsymbol{\lambda}^0 & \ldots & \boldsymbol{\lambda}^q \\ | & & | \end{pmatrix}. \tag{5.90}$$

Then we rewrite (5.88) into

$$\mathbf{M}_{\mathrm{H}} \mathbf{T} = \mathbf{I} \;\; \Leftrightarrow \;\; \mathbf{T} \mathbf{M}_{\mathrm{H}} = \mathbf{I}. \tag{5.91}$$

Left multiplication by $\boldsymbol{\Phi}(\mathbf{x})^\top$ yields

$$\boldsymbol{\Phi}(\mathbf{x})^\top \mathbf{T} \mathbf{M}_{\mathrm{H}} = \boldsymbol{\Phi}(\mathbf{x})^\top.$$

We apply (5.90)

$$\left(\phi_0(\mathbf{x}), \ldots, \phi_q(\mathbf{x})\right) \begin{pmatrix} | & & | \\ \boldsymbol{\lambda}^0 & \ldots & \boldsymbol{\lambda}^q \\ | & & | \end{pmatrix} \mathbf{M}_{\mathrm{H}} = \left(\phi_0(\mathbf{x}), \ldots, \phi_q(\mathbf{x})\right)$$

and (5.89)

$$\left(t^0(\mathbf{x}), \ldots, t^q(\mathbf{x})\right) \mathbf{M}_{\mathrm{H}} = \left(\phi_0(\mathbf{x}), \ldots, \phi_q(\mathbf{x})\right).$$

By transposing we obtain

$$\mathbf{M}_{\mathrm{H}}^\top \begin{pmatrix} t^0(\mathbf{x}) \\ \vdots \\ t^q(\mathbf{x}) \end{pmatrix} = \begin{pmatrix} \phi_0(\mathbf{x}) \\ \vdots \\ \phi_q(\mathbf{x}) \end{pmatrix},$$

which is per definition equivalent to

$$\mathbf{M}_{\mathrm{H}}^{\top}\mathbf{t}(\mathbf{x}) = \mathbf{\Phi}(\mathbf{x}).$$

Hence, $\mathbf{t}$ solves (5.87), i.e., $\mathbf{t}(\mathbf{x}) \equiv \mathbf{l}(\mathbf{x})$.

For the uniquely defined polynomial $\widetilde{m}_{\mathrm{H}}(\mathbf{x})$ solving the Hermite interpolation problem $\mathbf{M}_{\mathrm{H}}\mathbf{v} = \mathbf{b}_{\mathrm{H}}$ with $\mathbf{M}_{\mathrm{H}} \in \mathbb{R}^{q_1 \times q_1}$ and $\mathbf{b}_{\mathrm{H}} \in \mathbb{R}^{q_1}$ holds

$$\widetilde{m}_{\mathrm{H}}(\mathbf{x}) = \sum_{i=0}^{p} f(\mathbf{y}^i)t^i(\mathbf{x}) + \sum_{i=0}^{p}\sum_{j=1}^{n_{\mathrm{kd}}} \frac{\partial f}{\partial x_j}(\mathbf{y}^i)t^{jp_1-1+i}(\mathbf{x}), \tag{5.92}$$

where $p_1 = p + 1$ is the number of interpolation points and $|\Phi| = q_1 = (1 + n_{\mathrm{kd}})p_1$ holds.

We extend the concept above for Hermite interpolation to Hermite least squares, i.e., to the case $|\mathcal{T}_{\mathrm{H}}| > q_1$. In order to obtain the Lagrange-type polynomials for Hermite least squares, we solve

$$\mathbf{M}_{\mathrm{H}}\boldsymbol{\lambda}^i \overset{\text{l.s.}}{=} \mathbf{e}^{i+1}, \tag{5.93}$$

instead of (5.88). The interpolation set is updated based on Λ-poisedness in the Hermite least squares sense. This means that we choose the leaving interpolation point by maximizing (3.48) over the first p_1 Lagrange polynomials, where the Lagrange polynomials are obtained from (5.93). This ensures that the derivative information is considered when maintaining the interpolation set. In practice we observe that often the chosen leaving point would be the same without considering derivative information. However, there are several iterations where another leaving point is chosen when derivative information is considered. Once a leaving point is chosen, we replace this point with all its corresponding information (function value and derivative information) by the new data point.

Second Order Derivatives

In the beginning of Sect. 5.4 we mentioned that the usage of second order derivatives in the Hermite least squares method is straightforward. For the sake of completeness we formulate the resulting linear system analogously to (5.78)–(5.80). Let $\mathbf{M}_{\mathcal{D}_2,i}$, $i = 0, \ldots, p$, denote the matrix with second order derivative information for the i-th interpolation point $\mathbf{y}^i$, and $\mathbf{b}_{\mathcal{D}_2,i}$ the corresponding right hand side. Further, let $n_{2\mathrm{kd}} \leq n$ denote the number of known second order derivative directions. Then, $\mathbf{M}_{\mathcal{D}_2,i}$ and $\mathbf{b}_{\mathcal{D}_2,i}$ are given by

$$\mathbf{M}_{\mathcal{D}_2,i} = \begin{pmatrix} \frac{\partial^2}{\partial x_1 \partial x_1}\phi_1(\mathbf{y}^i - \mathbf{x}^{\text{opt}}) & \dots & \frac{\partial^2}{\partial x_1 \partial x_1}\phi_q(\mathbf{y}^i - \mathbf{x}^{\text{opt}}) \\ \frac{\partial^2}{\partial x_1 \partial x_2}\phi_1(\mathbf{y}^i - \mathbf{x}^{\text{opt}}) & \dots & \frac{\partial^2}{\partial x_1 \partial x_2}\phi_q(\mathbf{y}^i - \mathbf{x}^{\text{opt}}) \\ \vdots & & \vdots \\ \frac{\partial^2}{\partial x_1 \partial x_{n_{2\text{kd}}}}\phi_1(\mathbf{y}^i - \mathbf{x}^{\text{opt}}) & \dots & \frac{\partial^2}{\partial x_1 \partial x_{n_{2\text{kd}}}}\phi_q(\mathbf{y}^i - \mathbf{x}^{\text{opt}}) \\ \frac{\partial^2}{\partial x_2 \partial x_2}\phi_1(\mathbf{y}^i - \mathbf{x}^{\text{opt}}) & \dots & \frac{\partial^2}{\partial x_2 \partial x_2}\phi_q(\mathbf{y}^i - \mathbf{x}^{\text{opt}}) \\ \vdots & & \vdots \\ \frac{\partial^2}{\partial x_{n_{2\text{kd}}} \partial x_{n_{2\text{kd}}}}\phi_1(\mathbf{y}^i - \mathbf{x}^{\text{opt}}) & \dots & \frac{\partial^2}{\partial x_{n_{2\text{kd}}} \partial x_{n_{2\text{kd}}}}\phi_q(\mathbf{y}^i - \mathbf{x}^{\text{opt}}) \end{pmatrix} \in \mathbb{R}^{(n_{2\text{kd}}^2 + n_{2\text{kd}})/2 \times q} \tag{5.94}$$

and

$$\mathbf{b}_{\mathcal{D}_2,i} = \begin{pmatrix} \frac{\partial^2}{\partial x_1 \partial x_1} f(\mathbf{y}^i) \\ \frac{\partial^2}{\partial x_1 \partial x_2} f(\mathbf{y}^i) \\ \vdots \\ \frac{\partial^2}{\partial x_1 \partial x_{n_{2\text{kd}}}} f(\mathbf{y}^i) \\ \frac{\partial^2}{\partial x_2 \partial x_2} f(\mathbf{y}^i) \\ \vdots \\ \frac{\partial^2}{\partial x_{n_{2\text{kd}}} \partial x_{n_{2\text{kd}}}} f(\mathbf{y}^i) \end{pmatrix} \in \mathbb{R}^{(n_{2\text{kd}}^2 + n_{2\text{kd}})/2}. \tag{5.95}$$

Then, $\mathbf{M}_{\mathcal{D}_2,i}$ and $\mathbf{b}_{\mathcal{D}_2,i}$ can be added as additional lines to the system (5.80) and solved by least squares regression.

We assume that the second order derivatives are available for all directions, i.e., the index set of known second order derivative directions is given by $\mathcal{I}_{\mathcal{D}_2} = \{1, \dots, n\} \times \{1, \dots, n\}$. Further, utilizing that Φ is the monomial basis as defined in (5.55), then, the matrix $\mathbf{M}_{\mathcal{D}_2,i}$ can be simplified to

$$\mathbf{M}_{\mathcal{D}_2,i}^{\text{simpl.}} = \begin{pmatrix} 0 & \dots & 0 & 1 & & \\ \vdots & & \vdots & & \ddots & \\ 0 & \dots & 0 & & & 1 \end{pmatrix}, \tag{5.96}$$

i.e., the linear part vanishes and the quadratic part reduces to the identity matrix.

5.4.2 *Hermite BOBYQA Method*

A characterizing feature of the original BOBYQA method is, that it can handle underdetermined interpolation settings. Existence and uniqueness of the solution are guaranteed by minimizing the change between the current and the last Hessian matrix of the quadratic model (in the Frobenius norm), cf. Sect. 3.2.2. Thereby, the number of required interpolation points is reduced to $n + 2 \leq ||\mathcal{T}_{\text{B}} \equiv p_1 < (n+1)(n+2)/2$.

The Hermite BOBYQA approach extends this underdetermined interpolation system (5.70) with derivative information and applies least squares regression. The corresponding system matrix (5.71) and right hand side (5.75) need modification.

Taking the coefficients (5.76)–(5.77) obtained by solving the original underdetermined BOBYQA system (5.70), we obtain the quadratic BOBYQA model by

$$\widetilde{m}_{\mathrm{B}}^{(k)}(\mathbf{x}) = c_{\mathrm{B}}^{(k)} + \mathbf{g}_{\mathrm{B}}^{(k)\top}(\mathbf{x}-\mathbf{x}^{(k)}) + \frac{1}{2}(\mathbf{x}-\mathbf{x}^{(k)})^\top \mathbf{H}_{\mathrm{B}}^{(k)}(\mathbf{x}-\mathbf{x}^{(k)}), \tag{5.97}$$

where we write the subscript B to emphasize the correspondence to the original BOBYQA system. The gradient of $\widetilde{m}_{\mathrm{B}}^{(k)}(\mathbf{x})$ reads

$$\nabla_{\mathbf{x}}\widetilde{m}_{\mathrm{B}}^{(k)}(\mathbf{x}) = \mathbf{g}_{\mathrm{B}}^{(k)} + \mathbf{H}_{\mathrm{B}}^{(k)}(\mathbf{x}-\mathbf{x}^{(k)}). \tag{5.98}$$

We replace the Hessian $\mathbf{H}_{\mathrm{B}}^{(k)}$ by (5.77) and obtain

$$\nabla_{\mathbf{x}}\widetilde{m}_{\mathrm{B}}^{(k)}(\mathbf{x}) = \mathbf{g}_{\mathrm{B}}^{(k)} + \left(\mathbf{H}^{(k-1)} + \sum_{i=0}^{p-1} v_{\mathrm{B},i+1}^{(k)}\left((\mathbf{y}^i-\mathbf{x}^{\mathrm{opt}})(\mathbf{y}^i-\mathbf{x}^{\mathrm{opt}})^\top\right)\right)(\mathbf{x}-\mathbf{x}^{(k)}). \tag{5.99}$$

First, we assume that all first order derivatives are available, i.e., $n_{\mathrm{kd}} = n$. Additionally to the interpolation conditions (3.39) we have the conditions

$$\nabla\widetilde{m}^{(k)}(\mathbf{y}^j) = \nabla f(\mathbf{y}^j) \;\; \forall \mathbf{y}^j \in \mathcal{T}_{\mathrm{H}}. \tag{5.100}$$

We insert (5.99) into (5.100). For each $\mathbf{y}^j \in \mathcal{T}_{\mathrm{H}}$ we obtain

$$\sum_{i=0}^{p-1} v_{\mathrm{B},i+1}^{(k)} \underbrace{\left((\mathbf{y}^i-\mathbf{x}^{\mathrm{opt}})(\mathbf{y}^i-\mathbf{x}^{\mathrm{opt}})^\top\right)}_{=:\mathbf{C}^i}(\mathbf{y}^j-\mathbf{x}^{(k)}) + \mathbf{g}_{\mathrm{B}}^{(k)} = \nabla f(\mathbf{y}^j) - \mathbf{H}^{(k-1)}(\mathbf{y}^j-\mathbf{x}^{(k)}) \tag{5.101}$$

as condition. From $\mathbf{C}^i \in \mathbb{R}^{n\times n}$ follows that $\mathbf{C}^i(\mathbf{y}^j-\mathbf{x}^{(k)}) \in \mathbb{R}^{n\times 1}$ and thus on the left and right hand side of (5.101) we have vectors in $\mathbb{R}^n$. We rewrite (5.101) as matrix, which yields the linear system

$$\underbrace{\begin{pmatrix} | & & | & 1 & & 0 \\ \mathbf{C}^0(\mathbf{y}^j-\mathbf{x}^{(k)}) & \ldots & \mathbf{C}^{p-1}(\mathbf{y}^j-\mathbf{x}^{(k)}) & & \ddots & \\ | & & | & 0 & & 1 \end{pmatrix}}_{=:\mathbf{M}_+^j} \mathbf{v}^{(k)} = \underbrace{\begin{pmatrix} | \\ \nabla f(\mathbf{y}^j) - \mathbf{H}^{(k-1)}(\mathbf{y}^j-\mathbf{x}^{(k)}) \\ | \end{pmatrix}}_{=:\mathbf{b}_+^j} \tag{5.102}$$

for each $\mathbf{y}^j \in \mathcal{T}_{\mathrm{H}}$. This can be appended to the original BOBYQA system, to obtain the Hermite BOBYQA (HB) system matrix and right hand side

$$\mathbf{M}_{\text{HB}} = \begin{pmatrix} \mathbf{M}_{\text{B}} \\ \mathbf{M}_{+}^{0} \\ \vdots \\ \mathbf{M}_{+}^{p} \end{pmatrix} \quad \text{and} \quad \mathbf{b}_{\text{HB}} = \begin{pmatrix} \mathbf{b}_{\text{B}} \\ \mathbf{b}_{+}^{0} \\ \vdots \\ \mathbf{b}_{+}^{p} \end{pmatrix}. \tag{5.103}$$

Now we consider the general case that only some partial derivatives are available, i.e., $n_{\text{kd}} < n$. Then, only the first n_{kd} rows of each $\mathbf{M}_{+}^{j}$ and $\mathbf{b}_{+}^{j}$ are included. Since $\mathbf{b}_{\text{B}} \in \mathbb{R}^{p+n}$ and each $\mathbf{b}_{+}^{j} \in \mathbb{R}^{n_{\text{kd}}}$, we obtain $\mathbf{b}_{\text{HB}} \in \mathbb{R}^{p+n+p_1 n_{\text{kd}}}$, and $\mathbf{M}_{\text{HB}} \in \mathbb{R}^{p+n+p_1 n_{\text{kd}} \times p+n+p_1 n_{\text{kd}}}$ respectively. The overdetermined Hermite BOBYQA system is formulated as

$$\mathbf{M}_{\text{HB}} \mathbf{v}_{\text{HB}}^{(k)} \overset{\text{l.s.}}{=} \mathbf{b}_{\text{HB}}, \tag{5.104}$$

After solving (5.104) with least squares regression, the coefficients of the quadratic Hermite BOBYQA model are calculated analogously to the original BOBYQA method by (5.76)–(5.77).

Λ-Poisedness for Hermite BOBYQA

Again we discuss how the interpolation set is updated, with the aim of improving the Λ-poisedness. In the original BOBYQA method, the new and leaving data points are chosen as described in (3.48) and (3.49), based on the Lagrange polynomials in the minimum-norm sense and the corresponding Definition 3.3 for Λ-poisedness in the minimum-norm sense. Like in the Hermite least squares approach, we could apply the BOBYQA updating procedure also in the Hermite BOBYQA method, arguing that for a subset of the Hermite BOBYQA data set Λ-poisedness (in the minimum-norm sense) can be ensured. However, again the derivative information would not be considered. To do this, we solve

$$\mathbf{M}_{\text{HB}} \boldsymbol{\lambda}^{i} \overset{\text{l.s.}}{=} \mathbf{e}^{i+1} \tag{5.105}$$

and build polynomials using the entries of $\boldsymbol{\lambda}^{i}$ as coefficients, according to (5.76)–(5.77). Then we use these polynomials instead of the Lagrange polynomials when determining leaving and new data points with (3.48)–(3.49).

Second Order Derivatives

As the Hermite least squares method, the Hermite BOBYQA method can be extended to handle second order derivatives. Let $\mathbf{y}^{i} \in \mathcal{T}_{\text{H}}$ be the i-th interpolation point and

$$\mathbf{C}^{i} = \begin{pmatrix} c_{11}^{i} & \dots & c_{1n}^{i} \\ \vdots & & \vdots \\ c_{n1}^{i} & \dots & c_{nn}^{i} \end{pmatrix} \tag{5.106}$$

the corresponding matrix as defined in (5.101). The Hessian of the last iteration $\mathbf{H}^{(k-1)}$ can be written as

$$\mathbf{H}^{(k-1)} = \begin{pmatrix} h_{11}^{(k-1)} & \dots & h_{1n}^{(k-1)} \\ \vdots & & \vdots \\ h_{n1}^{(k-1)} & \dots & h_{nn}^{(k-1)} \end{pmatrix}. \tag{5.107}$$

Again, we denote the matrix with second order derivative information by $\mathbf{M}_{\mathcal{D}_2,i}$, $i = 0, \dots, p$, and the right hand side by $\mathbf{b}_{\mathcal{D}_2,i}$, respectively. Assuming that the second order derivatives are available for all directions, we obtain

$$\mathbf{M}_{\mathcal{D}_2,i} = \begin{pmatrix} c_{11}^i & \dots & c_{11}^i & 0 & \dots & 0 \\ \vdots & & \vdots & \vdots & & \vdots \\ c_{1n}^i & \dots & c_{1n}^i & 0 & \dots & 0 \\ c_{21}^i & \dots & c_{21}^i & 0 & \dots & 0 \\ \vdots & & \vdots & \vdots & & \vdots \\ c_{2n}^i & \dots & c_{2n}^i & 0 & \dots & 0 \\ c_{31}^i & \dots & c_{31}^i & 0 & \dots & 0 \\ \vdots & & \vdots & \vdots & & \vdots \\ c_{nn}^i & \dots & c_{nn}^i & 0 & \dots & 0 \end{pmatrix} \in \mathbb{R}^{((n^2+n)/2)\times(p+n)} \tag{5.108}$$

and

$$\mathbf{b}_{\mathcal{D}_2,i} = \begin{pmatrix} \frac{\partial^2}{\partial x_1 \partial x_1} f(\mathbf{y}^i) - h_{11}^{(k-1)} \\ \vdots \\ \frac{\partial^2}{\partial x_1 \partial x_n} f(\mathbf{y}^i) - h_{1n}^{(k-1)} \\ \frac{\partial^2}{\partial x_2 \partial x_2} f(\mathbf{y}^i) - h_{22}^{(k-1)} \\ \vdots \\ \frac{\partial^2}{\partial x_2 \partial x_n} f(\mathbf{y}^i) - h_{2n}^{(k-1)} \\ \frac{\partial^2}{\partial x_3 \partial x_3} f(\mathbf{y}^i) - h_{33}^{(k-1)} \\ \vdots \\ \frac{\partial^2}{\partial x_n \partial x_n} f(\mathbf{y}^i) - h_{nn}^{(k-1)} \end{pmatrix} \in \mathbb{R}^{(n^2+n)/2}. \tag{5.109}$$

In case that not all derivative directions are available, the corresponding rows in $\mathbf{M}_{\mathcal{D}_2,i}$ and $\mathbf{b}_{\mathcal{D}_2,i}$ must be removed.

5.4.3 Preconditioning for Hermite-Type Methods

In this section we discuss possible preconditioning techniques for solving the linear equations system.

Scaling

In PyBOBYQA, the Python implementation of BOBYQA [12], the system is scaled before it is solved. I.e., instead of solving

$$\mathbf{Mv} = \mathbf{b} \tag{5.110}$$

the system

$$\mathbf{LMRR}^{-1}\mathbf{v} = \mathbf{Lb} \tag{5.111}$$

is solved, where the matrices **L** and **R** are diagonal and of same size as **M**. We recall that Δ is the trust region radius. Note, for simplicity of notation we omit the index k for the current iteration. Each interpolation point $\mathbf{y}^i$, $i = 0, \ldots, p$, is scaled with the factor $1/\Delta$. In case of original BOBYQA with $p_1 = q_1$, this yields a multiplication of the columns of **M** representing the linear part with $1/\Delta$, and a multiplication of the columns representing the quadratic part with $1/\Delta^2$. Thus, **L** and **R** are given by

$$\mathbf{L} = \mathbf{I} \text{ and } \mathbf{R} = \operatorname{diag}\Bigg(\underbrace{\frac{1}{\Delta}, \ldots, \frac{1}{\Delta}}_{p}, \underbrace{\frac{1}{\Delta^2}, \ldots, \frac{1}{\Delta^2}}_{p-n}\Bigg). \tag{5.112}$$

Preserving the same scaling scheme in the Hermite least squares approach would leave the right scaling matrix **R** unchanged while for the left scaling matrix **L** we obtain

$$\mathbf{L} = \operatorname{diag}(\underbrace{1, \ldots, 1}_{p}, \underbrace{\Delta, \ldots, \Delta}_{p_1 n_{\mathrm{kd}}}). \tag{5.113}$$

In the original underdetermined BOBYQA setting ($p_1 < q_1$), the scaling of $\mathbf{y}^i$ with $1/\Delta$ yields the scaling matrices

$$\mathbf{L} = \mathbf{R} = \operatorname{diag}\Bigg(\underbrace{\frac{1}{\Delta^2}, \ldots, \frac{1}{\Delta^2}}_{p}, \underbrace{\Delta, \ldots, \Delta}_{n}\Bigg). \tag{5.114}$$

For Hermite BOBYQA follows

$$\mathbf{L} = \operatorname{diag}\Bigg(\underbrace{\frac{1}{\Delta^2}, \ldots, \frac{1}{\Delta^2}}_{p}, \underbrace{\Delta, \ldots, \Delta}_{n}, \underbrace{\frac{1}{\Delta}, \ldots, \frac{1}{\Delta}}_{p_1 n_{\mathrm{kd}}}\Bigg) \tag{5.115}$$

The right scaling matrix remains as defined in (5.114).

Weighting

An optional preconditioning step is a weighting of the information in the regression problem, cf. weighted regression [18]. This can be realized by left multiplication of a diagonal matrix **D** (of the same size as **M**), i.e.,

$$\mathbf{DMv} \overset{\text{l.s}}{=} \mathbf{Db}. \tag{5.116}$$

The idea is to give more weight to the interpolation points which are close to the current iterate solution, and less weight to those interpolation points that are far from the current iterate solution. In the Hermite least squares approach, this can be applied to function and derivative information, since each information is represented by one row. In the Hermite BOBYQA method, the weighting shall only be applied to the derivative information, since the first $p + n$ rows do not belong to one specific interpolation point. Thus, the first $p + n$ diagonal entries of **D** are set to 1.

5.5 Summary

In this chapter we formulated SOO and MOO problems for yield optimization. We calculated the first and second order derivatives of the yield function, with respect to uncertain Gaussian distributed parameters, uncertain non-Gaussian distributed parameters and deterministic parameters. We observed that the derivatives with respect to uncertain Gaussian distributed parameters can be obtained without any additional computing effort, while this is not the case for the other parameter types. This motivated the development of the Hermite-type modifications of the BOBYQA method. These methods are well suited for optimization problems with mixed gradient information and are not limited to the application of yield optimization. Under reasonable assumptions, global convergence was proven for the unconstrained case. For a yield optimized over uncertain Gaussian distributed parameters, we proposed the adaptive Newton-MC method, which achieves high efficiency by adaptively increasing the MC sample size. In combination with the hybrid approaches for yield estimation, all errors (MC, SC/GPR and FE) can be fully controlled. The adaptive MC sample size increase can also be transferred to the Hermite-type approaches, when they are used for yield optimization. In the next chapter we will investigate how the proposed methods perform in practice.

References

1. Gräb, H.E (2007) Analog design centering and sizing. Springer (2007)
2. Lacaze S, Brevault L, Missoum S, Balesdent M (2015) Probability of failure sensitivity with respect to decision variables. Struct Multidiscip Optim 52(2):375–381

3. Fuhrländer M, Georg N, Römer U, Schŭps S (2020) Yield optimization based on adaptive Newton-Monte Carlo and polynomial surrogates. Int J Uncert Quantific 10(4):351–373. https://doi.org/10.1615/Int.J.UncertaintyQuantification.2020033344
4. Geiersbach C, Wollner W (2020) A stochastic gradient method with mesh refinement for PDE constrained optimization under uncertainty. SIAM J Sci Comput 42(5):A2750–A2772. https://doi.org/10.1137/19M1263297
5. Feliot P, Bect J, Vazquez E (2017) A bayesian approach to constrained single- and multi-objective optimization. J Glob Optim 67(1):97–133
6. Wang M, Lv W, Yang F, Yan C, Cai W, Zhou D, Zeng X (2017) Efficient yield optimization for analog and SRAM circuits via Gaussian process regression and adaptive yield estimation. IEEE Trans Comput-Aided Design Integrat Circuits Syst 37(10):1929–1942
7. Mockus J, Tiesis V, Zilinskas A (1978) The application of Bayesian methods for seeking the extremum. Towards Glob Optim 2:117–129
8. Ulbrich M, Ulbrich S (2012) Nichtlineare optimierung. Birkhäuser (2012)
9. Fuhrländer, M, Schŏps, S (2022) Efficient yield optimization with limited gradient information. In: Progress in industrial mathematics at ECMI 2021, vol 39, pp 203–209. Springer (2022)
10. Fuhrländer M, Schöps S (2023) Hermite least squares optimization: a modification of BOBYQA for optimization with limited derivative information. Optim Eng. https://doi.org/10.1007/s11081-023-09795-y
11. Powell MJ (2009) The BOBYQA algorithm for bound constrained optimization without derivatives, vol 26. Cambridge NA Report NA2009/06. University of Cambridge, Cambridge, pp 26–46
12. Cartis, C, Fiala, J, Marteau, B, Roberts, L (2019) Improving the flexibility and robustness of model-based derivative-free optimization solvers. ACM Trans Math Softw 45(3). https://doi.org/10.1145/3338517
13. Conn AR, Scheinberg, K, Vicente LN (2009) Introduction to derivative-free optimization. SIAM (2009)
14. Conn AR, Scheinberg K, Vicente LN (2008) Geometry of interpolation sets in derivative free optimization. Math Program 111(1):141–172
15. Hermann, M (2011) Numerische Mathematik. Oldenbourg Wissenschaftsverlag (2011)
16. Sauer T, Xu Y (1995) On multivariate hermite interpolation. Adv Comput Math 4(1):207–259
17. Conn AR, Scheinberg K, Vicente LN (2008) Geometry of sample sets in derivative-free optimization: polynomial regression and underdetermined interpolation. IMA J Numer Anal 28(4):721–748
18. Björck Å (1996) Numerical methods for least squares problems. SIAM (1996)

Chapter 6
Numerical Applications and Results

In the previous two chapters methods for yield estimation and optimization were proposed. In this chapter these methods are numerically evaluated. In Chap. 2 we introduced two benchmark problems, a rectangular waveguide and a PMSM. In Sect. 6.1 we will specify these general applications to concrete model problems, which will be used for the numerical testing afterwards. The chapter is similarly structured as the theory chapters. We start with the evaluation of the yield estimation methods, first for the waveguide, then for the PMSM problem. As reference solution we consider a classic MC analysis. We continue with yield optimization. First, the adaptive Newton-MC method in combination with different yield estimation approaches is tested on the waveguide problem. Then, MOO problems for a simultaneous optimization of the reliability (yield) and performance (size / costs) of the waveguide and the PMSM are formulated and solved. Finally, the Hermite-type optimization methods are evaluated, first on a set of 30 nonlinear test problems, then on the waveguide yield problem.

6.1 Models

This section is dedicated to the detailed specification of the benchmark problems. We start with the waveguide problem and continue with the PMSM problem.

6.1.1 Waveguide

In Sect. 2.4, we formulated the waveguide problem for a rectangular waveguide. In this section we introduce a particular model of a rectangular waveguide with dielectric inlay, which is used to numerically test and evaluate the methods proposed in the previous chapters. This model has been introduced by [1] and used as a benchmark

M. Fuhrländer, *Design Methods for Reducing Failure Probabilities with Examples from Electrical Engineering*, Springer Theses,
https://doi.org/10.1007/978-3-031-37019-9_6

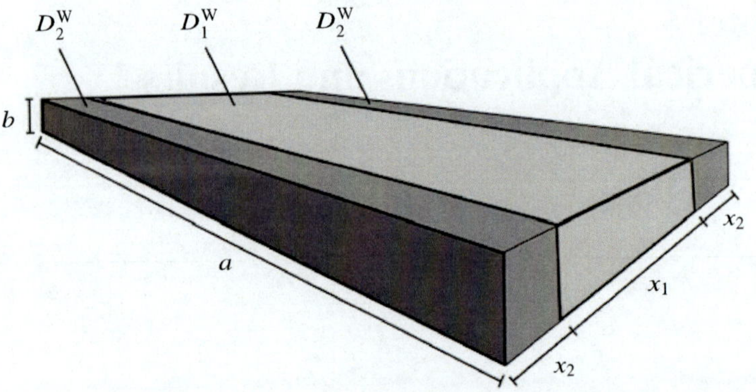

Fig. 6.1 Model of a rectangular waveguide with dielectric inlay. The waveguide ports are given in dark gray, the inlay in light gray and the offset in gray. Figure based on [5, Fig. 5]

problem e.g. in [2–4]. It is well suited for benchmarking since a closed form solution exists. The waveguide is illustrated in Fig. 6.1.

It consists of two ports $\partial D_{\mathrm{P1}}^{\mathrm{W}}$ and $\partial D_{\mathrm{P2}}^{\mathrm{W}}$, a dielectric inlay, and two vacuum offsets. The boundary conditions and the numerical model are as stated in Sect. 2.4. The computational domain for the inlay is denoted by D_1^{W}, the computational domain of the offsets is denoted by D_2^{W}. While for D_1^{W} different material properties will be assumed and later optimized, D_2^{W} is assumed to have vacuum permittivity ϵ_0 and permeability μ_0. The width a and the height b of the waveguide are assumed to be fixed. The geometry parameters indicating the length of the inlay and the length of the offset are changeable and are subject to optimization.

Depending on the test case, we consider four or twelve design parameters. Let

$$\mathbf{x}^{\{12\}} = (x_1, \ldots, x_{12})^\top \tag{6.1}$$

denote the design parameter vector in case of twelve and

$$\mathbf{x}^{\{4\}} = (x_1, x_2, x_{13}, x_{14})^\top \tag{6.2}$$

in case of four design parameters, respectively. The parameters x_1 and x_2 are geometry parameters (in mm), i.e., the length of the dielectric inlay and the length of the vacuum offset, see Fig. 6.1. The other parameters are material parameters with an effect on the relative permittivity $\epsilon_{\mathrm{r}}|_{D_1^{\mathrm{W}}}$ and permeability $\mu_{\mathrm{r}}|_{D_1^{\mathrm{W}}}$ on the dielectric inlay. In case of twelve design parameters the effects are represented by

$$\epsilon_{\mathrm{r}}^{\{12\}}|_{D_1^{\mathrm{W}}} = x_5 + (x_3 - x_5)\left(1 + \mathrm{i}\omega x_6 \tau\right)^{-1} + (x_4 - x_5)\left(1 + \mathrm{i}\omega x_7 \tau\right)^{-1}, \tag{6.3}$$

$$\mu_{\mathrm{r}}^{\{12\}}|_{D_1^{\mathrm{W}}} = x_{10} + (x_8 - x_{10})\left(1 + \mathrm{i}\omega x_{11} \tau\right)^{-1} + (x_9 - x_{10})\left(1 + \mathrm{i}\omega x_{12} \tau\right)^{-1}, \tag{6.4}$$

where

$$\tau = \frac{1}{2\pi\ (20\,\text{GHz})}. \tag{6.5}$$

In case of four design parameters, the relation is given by

$$\epsilon_{\mathrm{r}}^{\{4\}}|_{D_1^{\mathrm{w}}} = 1 + x_{13} + (1 - x_{13})\left(1 + \mathrm{i}\omega\left(2\pi 5 \cdot 10^9\right)^{-1}\right)^{-1}, \tag{6.6}$$

$$\mu_{\mathrm{r}}^{\{4\}}|_{D_1^{\mathrm{w}}} = 1 + x_{14} + (2 - x_{14})\left(1 + \mathrm{i}\omega\left(1.1 \cdot 2\pi 20 \cdot 10^9\right)^{-1}\right)^{-1}. \tag{6.7}$$

Depending on the considered test case, some of the design parameters are assumed to be uncertain, while others are assumed to be deterministic. This will be clarified in the respective section. The uncertain parameters are modeled as independent random variables. In order to represent the manufacturing uncertainties and avoid unphysical parameter values, they are assumed to follow a truncated Gaussian distribution, cf. (3.6). In Sect. 6.3.1 we discuss the consequences for gradient based optimization when assuming a truncated Gaussian distribution instead of a Gaussian distribution. For uncertain geometry parameters we write ξ_i instead of x_i and set

$$\xi_i \sim \mathcal{N}_{\mathcal{T}}(\overline{\xi}_i, 0.7^2, \overline{\xi}_i - 3, \overline{\xi}_i + 3), \quad i \in \{1, 2\}, \tag{6.8}$$

for uncertain material parameters we also write ξ_i instead of x_i and assume that

$$\xi_i \sim \mathcal{N}_{\mathcal{T}}(\overline{\xi}_i, 0.3^2, \overline{\xi}_i - 0.3, \overline{\xi}_i + 0.3), \quad i \in \{3, \ldots, 14\}. \tag{6.9}$$

If a design parameter x_i is considered as deterministic, we write d_i, $i \in \{1, \ldots, 14\}$.

As stated in Sect. 2.4, the S-parameter is considered as QoI. Please note, that the S-parameter depends on the deterministic design parameter $\mathbf{d}$, the uncertain design parameter $\boldsymbol{\xi}$ and the angular frequency ω. The performance feature specifications (PFS) are given by

$$Q_\omega(\boldsymbol{\xi}, \mathbf{d}) := |S_\omega(\boldsymbol{\xi}, \mathbf{d})| \overset{!}{\leq} S_{\text{pfs}} = -24\,\text{dB} \;\; \forall \omega \in T_{\mathrm{r}} = [2\pi 6.5, 2\pi 7.5] \text{ in GHz}. \tag{6.10}$$

This implies that the angular frequency ω is considered as a range parameter. In the computations we consider eleven equidistant frequency points $\omega^{(j)} \in T_{\mathrm{d}} \subset T_{\mathrm{r}}$ in the frequency range. The PFS have to be fulfilled in these points, i.e.,

$$|S_{\omega^{(j)}}(\boldsymbol{\xi}, \mathbf{d})| \overset{!}{\leq} -24\,\text{dB} \;\; \forall \omega^{(j)} \in T_{\mathrm{d}}. \tag{6.11}$$

In the proposed surrogate based approaches, for each $\omega^{(j)}$, and for the real and the imaginary part, separate surrogate models are built. To overcome this, approximations along the frequency could be built, e.g. using model order reduction [6], [7, Chap. 5.2] or a specific GPR kernel [4, Chap. 3].

We provide some details on the implementation in the programming language Python. For the calculation of the S-parameter (2.51) with FEM, we employ the FE library FEniCS [8]. Since complex numbers are not supported in FEniCS 2017.2.0, the real and the imaginary parts of the matrices are assembled separately and combined afterwards. The resulting linear system of equations is solved with a sparse LU decomposition [9], using scipy [10].

6.1.2 PMSM

The (multi-objective) design optimization under uncertainty has a long tradition in the field of electrical machines [11]. Common approaches are the Taguchi methods [12], worst case optimization [13] and six sigma design optimization [14]. In [15] a recent study on these methods is provided. We follow this tradition and also investigate the performance of the proposed yield estimation and optimization methods on the example of an electrical machine. In Sect. 2.5 we introduced the PMSM problem. In this section we will specify that problem and discuss the modeling and simulation details. The PMSM we will investigate is based on the benchmark problem introduced in [16] and used for example in [17]. The model is illustrated in Fig. 6.2. It has six PMs, i.e., $N_{\mathrm{pp}} = 3$ pole pairs, three phases and two slots per pole and phase, i.e., a total of 36 slots. The three phases are represented with different colors of the windings in Fig. 6.2. The radius of the entire machine, i.e., the outer stator radius, is $R_{\mathrm{s,o}} = 67.5\,\mathrm{mm}$, the length of the machine is $l_z = 100\,\mathrm{mm}$. The materials used are copper for the coils, laminated steel for the machine and rare-earth material for the PMs, e.g. NdFeB-magnets. Since rare-earth materials are expensive and their extraction pollutes the environment [18], it is of high interest to reduce the size of the PMs. A detailed description of the material and geometry parameters of the investigated PMSM is provided in Appendix A.

However, the parameters which are subject to optimization and the uncertain parameters will be discussed in the following. In Fig. 6.3a the deterministic geometry parameter

$$\mathbf{d} = (d_1, d_2, d_3, d_4)^{\top} \tag{6.12}$$

is illustrated. The length of d_1, d_2 and d_3 is given in mm, the angle of d_4 in degree. Later we will consider **d** as the optimization variable. The magnitude of the magnetic field induced by the i-th PMs B_i (in Tesla (T)) and the magnetization direction of the PMs ϕ_i (in degree) are considered uncertain, see Fig. 6.3b. Since these uncertainties are assumed to be caused by manufacturing imperfections they may differ for each PMs within the PMSM. Hence, for six PMs we obtain a total of twelve uncertain parameters. Following [17] and the *Laplace principle of unsufficient reasons* [20], they are modeled as uniformly distributed random variables with mean value

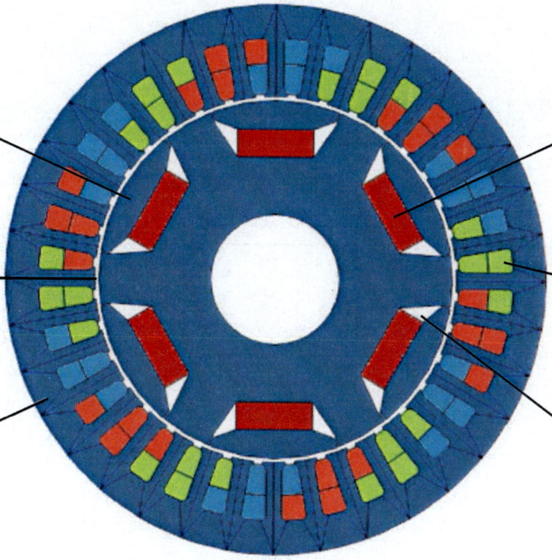

Fig. 6.2 PMSM model with six PMs and 36 slots. Figure taken from [19, Fig. 1], © 2022 IEEE. Reprinted, with permission, from [19]

$$\overline{\boldsymbol{\xi}} = \left(\overline{\xi}_1, \ldots, \overline{\xi}_{12}\right)^\top = \left(\overline{B}_1, \ldots, \overline{B}_6, \overline{\phi}_1, \ldots, \overline{\phi}_6\right)^\top = (0.94, \ldots, 0.94, 0, \ldots, 0)^\top \tag{6.13}$$

and a distribution range of ± 0.05 T and $\pm 3\,°$, respectively. We write

$$\begin{aligned} \xi_k &\sim \mathcal{U}(\overline{\xi}_k - 0.05, \overline{\xi}_k + 0.05) && \text{for } k = 1, \ldots, 6 \\ \xi_k &\sim \mathcal{U}(\overline{\xi}_k - 3, \overline{\xi}_k + 3) && \text{for } k = 7, \ldots, 12. \end{aligned} \tag{6.14}$$

The uncertain parameters are material properties which are not subject to optimization.

In this work, the average torque τ_{avg} is considered as QoI. In Sect. 2.5 the torque has been introduced as function of the (discretized) magnetic vector potential (MVP), which depends on the geometry and material properties of the PMs. For emphasizing this relation we can write

$$\tau_{\text{avg}} \equiv \tau_{\text{avg}}(\mathbf{a}(\boldsymbol{\xi}, \mathbf{d})). \tag{6.15}$$

We introduce a lower bound on the average torque for preventing possible machine failures. We obtain the PFS

$$Q(\boldsymbol{\xi}, \mathbf{d}) := \tau_{\text{avg}}(\mathbf{a}(\boldsymbol{\xi}, \mathbf{d})) \overset{!}{\geq} \tau_{\text{pfs}} = 10.8\,\text{Nm}. \tag{6.16}$$

We proceed with some details on the modeling and simulation of this model in CST Studio Suite® 2021 (CST). For cylindrical machines, having a length larger than or similar to the size of their diameter, sufficient accuracy can be achieved in

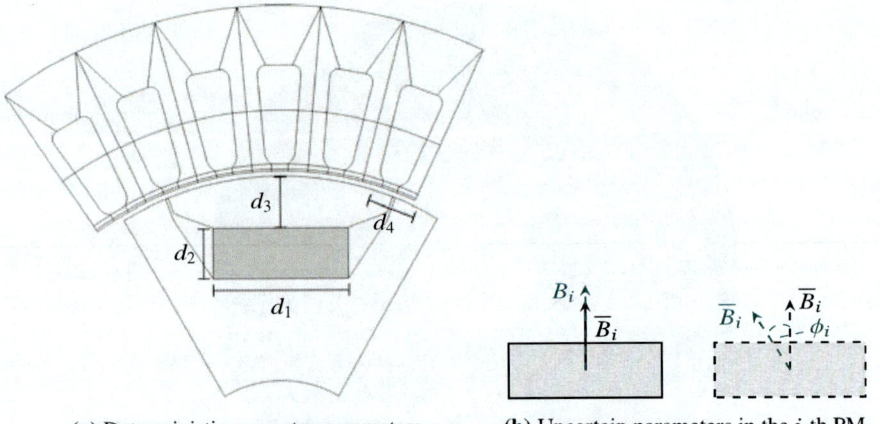

(a) Deterministic geometry parameters. **(b)** Uncertain parameters in the i-th PM.

Fig. 6.3 Illustration of deterministic geometry parameters and uncertain material properties in the PMSM. Figures taken from [19, Figs. 4, 5], © 2022 IEEE. Reprinted, with permission, from [19]

the computations by relying on 2D models [17]. For the investigated PMSM this is the case. We assume $\mathbf{B} = (B_x(x, y), B_y(x, y), 0)$, and thus $\mathbf{J}_s = (0, 0, J_z(x, y))$ and $\mathbf{A} = (0, 0, A_z(x, y))$. With this assumptions, the strong formulation (2.62) can be transformed into

$$-\nabla \cdot (\nu \nabla A_z(x, y)) = J_z(x, y) + (\nabla \times \mathbf{M}_{\mathrm{pm}})_z. \tag{6.17}$$

The homogeneous Dirichlet boundary conditions we assumed in (2.63) become

$$\mathbf{n} \times A_z(x, y) = 0 \ \text{ on } \partial D^{\mathrm{M}} = \partial D_{\mathrm{i}}^{\mathrm{M}} \cup \partial D_{\mathrm{o}}^{\mathrm{M}}. \tag{6.18}$$

Instead of modeling the Dirichlet boundary conditions (6.18) directly, we approximate them by modeling air with $\mu = \epsilon = 1$ in the origin and outside the machine. The magnetic flux density in the air outside the machine is approximately 0.5 mT compared to 0.94 T in the PMs. Hence, this approximation is justified. Within the CST FE solver, nodal basis functions of second order are applied.

In the simulation, as τ_{avg} we refer to the average torque value over one electrical period

$$t_{\mathrm{ep}} = \frac{1}{s_{\mathrm{m}} \cdot N_{\mathrm{pp}}} = \frac{60}{1930\,\mathrm{rpm} \cdot 3} \approx 0.01036\,\mathrm{s}, \tag{6.19}$$

where s_{m} denotes the speed of the machine in revolutions per minute (rpm). Let N_{per} denote the number of time steps used to resolve the electrical period and $\tau_k \equiv \tau(\mathbf{a}_k(\boldsymbol{\xi}, \mathbf{d}))$ the torque value in the k-th time step, then the average torque is given by

$$\tau_{\text{avg}} = \frac{1}{N_{\text{per}}} \sum_{k=1}^{N_{\text{per}}} \tau_k. \tag{6.20}$$

In the following computations we set $N_{\text{per}} = 60$ since it provides sufficient accuracy in the order of 10^{-4} within a reasonable simulation time of ≈ 85 s.

6.2 Yield Estimation

In this section we will evaluate the performance of the proposed yield estimation approaches, i.e., SC-Hybrid and GPR-Hybrid, on the two benchmark problems. We follow our work in [21, 22]. As the reference solution a classic MC analysis is considered. In the evaluation we are mainly interested in two aspects, the accuracy of the yield estimate and the efficiency of the method. For measuring the accuracy we consider the relative error compared to the reference solution given by

$$\varepsilon_{\text{rel}} = \frac{\left|Y_{\text{ref}} - \widetilde{Y}\right|}{Y_{\text{ref}}}, \tag{6.21}$$

where Y_{ref} denotes the value of the reference solution and $\widetilde{Y}$ the value of the considered yield estimator. For measuring the efficiency or the computational costs, we consider the number of high fidelity function evaluations in order to obtain the estimate, i.e.,

$$\#\text{HF} \equiv \#\text{HF}_h = \#\text{ high fidelity evaluations}. \tag{6.22}$$

In our case, the term high fidelity evaluation typically refers to the FEM evaluation of the problem with mesh size h. However, in case of the SC-Hybrid approach with model refinement as proposed in Sect. 4.3, different levels of high fidelity evaluations are distinguished. In this work, the refinement is achieved by refining the mesh size h by division by two. The resulting difference in the computing effort depends on the computational model and the solver. In the following, we assume an optimal solver which is linear in the number of degrees of freedom (DoFs). This implies that each time the mesh size is halved, the computational effort increases by a factor of four in case of a 2D model and by eight in case of a 3D model. The electric field of the considered waveguide problem is constant in y-direction, thus, the refinement is only applied to x- and z-directions. Hence, the factors for a 2D model apply. The measure for the computational effort under consideration of three different refinement levels is then given by

$$\#\text{HF}_{\text{L}} = \#\text{HF}_h + 4\,\#\text{HF}_{h/2} + 16\,\#\text{HF}_{h/4}. \tag{6.23}$$

This measure is only relevant for the SC-Hybrid approach. On the other hand, in the GPR-Hybrid approach parallelization is enabled, cf. Sect. 4.4. The batch size N_{B} indicates the number of high fidelity evaluations performed possibly in parallel.

Assuming that sufficient parallel processors are available to realize this parallelization, the efficient costs are given by

$$\#\mathrm{HF}_{\mathrm{eff}} = \left\lceil \frac{\#\mathrm{HF}}{N_{\mathrm{B}}} \right\rceil . \tag{6.24}$$

In the hybrid approaches we further distinguish between offline and online high fidelity evaluations. Offline evaluations are the evaluations of the training data in order to build the initial surrogate model. Online evaluations are the reevaluations of critical sample points on the high fidelity model. Note that in the GPR-Hybrid approach online evaluations are used for surrogate model updates.

In both hybrid methods a safety factor $s \geq 1$ for determining critical sample points has to be chosen, cf. (4.13) and (4.16). In the following, this parameter is set $s = 2$, which is a rather conservative choice. This might lead to more sample points classified as critical, i.e., higher computational costs, but also to higher accuracy by avoiding misclassification of sample points, cf. Sect. 4.3.

6.2.1 Waveguide

In the following computations all design parameters of the waveguide introduced in Sect. 6.1.1 are considered as uncertain parameters, i.e., we write ξ_i instead of x_i and set the mean values to

$$\overline{\boldsymbol{\xi}}^{\{4\}} = (10.36, 4.76, 0.58, 0.64)^{\top} \tag{6.25}$$

and

$$\overline{\boldsymbol{\xi}}^{\{12\}} = (8.6, 3.8, 2, 0.5, 0.7, 0.6, 1.4, 2.8, 1.7, 0.8, 0.3, 1.4)^{\top}, \tag{6.26}$$

respectively. The size of the MC sample set N_{MC} is chosen based on the MC error estimator (4.7). In the following we allow an MC error of $\hat{\varepsilon}_{\mathrm{MC},Y} = 0.01$ which yields a sample size of $N_{\mathrm{MC}} = 2500$. For the waveguide problem with four uncertain parameters, in Fig. 6.4 the results of the MC yield estimators for different sample sizes are given. The black line indicates the value of the yield estimator for $N_{\mathrm{MC}} = 10000$, which is the most accurate solution calculated. The gray shaded area indicates the 0.01-sigma level for each sample size. We observe that for sample sizes below 1500 the sigma interval is rather large, then it decreases with increasing sample sizes, first rapidly, then slowly.

The reference solutions for the yield estimation, i.e., the yield values obtained with classic MC analysis with $N_{\mathrm{MC}} = 2500$ sample points on the closed form solution of the waveguide, are given by

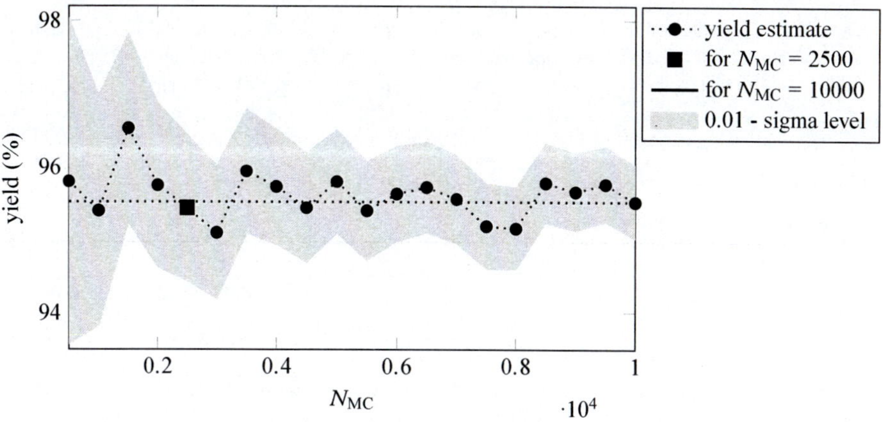

Fig. 6.4 Comparison of MC yield estimates with different sample sizes N_{MC} for the waveguide problem with four uncertain parameters. Figure taken from [22, Fig. 2]

$$Y_{\mathrm{ref}}^{\{4\}} = 95.44 \tag{6.27}$$

and

$$Y_{\mathrm{ref}}^{\{12\}} = 74.60\,\%. \tag{6.28}$$

SC-Hybrid Approach

We present the results for the SC-Hybrid approach from Sect. 4.3 applied to the waveguide problem. We follow our work in [21]. The results will be compared to classic MC and to the SC approach without hybridization.

For a fair evaluation (same final accuracy) of the computational effort of the SC-Hybrid approach compared to classic MC, the classic MC must be conducted on the finest mesh used within the SC-Hybrid refinement process, i.e., on a mesh with size $h/4$. We will refer to this approach as $\mathrm{MC}_{\mathrm{fine}}$. In order to evaluate the SC-Hybrid strategy, leaving out the effect of model refinement, we also apply the refinement to the MC approach, called MC_{L} to indicate the usage of different levels of high fidelity. For the SC surrogate model, the Leja nodes are evaluated on the first grid, i.e., with mesh size h. Since the range parameter interval is discretized in eleven equidistant range parameter points and for each of them separate surrogate models are built, cf. Sect. 6.1.1, the number of Leja nodes is multiplied with eleven to obtain the offline costs. The separate surrogate models for the imaginary part and the real part use the same Leja nodes. We investigate two different accuracy settings for the SC approach, which is reflected in the number of Leja nodes employed. In the four parameter example we use 30 and 120 Leja nodes per surrogate model, in the twelve parameter example 90 and 1600, respectively. For the SC-Hybrid approach the SC model with lower accuracy is employed as the initial surrogate model. For each approach the same MC sample set as for the reference solution is used.

Table 6.1 Computational effort and accuracy for yield estimation with SC-Hybrid approach compared to classic MC and pure SC, applied to waveguide problem with four uncertain parameters

Approach	# Leja	$\#\mathrm{HF}_h^{\mathrm{offline}}$	$\#\mathrm{HF}_h^{\mathrm{online}}$	$\#\mathrm{HF}_{h/2}^{\mathrm{online}}$	$\#\mathrm{HF}_{h/4}^{\mathrm{online}}$	$\#\mathrm{HF}_{\mathrm{L}}$	$\varepsilon_{\mathrm{rel}}$ (%)
$\mathrm{MC}_{\mathrm{fine}}$	–	–	–	–	26360	421760	0.0000
MC_{L}	–	–	26360	5	1	26396	0.0000
SC-Hybrid	30	330	165	5	1	**531**	**0.0000**
SC	30	330	–	–	–	330	0.1257
SC	120	1320	–	–	–	1320	0.0000

Table 6.2 Computational effort and accuracy for yield estimation with SC-Hybrid approach compared to classic MC and pure SC, applied to waveguide problem with twelve uncertain parameters

Approach	# Leja	$\#\mathrm{HF}_h^{\mathrm{offline}}$	$\#\mathrm{HF}_h^{\mathrm{online}}$	$\#\mathrm{HF}_{h/2}^{\mathrm{online}}$	$\#\mathrm{HF}_{h/4}^{\mathrm{online}}$	$\#\mathrm{HF}_{\mathrm{L}}$	$\varepsilon_{\mathrm{rel}}$ (%)
$\mathrm{MC}_{\mathrm{fine}}$	–	–	–	–	22705	363280	0.0000
MC_{L}	–	–	22705	25	6	22901	0.0000
SC-Hybrid	90	990	4812	25	6	**5998**	**0.0000**
SC	90	990	–	–	–	990	6.2235
SC	1600	17600	–	–	–	17600	0.4290

The results for the four parameter setting are summarized in Table 6.1, the results for the twelve parameter setting in Table 6.2.

In both tables we see the number of offline and online high fidelity evaluations. For the online evaluations we distinguish the three different refinement levels. In the last two columns we see the key indicators for each approach: the costs according to (6.23) and the relative error compared to the reference solution according to (6.21).

In the test case with twelve uncertain parameters (see Table 6.2), both MC approaches and the SC-Hybrid approach achieve the same yield estimate as the reference solution. Thereby, the SC-Hybrid approach requires the least computational effort. Compared to MC_{L} more than 73 % of the computing costs can be saved, compared to $\mathrm{MC}_{\mathrm{fine}}$ even 98 %. We observe that in both approaches applying the refinement strategy, i.e., SC-Hybrid approach and MC_{L}, most of the MC sample points are evaluated on the first mesh (h). The refinement $h/2$ is applied for 25 sample points, the refinement $h/4$ for only 6 sample points. The pure SC approach performs worse. Using the same surrogate model as for SC-Hybrid, the computational effort is reduced by more than 80 %, but at the cost of a relative error of more than 6 %. Increasing the accuracy of the surrogate model by increasing the number of Leja nodes reduces this error. However, even with 1600 Leja nodes per surrogate model, resulting in a computational effort three times higher than in the SC-Hybrid approach, the error is still larger than 0.4 %.

In the test case with four uncertain parameters, the results are similar. The MC approaches and the SC-Hybrid approach achieve the same yield estimate as the reference solution, while the SC approach employing 30 Leja nodes, but without hybridization, has an error of more than 0.1 %. In this setting the error of the SC approach can be reduced to zero by increasing the number of Leja nodes to 120. However, with 120 Leja nodes the computing effort of the SC method is more than twice the effort of the hybrid approach. Considering the computational effort of the MC approaches and the SC-Hybrid approach, we observe that the hybrid approach saves almost 98 % compared to $\mathrm{MC_L}$, and even 99.8 % compared to $\mathrm{MC_{fine}}$.

We conclude, in both settings the SC-Hybrid approach performs best, achieving the reference solution for the yield estimate at the lowest computational costs. Comparing the test case with twelve and four uncertain parameters, we observe that the advantage of the SC-Hybrid approach over classic MC decreases for an increasing number of uncertain parameters. Nevertheless, by construction we know that the number of high fidelity evaluations in the SC-Hybrid approach can never exceed the number of high fidelity evaluations in classic MC, excluding the effort for calculating the error indicator and for the offline high fidelity evaluations (which might scale poorly for many uncertain parameters). Further, we observe that the higher the number of uncertain parameters, the higher the benefit which can be expected from the SC-Hybrid approach compared to pure SC. This can also be explained by the scaling of collocation nodes and dimension in the SC approach. The hybrid approach allows lower accuracy in the surrogate model since critical sample points are reevaluated on the high fidelity model before classification.

Since the GPR-Hybrid approach, which will be investigated next, is not implemented with model refinement, we perform one calculation of the SC-Hybrid approach without refinement to allow a valid comparison. For the waveguide problem with four uncertain parameters and 30 Leja nodes we perform all high fidelity evaluations on a mesh with size h. The reference value of the yield can be achieved while the total number of high fidelity evaluations is

$$\#\mathrm{HF}^{\text{SC-H}} = \#\mathrm{HF}_h^{\text{offline}} + \#\mathrm{HF}_h^{\text{online}} = 330 + 165 = 495. \tag{6.29}$$

GPR-Hybrid Approach

For the GPR-Hybrid approach we investigate the waveguide problem with four uncertain parameters. For this benchmark problem a closed form solution exists and will be used here. However, we refer to this evaluation as a high fidelity evaluation since in practice the problem would be solved using a computationally expensive numerical method, e.g. FEM. The GPR is realized in Python with the package scikit-learn [23]. In the beginning of GPR a kernel and a mean function have to be chosen, cf. Sect. 3.3.3. The kernel applied to the waveguide problem is a product of a constant kernel and a squared exponential kernel, i.e.,

$$k(\boldsymbol{\xi}, \boldsymbol{\xi}') = \zeta \mathrm{e}^{\frac{|\xi - \xi'|}{2l^2}}, \tag{6.30}$$

where $\zeta \in \mathbb{R}$ and $l > 0$ are hyperparameters. For the hyperparameters we provide the starting points $\zeta_0 = 0.1$ and $l_0 = 1$ and ranges $\zeta \in [10^{-5}, 10^{-1}]$ and $l \in [10^{-5}, 10^5]$,

Table 6.3 Number of high fidelity evaluations for different sizes of the initial training data set in the GPR-Hybrid approach applied to the waveguide problem. In all cases batch size $N_B = 50$ is used

	$\lvert\mathcal{T}_{\mathrm{GPR}}\rvert = 5$	$\lvert\mathcal{T}_{\mathrm{GPR}}\rvert = 10$	$\lvert\mathcal{T}_{\mathrm{GPR}}\rvert = 30$
$\#\mathrm{HF}^{\mathrm{offline}}$	55	**110**	330
$\#\mathrm{HF}^{\mathrm{online}}$	306	**226**	179
#HF	361	**336**	509

then they are optimized within scikit-learn. For the hyperparameter optimization we allowed a maximum of 10 iterations. As the mean function the mean value of the training data evaluations is employed. The noise factor in the GPR is set to $\alpha = 10^{-5}$. This is recommended in order to avoid numerical issues, e.g. due to mesh noise. For more details on hyperparameters we refer to [24, Chap. 2.3].

As in the SC-Hybrid approach, for each frequency range point a separate surrogate model is built with its own training data points. Further, we build separate surrogate models for the imaginary part and the real part. Hence, the square root is avoided, which ensures (affin-)linearity of the QoI [22]. Since the same training data can be used for both and the computational costs for building the GP is negligible compared to the high fidelity evaluations of each sample point (cf. Sect. 3.3.3), the resulting computational effort for building these two surrogate models separately is also negligible.

The initial set of training data points is generated according to their probability density function, different sizes of the training data set $|\mathcal{T}_{\mathrm{GPR}}|$ are investigated. The batch size N_B defines the number of critical sample points evaluated before a model update is considered. It can be chosen according to the number of parallel processors available in order to allow parallel computing. However, a higher value also has the effect that the GPR model is updated less frequently, which might result in more sample points classified as critical due to evaluation on a less accurate surrogate model. Different values for N_B are compared.

For batch size $N_B = 50$ we investigate the best performing size of the training data set. The results are summarized in Table 6.3. In all following tests we set $|\mathcal{T}_{\mathrm{GPR}}| = 10$, since the sum of offline and online evaluations is minimal in this case. For smaller training data sets the number of offline evaluations decreases, while the number of online evaluations increases. For larger values we observe the opposite.

We compare different batch sizes $N_B \in \{1, 20, 50\}$ and apply either no sorting strategy, sorting based on the EGL criterion (4.18) or sorting based on the hybrid criterion (4.20). In all tests the reference solution of the yield estimate is achieved. Since the offline costs are the same for each of these test cases, in Table 6.4 only the online costs are compared. We distinguish between the number of online high fidelity evaluations $\#\mathrm{HF}^{\mathrm{online}}$ and the number of effective online evaluations $\#\mathrm{HF}_{\mathrm{eff}}^{\mathrm{online}}$ according to (6.24). The total costs can be obtained by adding the offline costs $\#\mathrm{HF}^{\mathrm{offline}} = 110$, or $\#\mathrm{HF}_{\mathrm{eff}}^{\mathrm{offline}} = \lceil 110/N_B \rceil$, respectively, to the online costs reported in Table 6.4.

Table 6.4 Online computational costs of different batch sizes and updating strategies within the GPR-Hybrid approach, applied to the waveguide problem. We distinguish high fidelity and effective evaluations

	No sorting		EGL criterion		Hybrid criterion	
	$\#\text{HF}^{\text{online}}$	$\#\text{HF}^{\text{online}}_{\text{eff}}$	$\#\text{HF}^{\text{online}}$	$\#\text{HF}^{\text{online}}_{\text{eff}}$	$\#\text{HF}^{\text{online}}$	$\#\text{HF}^{\text{online}}_{\text{eff}}$
$N_\text{B} = 1$	178	178	146	146	127	127
$N_\text{B} = 20$	197	10	163	9	160	8
$N_\text{B} = 50$	226	5	201	5	209	5

We evaluate the effect of using batches, i.e., of setting $N_\text{B} > 1$. For both sorting strategies and without sorting larger batches lead to higher numbers of high fidelity evaluations. However, the number of effective evaluations decreases in all cases significantly. This observation coincides with the expectation described above. It follows that the use of batches is only recommended when sufficient processors are available and parallelization is realized.

We observe that for smaller batch sizes the effect of sorting is stronger. Without batches ($N_\text{B} = 1$), sorting with the EGL criterion reduces the number of high fidelity evaluations by 18 %, sorting with the hybrid criterion by 29 %, compared to the strategy without sorting. In Fig. 6.5 the effect of sorting the sample points is illustrated. By sorting the sample points, the most critical sample points are evaluated first. This leads to faster improvement of the GPR model. For batch size $N_\text{B} = 50$, the plot shows the number of high fidelity evaluations over the number of MC sample points which have already been considered for classification. For the 0-th MC sample point the offline costs are plotted. The marks indicate when one batch is completed. It shows that using a sorting strategy, first the critical sample points are evaluated on the high fidelity model, then the non-critical sample points on the GPR model. Without sorting the critical sample points are more spread over the MC sample, although this curve is also flattening due to the improvement of the GPR model. After evaluating the whole MC sample set, the total number of high fidelity evaluations is similar for all strategies.

Finally, we compare the computational effort with the reference solution. Classic MC without batches (i.e., $N_\text{B} = 1$) requires 26360 high fidelity function evaluations. Enabling parallel computing for classic MC with N_B parallel processors yields 1318 effective evaluations for $N_\text{B} = 20$, 528 for $N_\text{B} = 50$, respectively. Depending on the sorting strategy and the batch size, using the GPR-Hybrid approach reduces the computational effort by 99.0 – 99.5 % compared to classic MC.

Linearization Approach

Linearization is a standard method for yield estimation in industry [25]. Assuming that the variations of the uncertain parameters are small enough to achieve valid results, the QoI is approximated by a simple linearization approach and then a MC analysis is applied. This paragraph is dedicated to the evaluation of the validity of

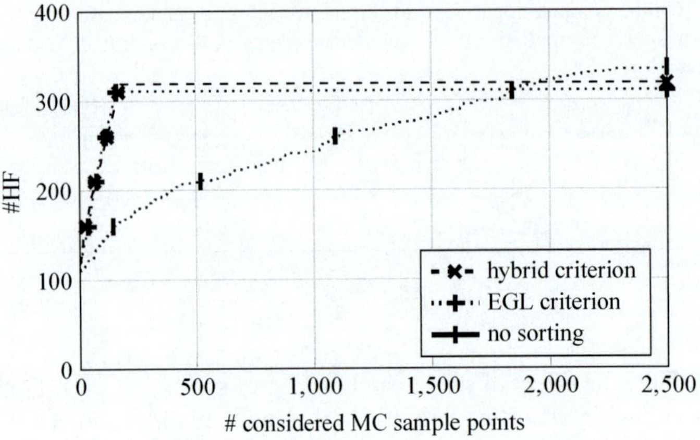

Fig. 6.5 Comparison of different sorting strategies within the GPR-Hybrid approach with batch size $N_B = 50$ applied to the waveguide problem. Number of high fidelity evaluations over the number of MC sample points which have already been considered for classification. The marks indicate completed batches. Figure taken from [22, Fig. 3]

this approach and the comparison of the results with the hybrid methods proposed in this work. We follow our work in [22].

We briefly describe the linearization approach. Let $\boldsymbol{\xi}^0$ denote the point for which a linearization shall be realized, probably $\overline{\boldsymbol{\xi}}$. For each direction a new point is generated by

$$\boldsymbol{\xi}^k = \boldsymbol{\xi}^0 + \delta \mathbf{e}^k, \quad k = 1, \dots, n_\xi, \tag{6.31}$$

and evaluated on the high fidelity model. In (6.31), $\mathbf{e}^k$ denotes the k-th unit vector and $\delta > 0$ the step size (if interpreted in the sense of finite differences). Having now two points per direction, we create a linear equations system with

$$Q(\boldsymbol{\xi}^k) = \sum_{i=1}^{n_\xi} (\alpha_i \xi_i^k) + \alpha_{n_\xi+1}, \quad k = 0, \dots, n_\xi, \tag{6.32}$$

where α_i, $i = 1, \dots, n_\xi$, denote the interpolation coefficients. By solving this system, the linear interpolation model is built. As in the GPR-Hybrid method, separate surrogate models are built for each range parameter point and for the real and the imaginary part. As above, for the surrogates of the real and the imaginary part the same interpolation points can be used. The number of required high fidelity evaluations per range parameter point is $n_\xi + 1$. In case of the waveguide problem with four uncertain parameters and eleven range parameter points, $5 \cdot 11 = 55$ high fidelity evaluations are required. If derivative information is available, this could be utilized alternatively.

We introduce the parameter $\upsilon \in [0, 1]$ to describe a varying magnitude of uncertainty. For the uncertain parameter we write $\boldsymbol{\xi} \sim \mathcal{N}_{\mathcal{T}}(\overline{\boldsymbol{\xi}}, \upsilon\boldsymbol{\Sigma}, \mathbf{a}, \mathbf{b})$, i.e., the covariance

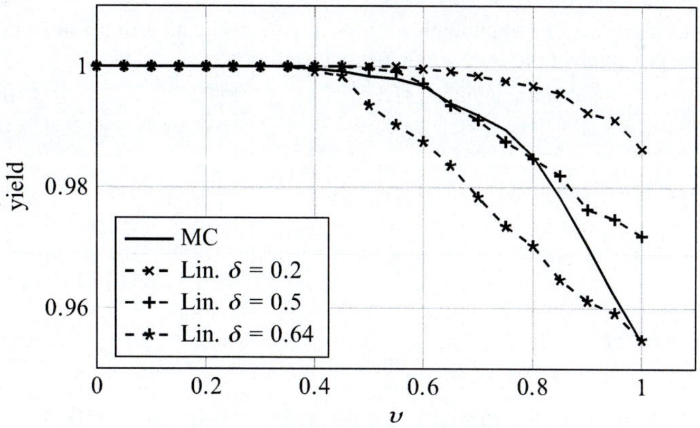

Fig. 6.6 Comparison of yield estimates by linearization (with step size δ) over increasing magnitudes of uncertainty υ for the waveguide problem with four uncertain parameters. Reference solution is from classic MC analysis. Figure taken from [22, Fig. 4]

matrix is multiplied with different values of υ. For $\upsilon = 0$ there is no uncertainty, for $\upsilon = 1$ we obtain the case as described in Sect. 6.1.1 and evaluated previously in this section. In Fig. 6.6 the yield estimates with increasing values of υ are illustrated. We compare the linearization approach with different values for the step size δ with the classic MC analysis. The hybrid approaches coincide with the MC yield estimates for all magnitudes of uncertainty and are not visualized here. For $\upsilon > 0.5$ the deviations of the linearization approach are significantly for any δ.

Comparison

In this section the proposed hybrid approaches have been applied to the waveguide problem. We evaluate their performance against each other and in comparison to the classic MC analysis. Both hybrid methods demonstrate high accuracy by coinciding with the MC results for all test cases. In order to analyze the computational effort under comparable settings, we consider the GPR-Hybrid solution without batches (i.e., $N_B = 1$) and without sorting and the SC-Hybrid solution without high fidelity model refinement. The computational costs for the waveguide problem with four uncertain parameters are summarized in Table 6.5. Compared to classic MC the computational effort is reduced by 98 % with the SC-Hybrid approach and even 99 % with the GPR-Hybrid approach. This significant reduction of computing time allows yield analysis in industrial applications. Comparing the two hybrid methods, the GPR-Hybrid approach is the recommended choice. It requires 40 % less high fidelity evaluations than the SC-Hybrid approach. Further, for GPR no detailed information of the FEM solution is required which allows the usage of proprietary software as a blackbox.

Table 6.5 Comparison of computational effort for different yield estimation approaches, applied to waveguide problem with four uncertain parameters

Approach	# Training points	$\#\mathrm{HF}^{\mathrm{offline}}$	$\#\mathrm{HF}^{\mathrm{online}}$	#HF
MC	–	–	26360	26360
SC-Hybrid	30	330	165	495
GPR-Hybrid	10	110	178	288

6.2.2 PMSM

In this section we estimate the yield for the PMSM described in Sect. 6.1.2. Since the machine is modeled and simulated in CST, details on the FEM matrices and solutions are not available. It follows that only classic MC and the GPR-Hybrid approach are employed. We discuss the yield estimation in the starting point of the optimization which follows in Sect. 6.4.2, i.e.,

$$\mathbf{d}^{(0)} = (19, 7, 7, 0)^\top. \tag{6.33}$$

The mean of the uncertain parameter $\boldsymbol{\xi}$ is as defined in (6.13). Equivalently to the waveguide problem, the size of the MC sample set is $N_{\mathrm{MC}} = 2500$. Due to the absence of a range parameter, only one surrogate model is required. For the reference solution, which is MC on the high fidelity (i.e., CST) model, this results in a computational effort of 2500 high fidelity evaluations. The MC yield estimate is

$$Y_{\mathrm{ref}} = 4.28\,\%. \tag{6.34}$$

Parallel computing was not available, thus, the batch size is set to $N_{\mathrm{B}} = 1$. Sorting is not applied. This section follows our work in [19].

Generation of the Training Data Set

This paragraph is dedicated to the generation of the training data set for the initial GPR surrogate model. In case of pure yield estimation, the deterministic parameter $\mathbf{d}$ is set according to (6.33) and remains unchanged. The training data set consists of random sample points drawn from the probability distribution of the uncertain parameter. We refer to this setting as *Case Est*.

Later we aim to optimize the yield. In contrast to the waveguide problem, where the mean of the uncertain parameter is the optimization variable, here we optimize over the deterministic parameter. Since the optimization is also using the surrogate model, for the purpose of optimization the training data set should also include variations of the deterministic parameter. We generate training data points $(\overline{\boldsymbol{\xi}}^{(i)\top}, \mathbf{d}^{(i)\top})^\top$, $i = 1, \dots, |\mathcal{T}_{\mathrm{GPR}}|$. The realizations for the uncertain parameter are distributed as in *Case Est*. The realizations of the deterministic parameters d_1, d_2 and d_3 are uniformly

Table 6.6 Comparison of computational costs and accuracy for different sizes of the initial training data set in the GPR-Hybrid approach applied to the PMSM problem

Case	$\lvert\mathcal{T}_{\mathrm{GPR}}\rvert$	#HF^{offline}	#HF^{online}	$\widetilde{Y}$ (%)	#HF	$\varepsilon_{\mathrm{rel}}$ (%)
Case Est	**10**	10	26	4.28	**36**	**0.0000**
	20	20	23	4.28	43	0.0000
	50	50	21	4.28	71	0.0000
Case Opt	15	15	34	3.92	49	8.4112
	20	20	40	4.36	**60**	**1.8692**
	30	30	43	4.32	73	0.9246

distributed in a $\pm 10\,\%$ range around the values defined in (6.33). The deterministic parameter d_4 is uniformly distributed in $[-3, 3]$. We refer to this setting as *Case Opt*.

Size of the Training Data Set

Once the generation of the training data points is clarified, a suitable size of the training data set is determined. Different values are tested, the results are summarized in Table 6.6. Please note that the offline costs coincide with the training data size, since there is no range parameter.

In *Case Est* ten initial training data points are sufficient to achieve a highly accurate yield estimate. By increasing the number of training data points the number of critical sample points (online evaluations) decreases only slightly. With 36 high fidelity evaluations the total costs are minimum for $\lvert\mathcal{T}_{\mathrm{GPR}}\rvert = 10$.

Based on the sample size $N_{\mathrm{MC}} = 2500$, the reference solution for the yield estimate and the equation for the error tolerance of the MC method (4.7) we obtain a MC error indicator of

$$\varepsilon_{\mathrm{MC},Y}(Y_{\mathrm{ref}}, N_{\mathrm{MC}}) = \sqrt{\frac{Y_{\mathrm{ref}}(1 - Y_{\mathrm{ref}})}{N_{\mathrm{MC}}}} \approx 0.00405. \tag{6.35}$$

Hence, all yield estimates in the region $Y_{\mathrm{ref}} \pm 0.405\,\%$ are valid. For the *Case Opt* at least 15 training data points are required to achieve this accuracy. However, using only five more data points yields 20 % higher total costs, but reduces the relative error from more than 8 % to less than 1.9 %. Only two MC sample points are classified differently as in the reference solution then. In the following optimization part, we apply *Case Opt* with $\lvert\mathcal{T}_{\mathrm{GPR}}\rvert = 20$ for training the initial surrogate model.

Figure 6.7 shows the standard deviation, i.e., the error, of the GPR models for *Case Est* and *Case Opt* during the procedure of one yield estimation. The observation from the waveguide model applies also to the PMSM: The number of critical sample points evaluated on the high fidelity model decreases over time, while the accuracy of the GPR model increases. The GP in the *Case Est* is already very accurate after the initial training. It is evident that for yield estimation *Case Est* performs slightly better than *Case Opt*, since the training data set solely includes points resembling the sample points drawn for the MC analysis, but not deterministic optimization

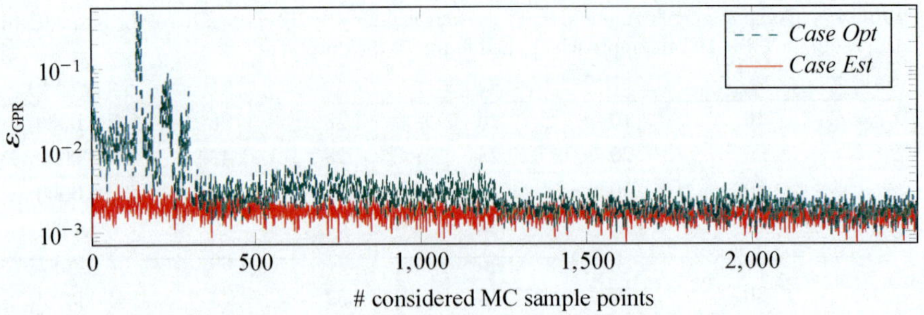

Fig. 6.7 Comparison of the GPR model error for *Case Est* and *Case Opt*. Figure based on [19, Fig. 6], © 2022 IEEE. Reprinted, with permission, from [19]

variables. However, comparing the effort to standard MC, the difference between *Case Est* and *Case Opt* is not significant.

Comparison

For pure yield estimation *Case Est* is the natural choice, since the deterministic parameter can just be considered as a constant model parameter. For yield optimization *Case Opt* is recommended, to account for the change of the deterministic parameters during the optimization process. Compared to classic MC the computational effort of yield estimation can be reduced by 98.56 % for *Case Est*, and 97.6 % for *Case Opt*, while achieving the same yield estimate.

6.3 Adaptive Newton-MC for Yield Optimization

In this section we apply the adaptive Newton-MC method for yield optimization proposed in Sect. 5.3 to the waveguide problem. For the yield estimation during the optimization procedure, classic MC, SC-Hybrid and GPR-Hybrid methods are employed. The results are compared to optimization with a standard globalized Newton method. The content of this section follows our work in [21, 26].

As in the previous section, the safety factor is set to $s = 2$ and the MC error tolerance to $\hat{\varepsilon}_{\mathrm{MC},Y} = 0.01$. For the standard Newton method, this error tolerance implies a sample set size of $N_{\mathrm{MC}} = 2500$. In the adaptive Newton-MC method we set the initial MC sample set size $N_{\mathrm{MC}}^{\mathrm{start}} = 100$ and the incremental factor $\mathrm{inc} = 1$. In both methods, adaptive Newton-MC and standard Newton, we limit the number of Armijo backward steps. If the inequality in line 10 of Algorithm 5.1 is not fulfilled after three steps, the step size is set to $\sigma^{(k)} = \beta^3$ and the next iteration starts.

6.3.1 *Waveguide*

We consider the waveguide problem as introduced in Sect. 6.1.1. As starting point for the optimization we set

$$\overline{\xi}^{\{4\},(0)} = (9, 5, 1, 1)^\top \tag{6.36}$$

for the waveguide problem with four uncertain parameters and

$$\overline{\xi}^{\{12\},(0)} = (9, 5, 2, 0.5, 1, 1, 1.1, 2.5, 1, 1, 1, 2)^\top \tag{6.37}$$

for the setting with twelve uncertain parameters. The initial yield estimates are

$$\widetilde{Y}^{\{4\},(0)} \approx 41\,\% \quad \text{and} \quad \widetilde{Y}^{\{12\},(0)} \approx 15\,\%. \tag{6.38}$$

Validity of Assuming a Truncated Gaussian Distribution

The adaptive Newton-MC method uses the derivatives of the yield, which can be easily computed in case of Gaussian distributed uncertain parameters, cf. Sect. 5.3. However, according to Sect. 6.1.1, we modeled the uncertain parameter as truncated Gaussian distributed, in order to avoid unphysical values. This implies, that the gradient used for optimization deviates from the exact gradient corresponding to the MC sample set. This can be interpreted as an inexact Newton method [27], with approximations in the root-finding problem, i.e., here inexact gradients and Hessians.

In order to avoid these inexact derivatives, alternatively, one could calculate the gradient with finite differences. However, the computational effort would increase significantly. While the gradient based on the Gaussian distribution can be calculated without any additional computing effort, the difference quotient requires at least one additional yield estimation per dimension. To distinguish the two types of gradients, in this paragraph we write $\nabla_{\overline{\xi}_1} Y_{\mathrm{G}}$ for the gradient based on the Gaussian distribution and calculated with (5.13), and $\nabla_{\overline{\xi}_1} Y_{\mathrm{DQ}}$ for the gradient based on the truncated Gaussian distribution and calculated with the difference quotient. We propose a simple modification of the adaptive Newton-MC method to ensure the optimality of the final yield solution with reduced additional computing costs. Once the algorithm has terminated, the gradient can be calculated with the difference quotient. If the error is small enough, i.e.,

$$\left|\nabla_{\overline{\xi}_1} Y_{\mathrm{G}} - \nabla_{\overline{\xi}_1} Y_{\mathrm{DQ}}\right| < \varepsilon_{\mathrm{grad}} \tag{6.39}$$

for some $\varepsilon_{\mathrm{grad}} > 0$, the solution is accepted. Otherwise, the next iteration is started using the difference quotient gradient from now. Please note that also this modification may increase the computational effort significantly, which could lead to a computationally prohibitive problem in practice.

In [21] the difference between these two gradients is investigated for a simplified version of the waveguide problem with only one uncertain parameter. Although there are differences, it is shown that they have very similar behavior. Especially in the important areas close to the optima both gradients agree well. It is also shown that the gradients converge towards each other for large N_{MC}. Hence, an increase of N_{MC} could also be applied in the modification mentioned above. However, we conclude that $\nabla_{\overline{\xi}} Y_{\mathrm{G}}$ is a good approximation for the yield based on truncated Gaussian distributed uncertain parameters. We use this approximation for an inexact Newton method, see e.g. [27]. This is also validated by the optimization results in the remainder of this section.

Adaptive Newton-MC with the SC-Hybrid Approach

We apply the adaptive Newton-MC and the standard Newton method to the waveguide problem with four and twelve uncertain parameters. The number of Leja nodes for building the surrogate model was chosen according to the estimation, i.e., 30 for four uncertain parameters, and 90 for twelve uncertain parameters, respectively. The refinement strategy in three mesh size levels (h,$h/2$,$h/4$) is also applied. Note that the surrogate model is only built once before the optimization starts. Alternatively, one could build a new surrogate model in each iteration or generate Leja nodes in a larger interval to cover a larger space. In Table 6.7 the results are summarized. Evaluating the performance of the optimization methods, again, we focus on two key indicators. First, the computational effort according to (6.23), i.e., under consideration of varying effort for different refinement levels. The offline costs are also included. Second, the optimal yield value achieved (Y^{opt}). In order to obtain comparable results the optimal yield value is always estimated with $N_{\mathrm{MC}} = 2500$, although the adaptive approach may require less sample points in the last iteration. Besides that, also the number of Newton iterations (#It) is provided in Table 6.7.

For the case with four uncertain parameters the yield can be improved to more than 95 %. Applying the adaptive method, the computational costs are 90 % lower than with the standard Newton method. We evaluate the proposed method, i.e., the adaptive Newton-MC method for optimization and the SC-Hybrid approach with model refinement for yield estimation against existing standard methods, i.e., non-adaptive Newton with classic MC on the finest mesh ($h/4$). The costs for this standard

Table 6.7 Comparison of adaptive Newton-MC and standard Newton method for yield optimization with SC-Hybrid estimation applied to the waveguide problem

Dim	Approach	#Leja	#It	#HF$_{\mathrm{L}}$	Y^{opt} (%)
4	Adapt. Newton-MC	30	12	**13716**	**95.44**
4	Standard Newton	30	30	138158	97.92
12	Adapt. Newton-MC	90	33	**376073**	**74.84**
12	Standard Newton	90	37	682745	78.20

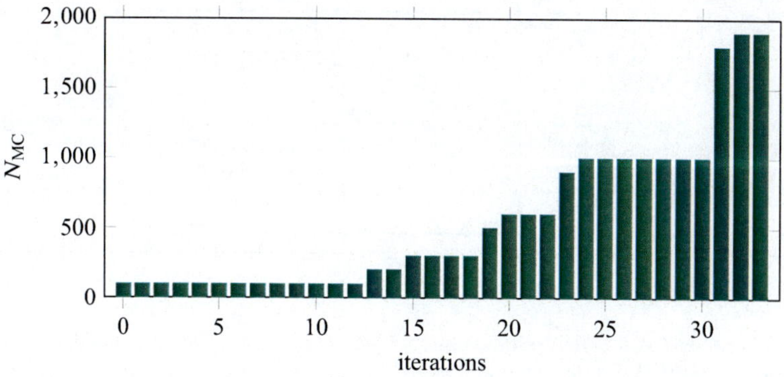

Fig. 6.8 Increase of the MC sample size N_{MC} in the adaptive Newton-MC method for yield optimization, applied to the waveguide problem with twelve uncertain parameters

approach would be $\#\text{HF}_{\text{L}} \approx 13 \cdot 10^6$. Hence, the proposed method can save 99.9 % of the computing costs.

In the case with twelve uncertain parameters, the adaptive Newton-MC and the non-adaptive standard Newton method terminate in different local optima with similar yield values. While the number of iterations is similar for both, the computational costs can be halved with the adaptive method. Figure 6.8 shows the size of the MC sample set in each iteration of the adaptive Newton-MC method. We observe that most iterations require only small sample sets. Only in the end does the number increase to reach the defined MC error tolerance.

Adaptive Newton-MC with the GPR-Hybrid Approach

We optimize the waveguide with four uncertain parameters using the GPR-Hybrid approach for yield estimation. The GPR model is updated after each evaluation of a critical sample point. Alternatively, one could set the batch size $N_{\text{B}} > 1$ or update after each iteration. In Fig. 6.9 the progress of the yield is plotted over the iterations of the optimization. Again, we consult the standard Newton method as reference algorithm. The yield increases very similarly. Both methods find the same optimum. While the adaptive method requires only 681 high fidelity evaluations, including offline costs, the non-adaptive method requires 2753 evaluations. Hence, using the adaptive Newton-MC optimization, the effort can be reduced by 75 %.

Comparison

We summarize the results for yield optimization with the adaptive Newton-MC method using different estimation approaches and compare it with the standard Newton method. Therefore, we consider the SC-Hybrid approach without model refinement, i.e., all high fidelity evaluations originate from FEM with mesh size h. In Table 6.8 the optimal yield values and the number of high fidelity evaluations are summarized. Further, we introduce the row *Saving*. It provides the percentage of high fidelity evaluations which can be saved compared to the reference solution, i.e., compared to standard Newton with classic MC.

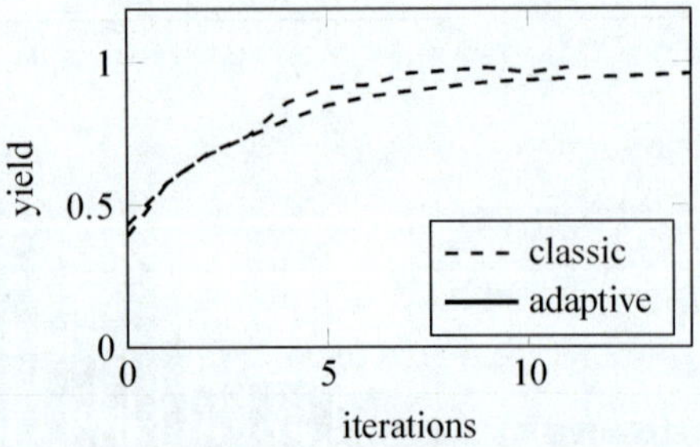

Fig. 6.9 Progress of the yield estimation with adaptive Newton-MC and standard Newton optimization and GPR-Hybrid estimation, applied to the waveguide with four uncertain parameters. Figure based on [26, Fig. 6]

Table 6.8 Comparison of the adaptive Newton-MC method and the standard Newton method with different yield estimation approaches, applied to the waveguide problem with four uncertain parameters

	Standard Newton			Adaptive Newton-MC		
	MC	SC-H	GPR-H	MC	SC-H	GPR-H
Y^{opt} (%)	97.92	97.92	98.32	95.40	95.44	**98.32**
#HF	863792	136556	2753	56652	13616	**681**
Saving (%)	ref	84.19	99.68	93.44	98.42	**99.92**

In all cases the yield is maximized significantly from 41 % to more than 95 %. The small differences do not allow any statement about the quality of the respective method. With all estimation techniques the adaptive optimization method reduces the computational effort significantly. Even with classic MC we record a saving of more than 93 %. However, the combination of the adaptive Newton-MC method for yield optimization and the GPR-Hybrid approach for yield estimation performs best. The computational costs are minimal, with a saving of 99.92 % compared to the reference method, while the optimal yield value is maximal with 98.32 % (although the differences in the yield values are not significant). The better performance of the GPR-Hybrid approach compared to the SC-Hybrid approach probably results from the fact that the GPR model can be updated during the optimization process, while the SC model cannot adapt to the advancing path of optimization, which may lead far away from the starting point. The initial model remains unchanged or completely new models are built. Due to the required structure of the interpolation points, the critical sample points cannot be used for potential new models.

6.4 MOO for Yield Optimization

In this section we formulate MOO problems according to (5.46) for the waveguide and the PMSM and solve them with the methods described in Sect. 3.2.2.

6.4.1 *Waveguide*

We consider the waveguide problem with four uncertain parameters and no deterministic ones, and follow our work in [26]. Besides maximizing the yield, we aim to minimize the width of the waveguide. The width of the waveguide is given by the length of the inlay and the offsets. Since these quantities are uncertain, we require the expectation value, i.e., $\mathbb{E}[\xi_1 + 2\xi_2]$. It is obtained by sampling the uncertain parameters. Note, the computing costs of this sum is negligible compared to the high fidelity evaluations required for the yield estimation. The MOO problem will be solved directly with a genetic algorithm. This implies a trade-off between the two objective functions. In order to ensure a minimum of reliability of the manufactured device, we introduce a lower bound for the yield estimate $Y_{\text{min}} = 80\,\%$. Further, we introduce lower and upper bounds for the optimization variable, i.e., the mean value of the uncertain parameter. We obtain the optimization problem

$$\begin{aligned} &\max_{\overline{\boldsymbol{\xi}}}\ Y(\overline{\boldsymbol{\xi}}) \\ &\min_{\overline{\boldsymbol{\xi}}}\ \mathbb{E}[\xi_1 + 2\xi_2] \\ &\text{s.t. } Y(\overline{\boldsymbol{\xi}}) \geq Y_{\text{min}} = 0.8 \\ &\qquad (5, 3, 0.5, 0.5)^\top \leq \overline{\boldsymbol{\xi}} \leq (25, 15, 1.5, 1.5)^\top. \end{aligned} \tag{6.40}$$

We apply the NSGAII (Non-dominated Sorting Genetic Algorithm II) algorithm [28] implemented in the Python package pymoo [29]. The initial population size is set to 200, the number of offsprings per generation to 100 and the maximum number of generations to 30. For the remaining parameters the default settings are used. After 30 generations, i.e., $200 + 30 \cdot 100 = 3200$ evaluations of the objective functions, we obtain the pareto front in Fig. 6.10. Depending on the weighting of the two objective functions, one of the pareto optimal solutions can be chosen.

6.4.2 *PMSM*

In the literature, many different aspects of electrical machines have been optimized, for example the shape of the rotor [30], the stator slots [31], or the PMs [13, 17]. Additionally, freeform [32] and topology [30] optimization is investigated. In this

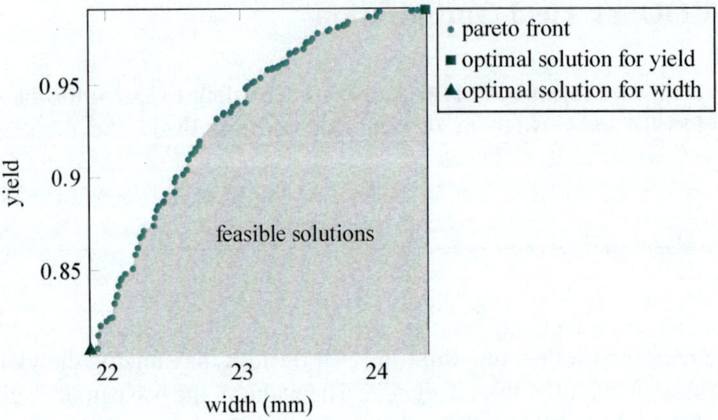

Fig. 6.10 Pareto front for the simultaneous maximization of the yield and minimization of the width of the waveguide with four uncertain parameters, see (6.40). Figure based on [26, Fig. 6]

work the shape and the position of the PMs will be subject to parametric optimization. In Sect. 6.1.2 we mentioned, that the material for the PMs is limited, expensive and its extraction harms the environment. It is only reasonable to reduce the amount of this material in the PMSM design, i.e., reduce the size of the PMs [16, 17]. In the 2D model, the size of the magnet is described by its surface. We define the objective function $C(\mathbf{d}) = d_1\, d_2$ (in mm^2), which is proportional to the costs of the PMs. We introduce bound constraints and linear constraints on the geometry parameters to ensure that the PMs fits into its position. The resulting MOO problem reads

$$\begin{aligned}
&\max_{\mathbf{d}}\ Y(\overline{\boldsymbol{\xi}}, \mathbf{d}) && \\
&\min_{\mathbf{d}}\ C(\mathbf{d}) && = d_1\, d_2 \\
&\text{s.t. } (d_1, d_2, d_3) && \geq (1, 1, 5)^\top \\
&\qquad d_3 && \leq 14 \\
&\qquad d_2 + d_3 && \leq 15 \\
&\qquad 3d_1 - 2d_3 && \leq 50 \\
&\qquad d_4 && \leq 8.5.
\end{aligned} \tag{6.41}$$

Problem (6.41) will be solved with a genetic algorithm and with the scalarization methods described in Sect. 3.2.2. We will also investigate a multi-start procedure for the weighted sum method. The content and the structure of this section follows our work in [19].

Reference Solution

Before we compare the different MOO approaches considering uncertainties, we calculate a reference solution. I.e., we minimize the surface dimension without consideration of material uncertainties. In (6.41) we neglect the first objective function (i.e., the maximization of the yield), therefore we introduce an additional constraint $\tau_{avg} \geq 10.8\,\text{Nm}$, cf. (6.16). Evaluating the QoI always on the high fidelity model, we obtain the optimal solution

$$\mathbf{d}_{\text{ref}} = (20.0, 4.69, 5.0, -0.09) \tag{6.42}$$

with optimal value $C_{\text{ref}} = 93.7759\,\text{mm}^2$. Now we assume that we have uncertainties as specified in Sect. 6.1.2. We estimate the yield for this solution and obtain a value of $Y_{\text{ref}} = 0.4696$.

In the following we show that significantly better yield values (around 0.99) can be achieved when uncertainties are already considered and optimized during the optimization process. In return, the surface dimension will be reduced a bit less than in the reference optimization run (to $108\,\text{mm}^2$).

Genetic Algorithm

We apply the same genetic algorithm implementation as for the waveguide problem, i.e., NSGAII as part of the python package pymoo [28, 29]. We set the initial population size to 100, the number of offsprings per generation to 50 and the maximum number of generations to 18. Further we define lower and upper bounds for the optimization variables by $d_1 \in [18, 21]$, $d_2 \in [6, 7]$, $d_3 \in [6, 8]$, $d_4 \in [-2, 2]$. After eight generations the pareto front is as illustrated in Fig. 6.11. The gray shaded area indicates the set of feasible solutions. After 18 generations, the algorithm has converged to a single point (indicated as square in Fig. 6.11) with yield value 100 % and surface dimension $108.146\,\text{mm}^2$. Please note that the reference solution (6.42) lies not within the bounds specified above.

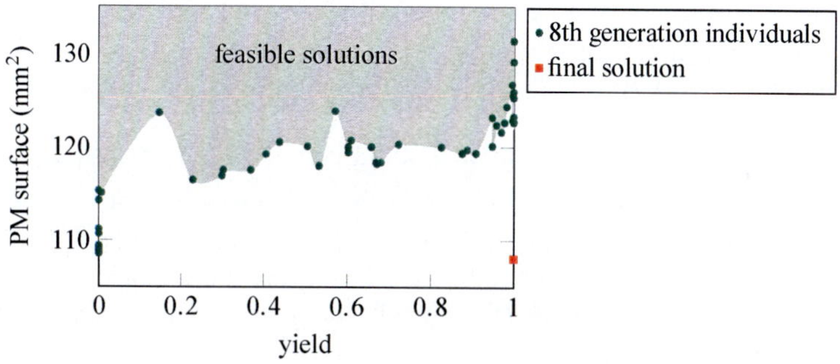

Fig. 6.11 Pareto front after eight generations for the simultaneous maximization of the yield and minimization of the size of the PMs, see (6.41). Optimization converges to single solution after 18 generations (red square). Figure based on [19, Fig. 15]

One advantage of genetic algorithms is their blackbox character. However, the computational costs are high. After 18 generations, the objective functions are evaluated $100 + 18 \cdot 50 = 1000$ times. Each time the objective functions are evaluated, a complete yield estimation is executed. Again the computational effort for the second objective function is negligible. Applying the GPR-Hybrid approach for yield estimation, about 11040 high fidelity evaluations are performed, including 20 offline evaluations. One high fidelity evaluation of the investigated PMSM in CST takes approximately 85 seconds. Accordingly, the entire optimization lasts eleven days. A calculation with classic MC would last almost seven years.

As explained in Sect. 3.2.2, the methods we discuss in the following require more previous knowledge about the problem than a genetic algorithm and return only one pareto optimal solution at a time, but are computationally much less expensive.

ε-constraint Method

In the ε-constraint scalarization method one objective function is transformed into a constraint. We keep the yield as the objective function and introduce an upper bound $C_{\max}$ for the PMs size. The resulting single-objective optimization (SOO) problem reads

$$\begin{aligned} \max_{\mathbf{d}} \; & Y(\overline{\boldsymbol{\xi}}, \mathbf{d}) & \\ \text{s.t.} \; (d_1, d_2, d_3) & \geq (1, 1, 5)^\top \\ d_3 & \leq 14 \\ d_2 + d_3 & \leq 15 \\ 3d_1 - 2d_3 & \leq 50 \\ d_4 & \leq 8.5 \\ C(\mathbf{d}) = d_1\, d_2 & \leq C_{\max}. \end{aligned} \tag{6.43}$$

Since the gradient of the yield with respect to a deterministic parameter is not available, cf. Sect. 5.2.1, and the constraint $C(\mathbf{d}) = d_1\, d_2$ is nonlinear, a derivative free solver for nonlinear constraints is required. We solve the optimization problem with COBYLA (Constrained Optimization BY Linear Approximation), which is part of the PDFO framework (Powell's Derivative Free Optimization solvers) [33] and able to handle arbitrary constraints.

Different values for $C_{\max}$ are compared. In Fig. 6.12a, b the progress of the yield estimate and the PMs surface is plotted for the different values of $C_{\max}$. The results are summarized in Table 6.9. The optimal objective function values Y^{opt} and C^{opt} are provided. For the computational effort we count the number of objective function calls, i.e., yield estimations (#YE), and the number of high fidelity evaluations, including 20 offline evaluations.

We observe that for most values of $C_{\max}$ the yield increases significantly, while the surface dimension remains close to the bound provided. But if the bound is chosen to low, the optimization routine fails to maximize the yield (see $C_{\max} = 100$). Hence, the choice of $C_{\max}$ is a trade-off between the two objective functions and is crucial for the optimization result. This is a drawback in practice, since previous knowledge

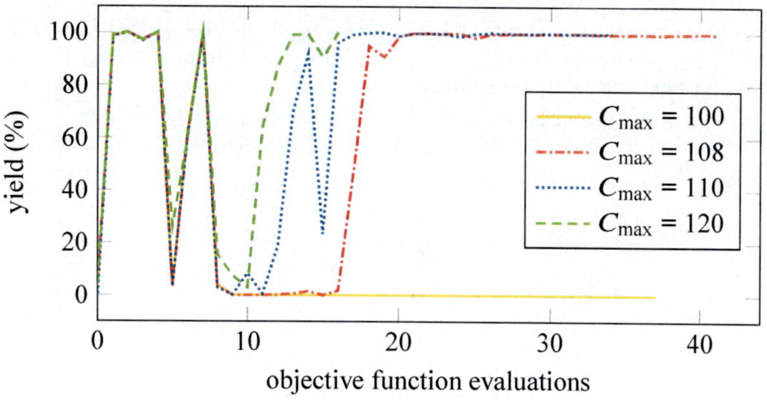

(a) Maximization of the yield estimate.

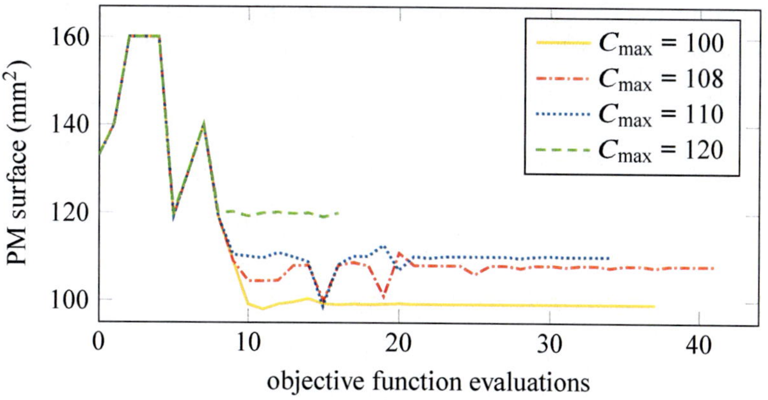

(b) Minimization of the PM surface.

Fig. 6.12 Progress of the two objective functions during the ε-constraint method applied to the PMSM. Figures based on [19, Figs. 7, 8], © 2022 IEEE. Reprinted, with permission, from [19]

Table 6.9 Comparison of different values for the upper bound C_{max} in the ε-constraint method applied to the PMSM

	Comp. effort		Object. fct. values	
C_{max}	#YE	#HF	Y^{opt} (%)	C^{opt} (mm^2)
120	17	370	100.0	119.938
110	35	382	99.8	110.000
108	**42**	**335**	**99.6**	**108.000**
100	38	182	0.0	99.092

Table 6.10 Comparison of different values for the weight w in the weighted sum method applied to the PMSM. Note that the test case $w = 5 \cdot 10^{-3}$ has terminated since the maximum number of objective function calls (100) was reached

	Comp. effort		Object. fct. values	
w	#YE	#HF	Y^{opt} (%)	C^{opt} (mm^2)
$1 \cdot 10^{-3}$	31	279	99.9	128.963
$2 \cdot 10^{-3}$	35	250	99.6	114.594
$3 \cdot 10^{-3}$	**59**	**282**	**99.5**	**109.702**
$5 \cdot 10^{-3}$	100*	754	97.8	105.174

is required. Later in this section we will discuss the multi-start procedure, a heuristic which addresses this drawback. For this example, we find that $C_{\max} = 108$ is the best choice. The computational effort is similar for all settings (362 high fidelity evaluations on average), except the failing setting ($C_{\max} = 100$).

Weighted Sum Method

In the weighted sum method all objective functions are merged into one single objective function and weighting factors are assigned. Since maximization and minimization are combined, the second objective function $C(\mathbf{d})$ is multiplied with -1, cf. (3.25). For (6.41) we obtain the scalarized problem

$$\begin{aligned} \max_{\mathbf{d}}\ f(\overline{\xi}, \mathbf{d}) &= Y(\overline{\xi}, \mathbf{d}) + w(-C(\mathbf{d})) \\ \text{s.t. } (d_1, d_2, d_3) &\geq (1, 1, 5)^{\top} \\ d_3 &\leq 14 \\ d_2 + d_3 &\leq 15 \\ 3d_1 - 2d_3 &\leq 50 \\ d_4 &\leq 8.5. \end{aligned} \tag{6.44}$$

The SOO problem has only linear constraints. Hence, we can apply the optimization algorithm LINCOA (LINearly Constrained Optimization Algorithm), which is also part of the PDFO framework [33]. Besides the trust region termination criteria implemented in LINCOA, we limit the number of objective function calls to 100 and terminate if there is no significant change of the objective function value, considering the last four iterations. I.e., the algorithm terminates if

$$\Delta_{f^{(j)}} = \left| \left(\frac{1}{4} \sum_{i=1}^{4} f^{(j-i)}(\overline{\xi}, \mathbf{d}) - f^{(j)}(\overline{\xi}, \mathbf{d}) \right) \right| < 10^{-4}. \tag{6.45}$$

Different values for the weight $w > 0$ are tested. In Fig. 6.13a, b the progress of the yield estimate and the PMs surface dimension employing the weighted sum method is visualized. In Table 6.10 the results are summarized. In all settings, the

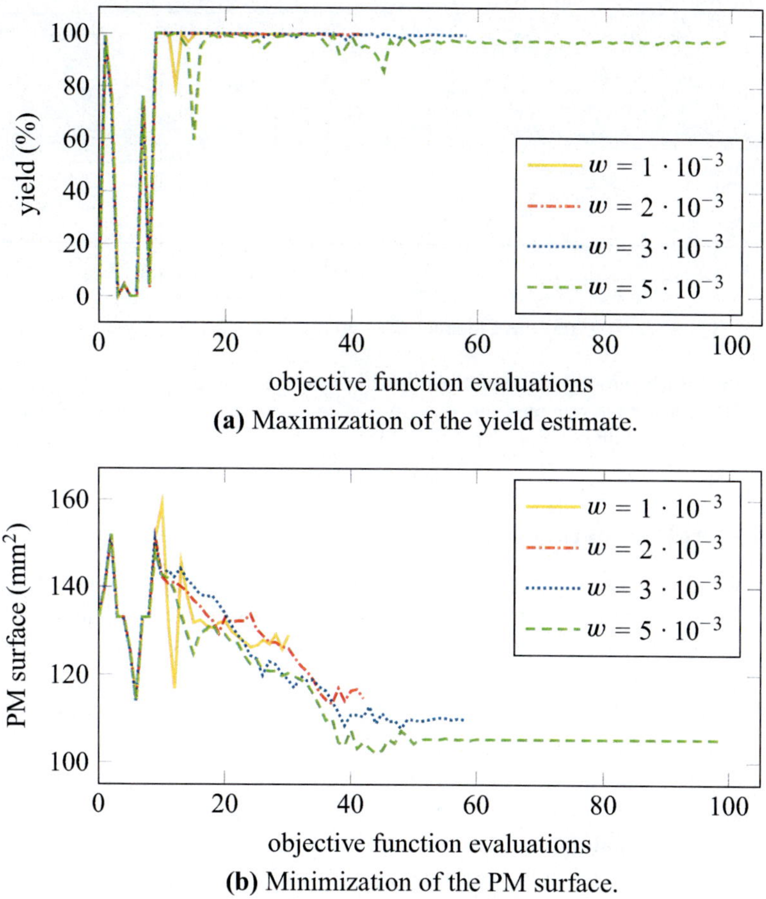

Fig. 6.13 Progress of the two objective functions during the weighted sum method applied to the PMSM. Figures based on [19, Figs. 9, 10], © 2022 IEEE. Reprinted, with permission, from [19]

yield estimate improves significantly. It is evident, that the influence of the surface dimension of the PMs increases for larger weight values. Consequently, the optimal value of the surface dimension decreases. On the other hand, the value of the optimal yield estimate deteriorates slightly for increasing w. Further, we observe that larger weights increase the demand on the weighted objective function which leads to higher computing costs. For $w = 5 \cdot 10^{-3}$, the algorithm terminates since the maximum of 100 objective function calls is reached. As the bound $C_{\max}$ in the ε-constraint method, the weight w is a trade-off between the two objective functions and is crucial for the optimization results. It should be chosen wisely, based on previous knowledge if possible. The best results are obtained with $w = 3 \cdot 10^{-3}$. The optimal objective function values are a good trade-off. The number of objective function calls is almost twice the number with lower weights, but thanks to the GPR-Hybrid approach, the number of high fidelity evaluations is only slightly higher.

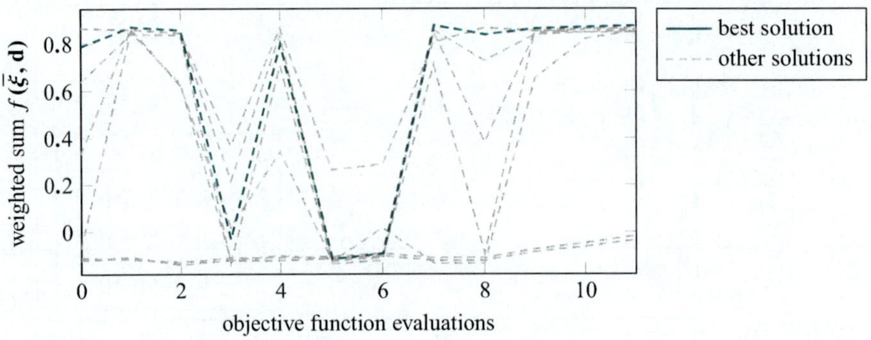

Fig. 6.14 Visualization of the exploration phase during the multi-start procedure applied to the PMSM. Progress of the weighted sum objective function $f(\bar{\xi}, \mathbf{d})$ from (6.44) for different starting points. Figure based on [19, Fig. 11], © 2022 IEEE. Reprinted, with permission, from [19]

Multi-start Procedure

A drawback of both scalarization methods is that a parameter has to be set, the bound C_{max} in the ε-constraint method or the weight w in the weighted sum method. Further, the solvers LINCOA and COBYLA are – as most deterministic optimization solvers – no global methods, i.e., they find a local optimum which depends on the starting point chosen. With the multi-start procedure we aim to cover the feasible set better and not being trapped in a local optimum.

Multi-start procedures are common heuristics, which are widely used to globalize local optimization solvers, cf. [34, Chap. 6]. We will briefly describe the strategy we apply here. We randomly generate a set of ten starting points. For each of them, the optimization is started with low accuracy, i.e., with sample size $N_{\mathrm{MC}} = 100$. This phase serves to explore the entire feasible set in order to determine the most promising region. For this purpose, high accuracy is not required, so we save computational effort by choosing a small sample set. Alternatively, one could use lower fidelity FEM models. Following the terminology of reinforcement learning [35], we call this phase the *exploration phase*. After twelve objective function calls per starting point the optimizations are stopped and the starting point is selected for which the objective function value is best (i.e., maximum) so far. For the corresponding starting point the optimization is continued with high accuracy, i.e., $N_{\mathrm{MC}} = 2500$. Now we seek for the best objective function value in this region, hence, we call this phase the *exploitation phase*.

We apply the multi-start procedure to the weighted sum problem (6.44). We choose the weight $w = 1 \cdot 10^{-3}$, since for this setting, the final objective function value C^{opt} is still very high. We investigate whether the multi-start procedure is able to compensate the poor result.

In Fig. 6.14 the trajectory for the weighted sum function $f(\bar{\xi}, \mathbf{d})$ in the exploration phase is plotted. Most of the starting points perform similarly well during that phase, while two improve barely. The last solution of the green curve is used for the exploitation phase. In Fig. 6.15a, b as well as Table 6.11, the results of the

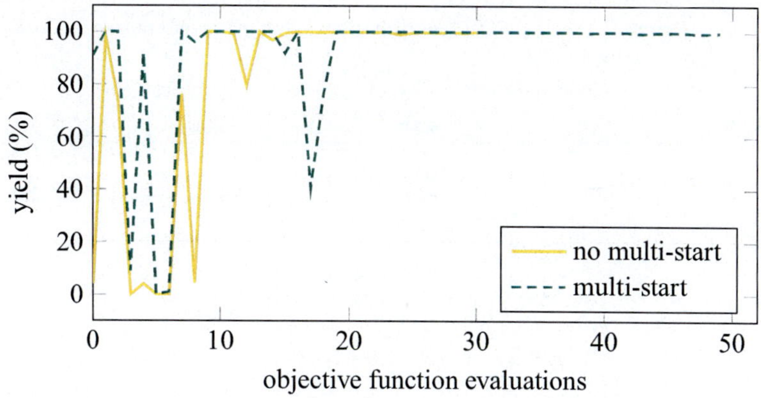

(a) Maximization of the yield estimate.

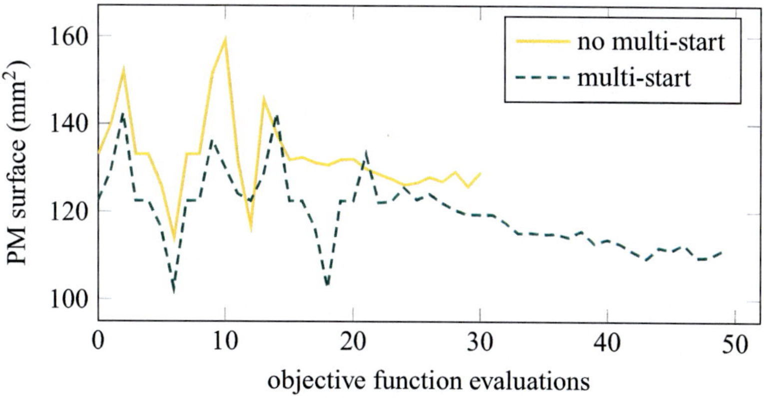

(b) Minimization of the PM surface.

Fig. 6.15 Progress of the two objective functions during the optimization of the PMSM. Comparison of weighted sum with and without multi-start procedure, for weight $w = 1 \cdot 10^{-3}$. Figures based on [19, Figs. 12, 13], © 2022 IEEE. Reprinted, with permission, from [19]

multi-start procedure are compared to the weighted sum results without multi-start. While the finite yield estimate does not differ significantly, the surface dimension of the PMs can be reduced to $\approx 111\,\text{mm}^2$, instead of $\approx 129\,\text{mm}^2$ without multi-start. The objective function is evaluated more than four times more often than without multi-starts, but most of the times with small MC sample sets. Thus, the number of high fidelity evaluations is only 50 % higher.

Comparison

With all MOO methods significant improvement of the two objective functions, i.e., the yield estimate and the surface dimension of the PMs, could be achieved. The yield estimate is increased from $\approx 4\,\%$ to at least 99.5 %, the surface dimension

Table 6.11 Comparison of weighted sum with and without multi-start procedure, for weight $w = 1 \cdot 10^{-3}$, applied to the PMSM

	Comp. effort		Object. fct. values	
Multi-start	#YE	#HF	Y^{opt} (%)	C^{opt} (mm^2)
no	31	279	99.9	128.963
yes	**135**	**414**	**99.8**	**111.203**

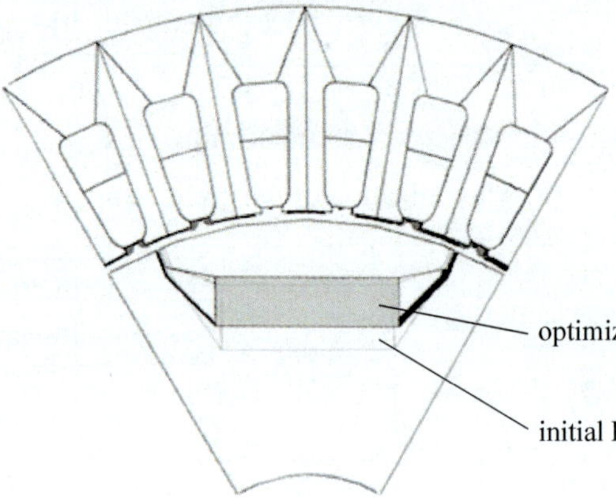

Fig. 6.16 Optimal design of the PMSM after optimizing with the weighted sum method employing the multi-start procedure. The gray shades area indicates the optimal design, the black contours the initial design. Figure taken from [19, Fig. 14], © 2022 IEEE. Reprinted, with permission, from [19]

is decreased from $\approx 133\,\mathrm{mm}^2$ to $\approx 110\,\mathrm{mm}^2$. In Fig. 6.16 the initial design of the PMSM and the optimal design achieved with the multi-start procedure are shown.

An advantage of the genetic algorithm is, that in contrast to the scalarization methods, it can be used as a blackbox, no parameters as weights or bounds have to be chosen. Further, it returns a set of pareto optimal solutions, not only one. On the other hand, with more than 11000 high fidelity evaluations it is computationally very expensive. The scalarization methods achieve equivalently good results, provided the weight w, or bound $C_{\max}$, respectively, is chosen well. But for this, previous knowledge is required. A set of pareto optimal solutions can only be achieved by solving the problem for many different values of weight and bound. A method to globalize the optimization and to reduce the impact of the weight, or bound, is the multi-start procedure. Although the solution found by the simple weighted sum method with the setting $w = 1 \cdot 10^{-3}$ performed poorly, the optimal solution after applying the multi-start procedure is almost as good as the best solutions found with the genetic algorithm or the scalarization methods after determining the most suitable value for w and $C_{\max}$, respectively. The computational effort is with 414 high fidelity evaluations much lower than with the genetic algorithm.

6.5 Hermite-Type Approaches for Mixed Gradient Optimization

This section is dedicated to the numerical evaluation of the Hermite least squares and the Hermite BOBYQA method and follows our work in [5]. As we emphasized in Sect. 5.4, these methods are not specifically for yield optimization, but in general for optimization of objective functions with mixed gradient information. Therefore, in Sect. 6.5.1 we will first evaluate the methods on a test set of almost 30 nonlinear, bound constrained problems of different dimensions, before they are applied to yield optimization of the waveguide problem in Sect. 6.5.2.

The PyBOBYQA implementation by [36] is considered as the reference solution. We use the default setting, i.e., the number of interpolation points is $p_1 = 2n + 1 < q_1$, and refer to it as ref_B in the following. Further, we compare the performance with SQP, approximating the unknown derivatives with finite differences. We apply the SLSQP algorithm from [37], implemented in SciPy [10]. The default settings of each method are used. Note that the proposed Hermite-type approaches are modifications of the PyBOBYQA implementation, which allows detailed comparisons of numerical tests. The SQP method, on the other hand, is a completely different method and implementation. For the Hermite BOBYQA method we use the same number of interpolation points as in ref_B, i.e., $p_1 = 2n + 1$. For the Hermite least squares method we set

$$p_1 = \max\left(2n + 1 - n_{\text{kd}}, \left\lceil \frac{(n+1)(n+2)}{2(1+n_{\text{kd}})} \right\rceil\right), \tag{6.46}$$

which showed best performance in numerical tests. Different constellations of available and unavailable derivatives are investigated. For an available derivative we always assume that it is available for all interpolation points.

As in the sections before, the key indicators to measure the performance of a method are the computational effort, measured by the number of objective function calls, and the ability to find an optimum. Please note that we consider only test problems which are solvable with the reference BOBYQA method.

6.5.1 Numerical Test Set

The test set consists of 29 problems commonly used for benchmarking and can be found on GitHub [38]. One of the test cases is the Rosenbrock function in $\mathbb{R}^2$, given by

$$f(\mathbf{x}) = 100(x_2 - x_1^2)^2 + (1 - x_1)^2, \tag{6.47}$$

where we assume that $\partial f/\partial x_2$ is available and $\partial f/\partial x_1$ is not available. We introduce this test case as it will be used for extensive testing in the following.

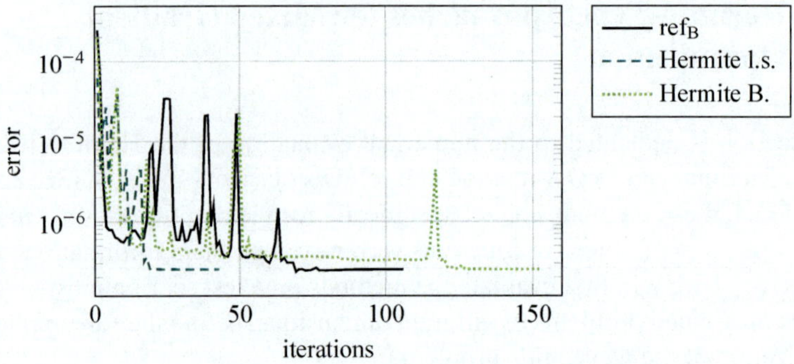

Fig. 6.17 Error of the quadratic model $\widetilde{m}^{(k)}(\mathbf{x})$ according to (6.48) with $\delta = 0.01$ for the Rosenbrock function (6.47). Figure based on [5, Fig. 1]

Accuracy of the Quadratic Model

Before we evaluate the optimization results for the complete test set, we introduce the approximation errors of the quadratic models $\widetilde{m}^{(k)}(\mathbf{x})$ of the various methods. Therefore, we consider the second order Taylor expansion as reference and calculate the error using the L^2-norm

$$\|\widetilde{m}^{(k)}(\mathbf{x}) - T_2 f(\mathbf{x}; \mathbf{x}^{(k)})\|_{L^2}^2 = \int_{\mathbf{x}^{(k)}-\delta}^{\mathbf{x}^{(k)}+\delta} ||\widetilde{m}^{(k)}(\mathbf{x}) - T_2 f(\mathbf{x}; \mathbf{x}^{(k)})^2 \, \mathrm{d}\mathbf{x}. \qquad (6.48)$$

In Fig. 6.17 the resulting errors are visualized for the Rosenbrock optimization problem (6.47). The error of the Hermite least squares model decreases first. After only 20 iterations it is below $2.5 \cdot 10^{-7}$. The ref_B model needs 65 iterations for reducing the error to this magnitude, the Hermite BOBYQA model 118 iterations, respectively. It should be noted that the Hermite BOBYQA model error is already below 10^{-6} after 50 iterations, but has an outlier in iteration 117 before it reduces to $2.5 \cdot 10^{-7}$.

Finally, all methods find the same optimum. In conclusion, the Hermite least squares method tends to approximate the objective function best, particularly at the beginning, i.e., in the first 40 iterations.

Evaluation of the Test Set

For each of the 29 problems from the test set we consider different cases. We vary the number of known derivate directions n_kd in the range $[\frac{1}{2}n, n]$ and compare the average number of objective function calls for each value of n_kd. For example, for a test problem with $n = 3$ and the assumption $n_\mathrm{kd} = 2$, we obtain three test cases: (1) $\partial f/\partial x_1$ and $\partial f/\partial x_2$ are known, (2) $\partial f/\partial x_1$ and $\partial f/\partial x_3$ are known and (3) $\partial f/\partial x_2$ and $\partial f/\partial x_3$ are known. For $n = 10$ three random permutations of known derivatives per n_kd are evaluated. The results are visualized in Fig. 6.18a–e, separately for each dimension n.

Applying the Hermite least squares optimization, the optimal solution is found in all test cases. Although there are single test cases for which the number of objective

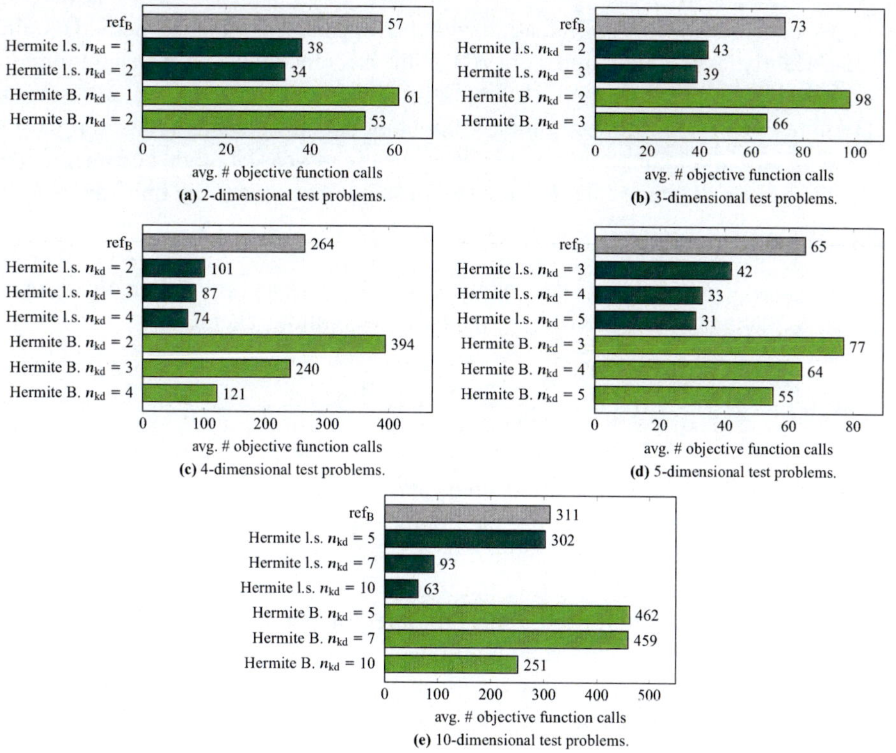

Fig. 6.18 Average number of objective function calls for all test problems with varying number of known derivatives n_{kd}. Figures based on [5, Fig. 2]

function calls is higher than with $\mathrm{ref_B}$, in average their number is significantly lower. The savings lie between 34 % and 80 %, depending on the dimension n and the number of known derivatives n_{kd}. The case $n = 10$ with $n_{\mathrm{kd}} = 5$ is an exception, there the saving is only 6 %. The Hermite BOBYQA finds the optimum for all test cases, except of three. These cases are excluded from the plots in Fig. 6.18, but are listed in [38]. For $n_{\mathrm{kd}} \approx \frac{1}{2}n$, the number of function evaluations is 8 % – 49 % higher compared to $\mathrm{ref_B}$, while for $n_{\mathrm{kd}} \approx n$ it is 7 % – 54 % lower. For both Hermite-type approaches we observe that the computational effort decreases, the more derivative information we can provide. When all derivatives are available, they conceptually approach SQP, see Sect. 3.2.2

Hence, all test problems are also evaluated with SQP. The unknown gradients are approximated with finite differences. In most test cases the number of objective function calls with SQP is lower than $\mathrm{ref_B}$, and often also lower than with Hermite least squares. Since the SQP implementation is completely different, for example how they solve the quadratic subproblem, the results are not visualized in the summary plots in Fig. 6.18.

We conclude that the optimization with the Hermite least squares method requires significantly less computational effort than the reference PyBOBYQA solution ref_B, provided that at least the half of the partial derivatives are available. By using the Hermite BOBYQA optimization the computational effort is only reduced if most of the derivative directions are available. However, in this case it might be rather recommended to use SQP and calculate the few missing derivatives with finite differences.

Weighted Regression

In Sect. 5.4.3 we mentioned the weighted regression as an optional preconditioning step. We define the weights of an exponential weighting scheme by

$$w(\mathbf{y}^i) = \frac{1}{\exp(\kappa)} \exp\left(\kappa - \kappa \frac{||\mathbf{y}^i - \mathbf{x}^{(k)}||_2}{\max_{i=0,\dots,p} ||\mathbf{y}^i - \mathbf{x}^{(k)}||_2}\right), \quad \forall i = 0, \dots, p, \tag{6.49}$$

where $\mathbf{y}^i$ is the i-th interpolation point and $\kappa > 0$ a scaling factor, set to $\kappa = 5$ in our tests. The intention is that data points far from the current iterate solution $\mathbf{x}^{(k)}$ are exponentially lower weighted in the regression than points close to $\mathbf{x}^{(k)}$. In the Hermite least squares method the weight $w(\mathbf{y}^i)$ is applied to each row of the linear equations system related to the i-th interpolation point, i.e., also to the rows containing the derivative information. In the Hermite BOBYQA method it can only be applied to the derivative rows, cf. Sect. 5.4.3. In the numerical tests there was no significant improvement observed. Therefore, the weighting is not applied per default.

Noisy Data

On the example of the Rosenbrock function (6.47) we compare the performance of the discussed optimization methods under noise. For that purpose, we follow [36] and model the noise by multiplying the objective function value and its derivative values by $1 + \zeta$, where ζ is a uniformly distributed random variable in the range $[-10^{-2}, 10^{-2}]$. As starting point for the optimization we set

$$\mathbf{x}^{\text{start}} = (1.2, 2) \text{ with } f\left(\mathbf{x}^{\text{start}}\right) = 31.4. \tag{6.50}$$

The optimal solution is

$$\mathbf{x}^{\text{opt}} = (1, 1) \text{ with } f\left(\mathbf{x}^{\text{opt}}\right) = 0. \tag{6.51}$$

We optimize the noisy Rosenbrock problem using the Hermite least squares method. After only 37 objective function calls (even 6 less than without noise), it terminates successfully with the optimal solution

$$\mathbf{x}^{\text{H.l.s.}} = (1, 1) \text{ with } f\left(\mathbf{x}^{\text{H.l.s.}}\right) = 1.02 \cdot 10^{-23}. \tag{6.52}$$

Applying the Hermite BOBYQA optimization, after 109 objective function calls we also find the optimal solution

Table 6.12 Number of objective function calls for the optimization of the Rosenbrock function (6.47) with the Hermite least squares method with and without using second order derivatives

	Known derivatives		
	1	2	1, 2
1st order	67	43	40
2nd order	62	40	38

$$\mathbf{x}^{\text{H.B.}} = (1, 1) \text{ with } f\left(\mathbf{x}^{\text{H.B.}}\right) = 2.97 \cdot 10^{-19}. \tag{6.53}$$

The reference solution ref_B fails in finding the optimal solution. After 43 objective function calls it returns

$$\mathbf{x}^{\text{B}} = (1.41, 1.98) \text{ with } f\left(\mathbf{x}^{\text{B}}\right) = 0.16. \tag{6.54}$$

In [36, Sect. 7], the authors propose a modification of the PyBOBYQA method for noisy data. The main differences compared to ref_B are: different default settings for the trust region radius adjustment, sample averaging and multiple restarts. We also test this modification and refer to it as ref_N. Even after 2000 objective function calls none of the termination criteria apply. To compare the performance with the Hermite least squares method, we set the budget to 37 function calls (which is the required number for Hermite least squares), refer to it as $\text{ref}_{\text{N},37}$ and obtain

$$\mathbf{x}^{\text{N},37} = (1.08, 1.17) \text{ with } f\left(\mathbf{x}^{\text{N},37}\right) = 0.01. \tag{6.55}$$

Hence, within that budget, the optimal solution has improved significantly, but is not sufficiently identified. The results are even worse for the SQP method. After 41 objective function calls the solution

$$\mathbf{x}^{\text{SQP}} = (0.21, -0.00) \text{ with } f\left(\mathbf{x}^{\text{SQP}}\right) = 0.31. \tag{6.56}$$

is returned.

The results are visualized in Fig. 6.19. In summary: as expected, the SQP method, dependent on finite differences approximations of the gradient, fails in case of noisy data. Also the standard PyBOBYQA method ref_B terminates without finding the optimal solution. The PyBOBYQA modification for noisy data ref_N approaches the optimum, but it does not terminate. The Hermite-type methods on the other hand show great robustness under noise. This can be intuitively explained by the least squares approach instead of interpolation.

Hermite Least Squares with Second Order Derivatives

Equations (5.108)–(5.109) show how second order derivatives can be used in the Hermite least squares approach. Since this information is rarely available in practice, it is only numerically evaluated for the Rosenbrock function (6.47). The results are

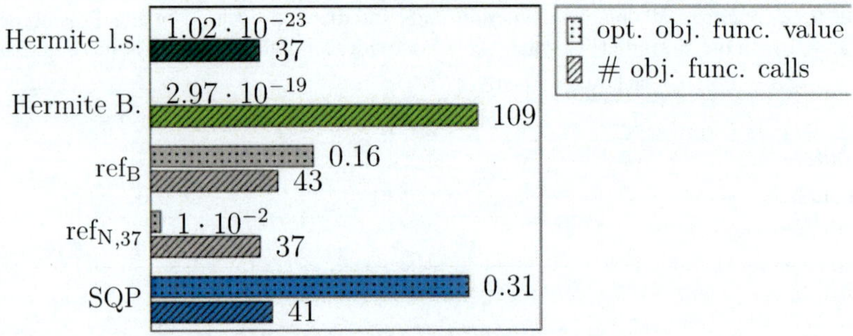

Fig. 6.19 Results for the optimization of the noisy Rosenbrock function (6.47). Optimal objective function values and number of objective function calls. Figure based on [5, Fig. 4]

summarized in Table 6.12. We observe that by using second order derivatives, the number of objective function calls can be slightly reduced in this example (Fig. 6.19).

6.5.2 Hermite-Type Approaches for Yield Optimization

In this section we apply the Hermite least squares and the Hermite BOBYQA method to the waveguide problem, again we compare the results with the reference solutions. We introduce a setting with two uncertain and two deterministic optimization variables. The geometry parameters ξ_1 and ξ_2 are uncertain, as defined in (6.8). The material parameters x_{13} and x_{14} from (6.2) are deterministic optimization variables. To emphasize that and in accordance with Sect. 5.1, they will be denoted with d_1 and d_2, respectively. The optimization variable $\mathbf{x} \in \mathbb{R}^4$ is then given by

$$\mathbf{x} = \begin{pmatrix} \overline{\boldsymbol{\xi}} \\ \mathbf{d} \end{pmatrix} = (\overline{\xi}_1, \overline{\xi}_2, d_1, d_2)^\top . \tag{6.57}$$

While the derivatives with respect to $\overline{\xi}_1$ and $\overline{\xi}_2$ are available without additional computing effort, cf. Sect. 5.1.1, the derivatives with respect to d_1 and d_2 are considered as unknown, cf. Sect. 5.1.2. The starting points are

$$\overline{\xi}_1^{\text{start}} = 9,\ \overline{\xi}_2^{\text{start}} = 5 \text{ and } d_1^{\text{start}} = d_2^{\text{start}} = 1. \tag{6.58}$$

The yield is estimated with MC, the initial yield value is $\widetilde{Y}^{\text{start}} \approx 43\,\%$.

In order to evaluate the yield optimization with and without noise we generate the MC sample set in two different ways. First, in each iteration a completely new random sample set is generated (noise), or second, the sample set is just shifted according to the movement of the mean value (no noise). The size of the sample set N_{MC} controls the accuracy and thus, the noise level. We investigate three settings:

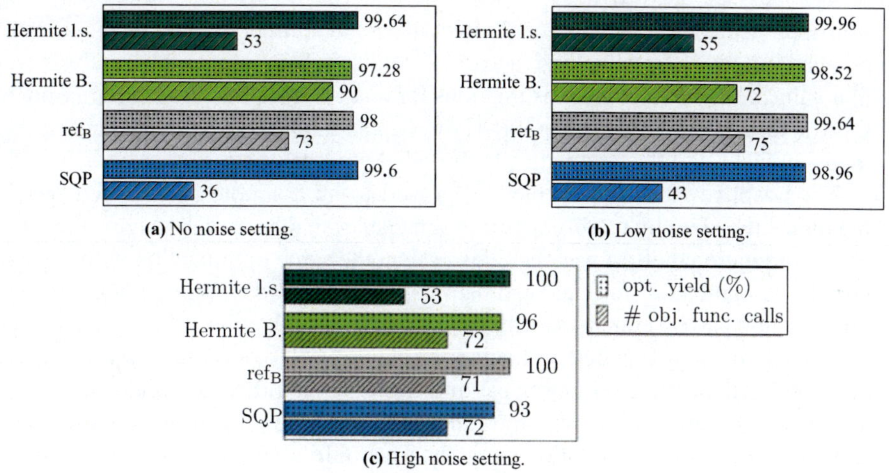

Fig. 6.20 Optimal yield value and number of objective function calls for the waveguide problem with two uncertain and two deterministic optimization variables. Figures based on [5, Fig. 7]

(a) no noise: shifted sample set and $N_{\mathrm{MC}} = 2500$; (b) low noise: new sample set and $N_{\mathrm{MC}} = 2500$; (c) high noise: new sample set and $N_{\mathrm{MC}} = 100$.

The results are summarized in Fig. 6.20. In the no noise setting (Fig. 6.20a) all approaches achieve optimal yield values above 97 %. The SQP method requires the least number of objective function evaluations, followed by Hermite least squares with 50 % more evaluations, $\mathrm{ref_B}$ with 100 % and finally Hermite BOBYQA with 150 % more evaluations. These results correspond to our findings in the previous section. Hermite least squares optimization outperforms the reference method $\mathrm{ref_B}$, when half of the derivative directions are available. For Hermite BOBYQA more derivative information is required to become competitive. SQP performs best, but is not directly comparable. In the noisy settings the results for $\mathrm{ref_B}$ and Hermite least squares do not change significantly, neither the optimal yield value nor the computing effort. For the Hermite BOBYQA method, the number of function evaluations decreases to the same level as for $\mathrm{ref_B}$. The optimal yield value in the high noise setting is slightly worse. The major difference can be observed for SQP. The number of objective function calls doubles from the no noise to the high noise setting, and thus, comes to the same level as $\mathrm{ref_B}$. At the same time, the optimal yield value found deteriorates from more than 99 % to only 93 %.

6.6 Summary

In the beginning of this chapter the two benchmark problems, i.e., the waveguide and the PMSM, were specified. Then, the yield estimation and optimization methods proposed in the previous chapters were evaluated. With the hybrid yield estimation approaches the computational effort of yield estimation could be significantly reduced (up to 99 %) while high accuracy standards from classic MC analysis were maintained. This allows yield estimations for real-world applications in a reasonable time. In direct comparison, the GPR-Hybrid approach performs better thanks to easy model updates and the automatically provided error estimate.

The adaptive Newton-MC method proved to be a simple and efficient way to maximize the yield by modifying uncertain Gaussian distributed design parameters. The computational effort was 93 % lower than the same optimization with the standard Newton method (for yield estimation with classic MC). With MOO, reliability optimization can be combined with performance optimization. While genetic algorithms require many function evaluations by nature, scalarization techniques find only one solution depending on chosen parameters, which requires expert knowledge. For achieving the same optima with the considered test cases, the effort was still significantly lower using scalarization methods. The multi-start procedure proved to be a solid globalization technique which also reduces the sensitivity of the scalarization parameter choice. We conclude that the application of the proposed efficient yield estimation techniques allows even computationally expensive tasks like MOO under uncertainty. The time scale of solving such a problem reduces from years to days. Hence, it becomes practically applicable.

The Hermite-type approaches were applied to a large set of nonlinear test problems with mixed gradient information. The Hermite least squares method outperformed the derivative free BOBYQA reference if at least half of the derivative directions were available. Both Hermite-type approaches showed high performance for noisy data. Tests on yield optimization with the waveguide problem confirmed these findings.

References

1. Loukrezis D (2019) Benchmark models for uncertainty quantification. https://github.com/dlouk/UQ_benchmark_models/tree/master/rectangular_waveguides/debye1.py
2. Loukrezis D, Römer U, De Gersem H (2019) Assessing the performance of Leja and Clenshaw-Curtis collocation for computational electromagnetics with random input data. Int J Uncertain Quantif 9(1)
3. Loukrezis D (2019) Adaptive approximations for high-dimensional uncertainty quantification in stochastic parametric electromagnetic field simulations. PhD thesis, Technische Universität Darmstadt
4. Georg N (2021) Surrogate modeling and uncertainty quantification for radio frequency and optical applications. PhD thesis, Technische Universität Darmstadt
5. Fuhrländer M, Schöps S (2023) Hermite least squares optimization: a modification of BOBYQA for optimization with limited derivative information. Optim Eng. https://doi.org/10.1007/s11081-023-09795-y
6. Hess MW, Benner P (2013) Fast evaluation of time-harmonic Maxwell's equations using the reduced basis method. IEEE Trans Microw Theory Tech 61(6):2265–2274. https://doi.org/10.1109/TMTT.2013.2258167
7. Eller M (2017) A low-frequency stable Maxwell formulation in frequency domain and industrial applications. PhD thesis, Technische Universität Darmstadt

8. Alnæs MS, Blechta J, Hake J, Johansson A, Kehlet B, Logg A, Richardson C, Ring J, Rognes ME, Wells GN (2015) The FEniCS project version 1.5. Arch Numer Softw 3(100). https://doi.org/10.11588/ans.2015.100.20553
9. Hackbusch W (2016) Iterative solution of large sparse systems of equations, 2nd edn. Springer
10. Virtanen P, Gommers R, Oliphant TE, Haberland M, Reddy T, Cournapeau D, Burovski E, Peterson P, Weckesser W, Bright J, van der Walt SJ, Brett M, Wilson J, Millman KJ, Mayorov N, Nelson ARJ, Jones E, Kern R, Larson E, Carey CJ, Polat İ, Feng Y, Moore EW, VanderPlas J, Laxalde D, Perktold J, Cimrman R, Henriksen I, Quintero EA, Harris CR, Archibald AM, Ribeiro AH, Pedregosa F, van Mulbregt P (2020) SciPy 1.0 contributors: SciPy 1.0: fundamental algorithms for scientific computing in python. Nat Methods 17:261–272. https://doi.org/10.1038/s41592-019-0686-2
11. Di Barba P (2010) Multiobjective shape design in electricity and magnetism. Lecture notes in electrical engineering. Springer
12. Taguchi G, Chowdhury S, Wu Y (2005) Taguchi's quality engineering handbook. Wiley. https://doi.org/10.1002/9780470258354
13. Bontinck Z, Lass O, Schöps S, De Gersem H, Ulbrich S, Rain O (2018) Robust optimisation formulations for the design of an electric machine. IET Sci Meas Technol 12(8):939–948
14. Xiao S, Li Y, Rotaru M, Sykulski JK (2015) Six sigma quality approach to robust optimization. IEEE Trans Magn 51(3):1–4
15. Lei G, Bramerdorfer G, Liu C, Guo Y, Zhu J (2020) Robust design optimization of electrical machines: a comparative study and space reduction strategy. IEEE Trans Energy Convers 36(1):300–313
16. Pahner U, Mertens R, De Gersem H, Belmans R, Hameyer K (1998) A parametric finite element environment tuned for numerical optimization. IEEE Trans Magn 34(5):2936–2939
17. Bontinck Z (2018) Simulation and robust optimization for electric devices with uncertainties. PhD thesis, Technische Universität Darmstadt
18. Binnemans K, Jones PT, Blanpain B, Van Gerven T, Yang Y, Walton A, Buchert M (2013) Recycling of rare earths: a critical review. J Clean Product 51:1–22
19. Huber M, Fuhrländer M, Schöps S (2023) Multi-objective yield optimization for electrical machines using Gaussian processes to learn faulty designs. IEEE Trans Ind Appl 59(2):1340–1350
20. Hlaváček I, Chleboun J, Babuška I (2004) Uncertain input data problems and the worst scenario method. Elsevier
21. Fuhrländer M, Georg N, Römer U, Schöps S (2020) Yield optimization based on adaptive Newton-Monte Carlo and polynomial surrogates. Int J Uncertain Quantif 10(4):351–373. https://doi.org/10.1615/Int.J.UncertaintyQuantification.2020033344
22. Fuhrländer M, Schöps S (2020) A blackbox yield estimation workflow with Gaussian process regression applied to the design of electromagnetic devices. J Math Ind 10(1):1–17
23. Pedregosa F, Varoquaux G, Gramfort A, Michel V, Thirion B, Grisel O, Blondel M, Prettenhofer P, Weiss R, Dubourg V, Vanderplas J, Passos A, Cournapeau D, Brucher M, Perrot M, Duchesnay E (2011) Scikit-learn: machine learning in Python. J Mach Learn Res 12:2825–2830
24. Rasmussen CE, Williams CK (2006) Gaussian processes for machine learning. The MIT Press
25. Dassault Systèmes Deutschland GmbH: CST Studio Suite® (2018). www.3ds.com
26. Fuhrländer M, Schöps S (2021) Yield optimization using hybrid Gaussian process regression and a genetic multi-objective approach. Adv Radio Sci 19(B):41–48
27. Dembo RS, Eisenstat SC, Steihaug T (1982) Inexact Newton methods. SIAM J Numer Anal 19(2):400–408. https://doi.org/10.1137/0719025
28. Deb K, Pratap A, Agarwal S, Meyarivan T (2002) A fast and elitist multiobjective genetic algorithm: NSGA-II. IEEE Trans Evol Comput 6(2):182–197
29. Blank J, Deb K (2020) pymoo: multi-objective optimization in python. IEEE Access
30. Gangl P, Amstutz S, Langer U (2015) Topology optimization of electric motor using topological derivative for nonlinear magnetostatics. IEEE Trans Magn 52(3):1–4
31. Arjona M, Hernandez C, Lara J (2021) Electromagnetic optimal design of a permanent magnet synchronous generator. ICS Newsl 28(3)

32. Gangl P, Köthe S, Mellak C, Cesarano A, Mütze A (2022) Multi-objective free-form shape optimization of a synchronous reluctance machine. COMPEL - Int J Comput Math Electr Electron Eng
33. Zhang, Ragonneau TM, Zaikun (2022) PDFO - powell's derivative-free optimization solvers. https://www.pdfo.net/docs.html. Accessed 25.06.2022
34. Marti RC, Pardalos PM, Resende MG (2018) Handbook of heuristics. Springer
35. Sutton RS, Barto AG (1998) Reinforcement learning: an introduction. The MIT Press
36. Cartis C, Fiala J, Marteau B, Roberts L (2019) Improving the flexibility and robustness of model-based derivative-free optimization solvers. ACM Trans Math Softw 45(3). https://doi.org/10.1145/3338517
37. Kraft D (1988) A software package for sequential quadratic programming
38. Fuhrländer M, Schöps S (2022). Hermite least squares based on BOBYQA. https://doi.org/10.5281/zenodo.7473363

Chapter 7
Conclusion and Outlook

In this work, different methods for efficient and accurate yield (or failure probability) estimation and optimization were investigated. The application to the design process of electrotechnical devices was discussed. However, the proposed methods are not limited to this application and can be applied in any manufacturing process affected by uncertainties, e.g. due to manufacturing imperfections or natural material deviations.

One main challenge of sampling based yield estimation methods is that they are computationally prohibitive, if the QoI involves to numerically solve a complex model. In the case of electrotechnical devices, this are typically PDEs originating from Maxwell's equations, solved for example with the FEM. In Chap. 4 we introduced yield estimation methods with significantly reduced computational effort while maintaining high accuracy. The SC-Hybrid approach is based on SC and adjoint error indicators for the SC error and the FE error. The GPR-Hybrid approach is a non-intrusive method using GPR and allows GPR model updates on the fly with negligible additional computing costs. Compared to classic MC methods, the computational effort can be reduced by 98 % with the SC-Hybrid approach, and 99 % with the GPR-Hybrid approach, respectively. By applying FE model refinement strategies, the savings increase to even 99.8 %. We showed that parallelization is possible and can further reduce the effective costs, also when using surrogate model based approaches with recurring model updates.

This reduction of computing time allows the estimation of the yield for industrial designs, where one evaluation of the QoI often takes several minutes or even hours. In practice, often not only yield estimation is requested, but also yield optimization, i.e., the maximization of the design reliability. Then, the objective function, i.e., the yield estimate, is evaluated many times during the optimization process. Hence, an efficient approach for yield estimation is particularly important.

Besides efficient yield estimation approaches which have been used within the optimization, we investigated how the computing costs of the optimization algorithm can be reduced directly. In Sect. 5.3 we proposed a strategy for an adaptive sample size increase within the estimations. In combination with a globalized Newton method, we named it adaptive Newton-MC optimization. We showed that gradients of the yield

M. Fuhrländer, *Design Methods for Reducing Failure Probabilities with Examples from Electrical Engineering*, Springer Theses,
https://doi.org/10.1007/978-3-031-37019-9_7

can only be obtained directly from the yield estimate, if all optimization variables are affected by uncertainties and are (almost) Gaussian distributed. Otherwise, they must be approximated by costly additional QoI evaluations. The Hermite-type approaches proposed in Sect. 5.4 are modifications of the derivative free BOBYQA (Bound constrained Optimization BY Quadratic Approximation) method, exploiting all (first and second order) derivative information available. Due to better local quadratic approximations than in the original BOBYQA method, the optimization converges faster, requiring less evaluations of the objective function. By utilizing derivative information for the half of the optimization variables, the number of yield estimations during the optimization procedure were reduced by 27 % with the Hermite least squares approach.

We demonstrated that the Hermite least squares and the Hermite BOBYQA methods are not limited to yield optimization, but are relevant for all optimization problems with mixed gradient information and computationally expensive objective functions. Especially for noisy problems, they show high robustness and outperform the original BOBYQA method or gradient based methods with derivative approximations. Further, global convergence of the Hermite-type approaches can be proven under suitable assumptions.

Besides reliability optimization, there are often other performance indicators to be optimized. The computing effort in MOO is particularly high, since a trade-off between different objectives has to be found. Using classic MC is prohibitive. Applying for example a genetic algorithm to the bi-objective design optimization of a PMSM took eleven days using the GPR-Hybrid approach for yield estimation. With classic MC it would have required up to seven years.

The main findings of this research can be summarized as follows. First, GPR based hybrid approaches are better suited for yield estimation than polynomial interpolation based hybrid approaches. This can be mainly explained by their blackbox character and their flexibility regarding model updates. Second, if at least half of the derivative information is available, the Hermite least squares method outperforms derivative free solvers. Additionally, (Hermite) least squares shows higher robustness than interpolation in case of noisy data. Particularly for noisy problems, the specialized Hermite least squares method is also superior to gradient based approaches using approximations of the derivatives. Third, expensive tasks in industrial design such as MOO under uncertainty become tractable by using the proposed hybrid approaches for yield estimation. The time scale to solve these problems reduces from years or months to days or hours.

We conclude that the investigations in this work are important steps to make uncertainty quantification and optimization *of* uncertainty and *under* uncertainty feasible in real-world applications. Reliable designs in manufacturing processes under uncertainty lead to an increasing number of fully functional and high performing devices. This allows to develop devices closer to their physical limits and it saves resources, time and money by reducing rejects due to malfunctioning.

Based on the theoretical and numerical findings in this treatise, new topics for further investigation arise.

- We mentioned that besides reducing the computational effort of one sample point evaluation, reducing the number of sample points evaluated is a commonly used technique to reduce the effort for yield estimation. The hybrid methods proposed in this work could be combined with such techniques, e.g. importance sampling [1, 2].
- We defined performance feature specifications (PFS) which have to be fulfilled in a whole range parameter interval. We modeled this with a discrete set of range parameter points and separate surrogate models, each corresponding to one range parameter point. An alternative approach is to build an approximation along the range parameter, e.g. by employing model order reduction [3, 4].
- One drawback of the SC-Hybrid approach is the requirement of structured data points which inhibits easy model updates during the estimation process. By using regression instead of interpolation this drawback could be circumvented, cf. Sect. 4.3.
- In Sect. 4.2 we discussed that the proposed yield estimation methods do not address the case of rare events, i.e., yield estimation where the number of failing realizations is very small. Common approaches to tackle that are importance sampling [1] or subset simulation [5–7]. Further research could investigate how one of these techniques can be combined with the GPR-Hybrid approach to enable rare event simulation within our framework.
- In the Hermite-type optimization approaches we focused on reducing the number of objective function calls, assuming that each call is computationally expensive. Using one of the hybrid approaches for yield estimation reduces the computational effort for each objective function call (in case of yield optimization). However, analogously to the adaptive Newton-MC optimization, we suggest to incorporate an adaptive sample size increase strategy. This would be a specialization of the (general) Hermite-type optimization framework only for yield optimization. In the MOO approaches one can also apply adaptive sample size increase.
- We proved global convergence of the Hermite least squares method for unconstrained problems in the case that the number of interpolation points is sufficient to uniquely solve the interpolation problem (without derivative information). However, we showed that a lower number of interpolation points performs better in practice, as long as the system is overdetermined when derivative information is included. For this, global convergence was not proven.

References

1. Gallimard L (2019) Adaptive reduced basis strategy for rare-event simulations. Int J Numer Methods Eng 1:1–20. https://doi.org/10.1002/nme.6135
2. Tyagi A, Jonsson X, Beelen T, Schilders W (2018) Hybrid importance sampling Monte Carlo approach for yield estimation in circuit design. J Math Ind 8. https://doi.org/10.1186/s13362-018-0053-4
3. Eller M (2017) A low-frequency stable Maxwell formulation in frequency domain and industrial applications. PhD thesis, Technische Universität Darmstadt

4. Hess MW, Benner P (2013) Fast evaluation of time-harmonic Maxwell's equations using the reduced basis method. IEEE Trans Microw Theory Tech 61(6):2265–2274. https://doi.org/10.1109/TMTT.2013.2258167
5. Au SK, Beck JL (2001) Estimation of small failure probabilities in high dimensions by subset simulation. Probab Eng Mech 16:263–277. https://doi.org/10.1016/S0266-8920(01)00019-4
6. Bect J, Li L, Vazquez E (2017) Bayesian subset simulation. SIAM/ASA J Uncertain Quantif 5(1):762–786. https://doi.org/10.1137/16m1078276
7. Dhulipala SL, Shields MD, Spencer BW, Bolisetti C, Slaughter AE, Laboure VM, Chakroborty P (2022) Active learning with multifidelity modeling for efficient rare event simulation. J Comput Phys

Appendix A
Geometry and Material Specifications for the PMSM

A specification of the material and geometry properties of the PMSM can be found in [1, Tables A.1, A.2], where this model has also been used for benchmarking. However, for the sake of completeness, in Tables A.1, A.2 and Fig. A.1 we provide a detailed description.

Table A.1 Geometry properties of the investigated PMSM

Geometry	
Inner rotor radius	16 mm
Outer rotor radius	44 mm
Model coupling radius	44.7 mm
Inner stator radius	45 mm
Outer stator radius	67.5 mm
Length	100 mm
Number of PMs	6
Number of pole pairs	3
Number of phases	3
Number of coils	36
Number of windings per coil	12
Skew angle	0.52
$c1$	8.2 mm
$c2$	5 mm
$c3$	5.4 mm
$c4$	4
$c5$	5.7
$c6$	7

M. Fuhrländer, *Design Methods for Reducing Failure Probabilities with Examples from Electrical Engineering*, Springer Theses,
https://doi.org/10.1007/978-3-031-37019-9

Table A.2 Material properties of the investigated PMSM

Material	
Iron conductivity	0 S/m
Copper conductivity	0 S/m
PM conductivity	0 S/m
Iron relative permeability	500
Copper relative permeability	1
PM relative permeability	1.05

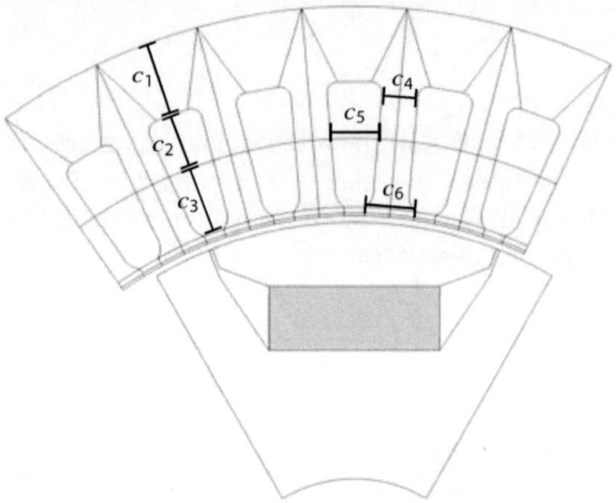

Fig. A.1 Geometry of the investigated PMSM

Reference

1. Bontinck Z (2018) Simulation and robust optimization for electric devices with uncertainties. PhD thesis, Technische Universität Darmstadt

About the Author

Dr.-Ing. Mona Fuhrländer is currently a Postdoctoral researcher at Technische Universität Darmstadt, in Prof. Dr. Sebastian Schöps' Computational Electromagnetics group. She is also part of the Institute for Accelerator Science and Electromagnetic Fields, the Centre for Computational Engineering, and the Graduate School Computational Engineering.

She was born in Frankfurt, Germany, in 1991. She obtained her Bachelor's and Master's degree in Mathematics from Technische Universität Darmstadt in 2015 and 2017, respectively. During her Bachelor studies she spend one year at the Universidad de Granada, Spain. In May 2018, she started her Ph.D. in Computational Engineering at the department of Electrical Engineering and Information Technology at Technische Universität Darmstadt. The Ph.D. was supervised by Prof. Dr. Sebastian Schöps and completed in December 2022. Her research focus is on uncertainty quantification, optimization and numerical analysis. In 2020, she was awarded the Young Scientist Award of the German branch of the International Union of Radio Science (U.R.S.I).

This book presents the results of her Ph.D. thesis. Motivated by uncertainties caused by manufacturing imperfections, natural material variations, or unpredictable environmental influences, her aim is to design electrotechnical devices with high reliability. Therefore, efficient simulation-based methods for estimating and minimizing the impact of uncertainty are investigated.

M. Fuhrländer, *Design Methods for Reducing Failure Probabilities with Examples from Electrical Engineering*, Springer Theses,
https://doi.org/10.1007/978-3-031-37019-9